5-20-77

INDUSTRIAL ENERGY CONSERVATION
A Handbook for Engineers and Managers

Other Pergamon Titles of Interest

ASHLEY et al.	Energy and the Environment—a Risk-Benefit Approach
BÖER	Sharing the Sun
BLAIR et al.	Aspects of Energy Conversion
BRATT	Have You Got the Energy?
DIAMANT	Total Energy
DUNN & REAY	Heat Pipes
HUNT	Fission, Fusion and the Energy Crisis
JONES	Energy and Housing
KARAM & MORGAN	Environmental Impact of Nuclear Power Plants
KARAM & MORGAN	Energy and the Environment: Cost-Benefit Analysis
KOVACH	Technology of Efficient Energy Utilization
McVEIGH	Sun Power, An Introduction to the Applications of Solar Energy
MESSEL & BUTLER	Solar Energy
MURRAY	Nuclear Energy
SIMON	Energy Resources
SMITH	The Technology of Efficient Electricity Use
SPORN	Energy in an Age of Limited Availability and Delimited Applicabil

INDUSTRIAL ENERGY CONSERVATION
A Handbook for Engineers and Managers

DAVID A. REAY

*International Research and Development Co. Ltd.,
Newcastle-upon-Tyne*

PERGAMON PRESS

OXFORD · NEW YORK · TORONTO · SYDNEY · PARIS · FRANKFURT

U.K.	Pergamon Press Ltd., Headington Hill Hall, Oxford OX3 0BW, England
U.S.A.	Pergamon Press Inc., Maxwell House, Fairview Park, Elmsford, New York 10523, U.S.A.
CANADA	Pergamon of Canada Ltd., 75 The East Mall, Toronto, Ontario, Canada
AUSTRALIA	Pergamon Press (Aust.) Pty. Ltd., 19a Boundary Street, Rushcutters Bay, N.S.W. 2011, Australia
FRANCE	Pergamon Press SARL, 24 rue des Ecoles, 75240 Paris, Cedex 05, France
WEST GERMANY	Pergamon Press GmbH, 6242 Kronberg-Taunus, Pferdstrasse 1, Frankfurt-am-Main, West Germany

Copyright © 1977 D. A. Reay

All Rights Reserved. No part of this publication may be reproduced, stored in a retrieval system or transmitted in any form or by any means: electronic, electrostatic, magnetic tape, mechanical, photocopying, recording or otherwise, without permission in writing from the publishers

First edition 1977

Library of Congress Cataloging in Publication Data
Reay, David A
Industrial energy conservation.

Bibliography: p.
Includes index.
1. Energy conservation--Handbooks, manuals, etc.
I. Title.
TJ163.3.R4 1977   621.4   76-56843
ISBN 0-08-020867-3
ISBN 0-08-021716-8 pbk.

In order to make this volume available as economically and rapidly as possible the author's typescript has been reproduced in its original form. This method unfortunately has its typographical limitations but it is hoped that they in no way distract the reader.

Printed in Great Britain by A. Wheaton & Co., Exeter

Contents

Foreword	by Lord Robens of Woldingham	ix
Preface		xi
Acknowledgements		xii
Introduction		1

Chapter 1. Current Primary Energy Resources
- 1.1 Coal — 4
- 1.2 Gas — 9
- 1.3 Oil — 12
- 1.4 Nuclear Fuel — 15
- 1.5 Industrial Implications of Energy Resources — 16

Chapter 2. Optimum Use of Prime Movers for Power Generation and Other Applications
- 2.1 Electricity Generation by National Authorities — 19
- 2.2 Generation of Power by Industry — 23
- 2.3 Steam Turbines — 24
- 2.4 Rankine Vapour Cycle Turbines Using Freons — 27
- 2.5 Gas Turbines — 28
- 2.6 Combined Cycles — 36
- 2.7 Diesel and Gas Engines — 38
- 2.8 Installation and Control Systems — 48
- 2.9 Conclusions — 49

Chapter 3. The Energy Intensive Industries - 1
- 3.1 Iron and Steel Production — 52
- 3.2 Aluminium — 59
- 3.3 The Chemical Industry — 62
- 3.4 Oil Refineries — 71

Chapter 4. The Energy Intensive Industries - 2
- 4.1 The Pulp and Paper Industry — 76
- 4.2 The Glass Manufacturing Industry — 85
- 4.3 The Food Processing Industry — 92
- 4.4 The Textile Industry — 99
- 4.5 Conclusions — 106

Chapter 5. Common Items of Plant - Good Housekeeping
- 5.1 Air Conditioning — 110
- 5.2 Air Conditioning Control Systems — 113
- 5.3 Boilers — 117
- 5.4 Chimneys — 118
- 5.5 Combustion Systems — 119

5.6	Compressed Air Systems	124
5.7	Cooling Towers	125
5.8	Dryers	126
5.9	Electricity Generation (In-Plant)	130
5.10	Electrical Equipment	131
5.11	Furnaces	132
5.12	General Process Heating	136
5.13	Incinerators	138
5.14	Lighting in Factories and Offices	139
5.15	Plant Buildings	143
5.16	Plant Domestic Hot Water Systems	145
5.17	Useful Aids to Energy Conservation Implementation	145
5.18	Steam Traps	146
5.19	Vats and Hot Storage Tanks	147
5.20	Fuel Additives	147
5.21	Sources of Other Data	148

Chapter 6. Common Items of Plant - Energy Recovery

6.1	Air Conditioning	152
6.2	Boilers	162
6.3	Burners - Heat Recovery to Improve Efficiency	169
6.4	Dryers - Heat Recovery and Fluidised Bed Systems	171
6.5	Furnaces	180
6.6	Incinerators	184
6.7	Thermal Fluid Heaters	191
6.8	'Non-Destructive' Waste Recovery Techniques	193

Chapter 7. Waste Heat Recovery Techniques

7.1	Heat Pipe Heat Exchangers	199
7.2	Liquid-Coupled Indirect Heat Exchangers	207
7.3	Gas-Coupled Indirect Heat Exchangers	211
7.4	Rotating Regenerators	211
7.5	Plate Heat Exchangers	219
7.6	Economisers	224
7.7	Waste Heat Boilers	227
7.8	Recuperators	232
7.9	The Heat Pump	238
7.10	Other Heat Exchanger Types	257

Chapter 8. Energy Storage

8.1	Thermal Storage Boilers and Accumulators	264
8.2	Thermal Insulation	267
8.3	Heat Storage Media	270

Chapter 9. New or Specialised Processes and Plant Having Energy-Saving Potential

9.1	Dielectric and Microwave Heating	276
9.2	Electron Beam Welding	279
9.3	Fluidised Bed Technology	282
9.4	Electric Foil Heating Elements	285
9.5	The Laser as a Welding Tool	286

Contents

Chapter 10. Alternative Sources of Energy

10.1	Wind Power	292
10.2	Geothermal Energy	294
10.3	Wave Power	296
10.4	The Hydrogen Economy Concept	298
10.5	Solar Energy	300
10.6	Conclusions	303

Appendices.

1	Energy Management and the Energy Audit	306
2	Financial Analyses for Evaluating Benefits of Energy Conservation Equipment	309
3	Organisations Offering Services to Industry in the Field of Energy Conservation	312
4	Manufacturers of Equipment for Energy Conservation	319
5	Bibliography	332
6	Useful Conversion Factors and Fuel Properties	342

Index.

344

Foreword

The basis of a prosperous economy lies in the availability of a cheap and uninterruptable supply of energy. Cheapness is of course a relative term, but higher priced fuels can be enormously cheapened in their use if every conceivable step is taken to cut out waste, unnecessary use and a careful economic energy survey in all commercial establishments.

Energy conservation is therefore of vital importance, as one sees in the years ahead a steady increase in primary energy costs.

As one who has had a fair amount to do with energy in the past thirty years, I welcome this book which will be of inestimable value to those who have any control over energy usage. Plant engineers will find it extremely useful, dealing as it does in some detail with the question of equipment reliability, heat recovery systems and energy conservation techniques, together with cost data.

The world demand for energy is rapidly mounting, and as more and more of the world's population moves from agricultural pursuits to industrialisation, the demand will continue to expand at an alarming rate.

Even if energy were for free, conservation is essential, if present resources are not to be run out more quickly than new resources can be found or new technologies introduced. But energy is not for free, and forms a quite substantial cost in manufacture and distribution.

Energy conservation is, therefore, an absolute must, not merely for this generation, but for all the generations to come whilst life lasts on this planet.

This book is, therefore, to be welcomed on all counts. My only hope is that the information will be well used. The author can only point the way, based on his experience and research. It is for the user to learn the lesson and take the right road now.

LORD ROBENS OF WOLDINGHAM
June 1976

Preface

The conservation of energy is a subject which is becoming of increasing interest to all sectors of society in the industrial nations. Government concern regarding dependence on imported energy, particularly oil, is typified by the European Economic Community proposal to reduce total energy consumption in 1985 by 10 per cent in relation to the amount initially estimated for that year, limiting to 40 per cent the dependence on energy from outside sources. In the United States vast funds are being allocated to the development of new energy resources, self-sufficiency in energy being the ultimate goal of 'Project Independence'.

Governments have a duty to encourage energy conservation, not least because a nation which is not self-sufficient must buy energy externally. The United Kingdom however, is fortunate in possessing adequate reserves of natural gas and, having a more recent impact, oil. The domestic consumer, faced with rising energy costs, is able to reduce losses by some form of insulation, or change to cheaper fuels by investment in new equipment, if financially acceptable. A direct reduction in energy consumption, at the possible expense of comfort or convenience, is an alternative method of balancing the domestic budget.

This book is concerned with the industrial consumer of energy. In the more energy-intensive industries energy conservation has always been recognised as a necessary component of developments in production technology. Now, however, the rising energy costs make most companies 'energy-intensive' in that the proportion of total expenditure attributable to energy is increasing, just as the labour bill rises in periods of wage inflation. There are now many opportunities for one sector of industry to benefit from the developments which are routinely applied to conserve energy in another sector: one main aim of this book is to identify these opportunities. Considerable data is available on energy conservation techniques and the cost of implementation: it is also the purpose of the book to present some of this data in a manner which will be of interest and value to the energy manager or his equivalent.

It would be presumptuous of me to claim that this is a 'Consumer's Guide to Industrial Energy Conservation'; however my desire to retain a degree of objectivity, while presenting data on specific products as well as on more general aspects of energy conservation, will, I hope, make this book a useful source of reference for energy managers and all those interested in conservation. This book will not provide all the answers, but if it is able to direct the reader, armed with the right questions, to the correct source, one of my aims will have been fulfilled.

David A. Reay
June 1976.

Acknowledgements

I am most grateful to the numerous companies and other organisations, both in the United Kingdom and the United States, who have assisted in providing data for use in this book.

Where possible, the appropriate credits are given within the main text, but separate acknowledgement is due to a number of contributors. I am grateful to Mr. W.H. Levers of the Standard Oil Company, in conjunction with the Institute of Electrical and Electronics Engineers, Inc. for the data on total energy systems included in Chapter 2. A considerable amount of statistical data is also presented with the permission of the Controller of Her Majesty's Stationery Office. Denis Csathy, President of Deltak Corporation, gave me much useful information on waste heat boilers, and Mr. E.C. Herrnicht of Cannon Air Engineering Ltd., Division of Fan Systems Group Ltd., provided valuable background information on the Rotary Air Jet dryer.

A second important source of statistics was the U.S. Environmental Protection Agency, and Howden Group Limited kindly provided papers on air preheaters.

In particular I would like to thank Lord Robens for writing what I believe to be a most pertinent Foreword; Mr. Alan Potts, Power Controller at C.A. Parsons & Co. Ltd., for his constructive comments on reading the draft, Pergamon Press for their encouragement, and last but not least, Pru Leach, for her assistance with editing, and for typing the manuscript.

David A. Reay
June, 1976.

Introduction

The need to conserve energy in industry is of considerable importance in view of rising costs and, at least in the short to medium term, potential shortages because of reducing reserves of fossil fuels. Cost is of more immediate interest to the manager in industry, and it is this factor which is likely to have the strongest influence on his attitude to conservation and increased efficiency.

The consumer, be he in industry or the domestic sector, does however, have a responsibility for improving the way in which he uses energy. This is expressed frankly by the European Economic Community:

"Special attention should be given to the role the consumers themselves can play. While the increased costs of energy may put a brake on demand in the short term, this affect may well tend to weaken in the long term. It is therefore essential that the consumers be confronted with their collective responsibility to make better use of energy resources, which from now on will be in shorter supply and will cost more than before."

Whether he be influenced by costs or by his social responsibility, the consumer in industry, through his energy manager or equivalent, needs accurate up-to-date information on techniques and equipment available to enable him to implement an energy conservation programme. The brief given to the energy manager will necessitate him answering a number of questions. These questions provided the basis for the synopsis of this book, and are as follows:

(i)   What items of plant use energy?

(ii)   How efficiently is the energy used?

(iii)   Can the efficiency of energy utilisation be improved?

(iv)   If so, how can improvements be implemented?
        (e.g. Good housekeeping, modified plant utilisation, different fuels, plant modification or plant replacement).

(v)   What energy savings will result?

(vi)   Who can advise on and/or implement the improvements needed?

(vii)   How much will it cost?

Briefly, the purpose of this book is to provide the answers to these questions and this is carried out in the following manner:

Chapter 1 discusses the availability of existing energy resources, by way of an introduction to energy utilisation and conservation. The cost of various forms of energy, and the efficiency of conversion are discussed, and the need for and advantages of energy conservation are detailed. The generation of power is responsible for a large proportion of the depletion of natural resources. However, as well as being provided by national authorities, power can be generated by industry itself. Chapter 2 discusses the techniques available for this, relative merits, and methods for maintaining a high load factor. Electricity generation will be discussed in some detail, but the use of prime movers for driving pumps and the like directly will also be considered. Total energy plant, a necessary part of Chapter 2, will also be discussed in subsequent Chapters on heat recovery techniques.

From a national point of view, energy conservation in what may be called the energy intensive industries (iron and steel, non-ferrous metals, chemicals, etc) is of great significance. Chapter 3 describes production methods used, and energy conservation techniques, with a view to isolating areas which can be of benefit to other smaller plant users. The expenditure on energy conservation in these high technology industries is considerable, and it is from here that 'technological fall-out' can be of great benefit to the smaller companies involved in manufacture.

Energy use in a number of industrial categories outside the major consumers remains significant. While the impetus to conserve energy in these industries, such as paper, textiles, food and allied products, etc., does not in many cases attract the serious attention of government agencies, energy costs are becoming increasingly important to the operators of plant in these areas. In Chapter 4, as in much of the remainder of the book, the emphasis is placed on this type of organisation. As well as providing the energy manager in these particular industries with a resumé of a variety of energy-conserving measures, many techniques developed in one industry may be applied in other areas. Heat recovery from effluents in the dairy industry, for example, can be carried out in similar ways in other processes. Therefore it is to be hoped that this book will stimulate the transfer of appropriate conservation techniques from one area of industry to another.

Chapters 3 and 4 are concerned largely with specific industries. In order to broaden the scope of the book, it is necessary to concentrate on particular items of plant common to other industries, as well as those discussed in the Chapters so far described. Chapter 5 details ways in which these common items of plant can be made to operate at high efficiency. As well as factors such as preventative maintenance, the use of control systems for temperature and air/fuel ratio selection, etc., many simple low-cost tasks can be undertaken to improve plant efficiency. In ovens, for example, modifications to the schedule to increase the hearth load factor are often possible. Pressure control as a means of reducing air in-leakage can also be beneficial. Each item of plant is considered from this point of view.

'Good housekeeping' is a form of energy conservation which the plant engineer is able to implement in a comparatively short time, but this may result in only modest savings. Chapter 6 contains information on the type of equipment in which the energy manager may invest as a further step to improve efficiency in particular processes. Opportunities exist for waste heat recovery, be it recirculated in the plant or used to supplement space heating. Waste water recovery and recirculation of other waste liquids are also among the topics discussed, and alternative heating methods, such as those using high temperature heat transfer fluids, are described.

Most engineers will agree that of the 'waste' in plant, the most significant is that lost in heat, be it in the form of exhaust gases or in hot liquid

Introduction

effluent. There are many ways in which heat may be recovered, ranging from comparatively simple regenerators and recuperators to multi-megawatt heat pump systems driven by gas turbines. Chapter 7 contains descriptions of these systems, case histories, and examples of the capital cost and return on investment accruing to the use of this equipment. Where a technique may not be sufficiently developed at present to meet all requirements, as is the case with the heat pump, current limitations are discussed.

In many instances energy may be available at low cost during 'off-peak' periods. Process waste heat may also be more profitably used at a later time. These situations can point to the beneficial use of energy storage techniques. Heat accumulators are familiar to many plant engineers, but the use of specific heat storage media and heat of fusion materials are still regarded by many as belonging to the realms of space technology. Chapter 8 discusses the status of thermal energy storage in the context of industry and buildings.

There is a number of processes, both under development and in current use which can be used for heating, welding, cutting, shaping etc., with less expenditure of energy than some commonly accepted techniques. What has often been considered as a research tool, or as being applicable to specialised tasks such as forming components for 'state-of-the-art' aircraft, must now become more widely applied. Features of these processes, and their current status, are described in Chapter 9.

Although the practical engineer may be justifiably sceptical about the relevance of 'alternative sources of energy' to his particular operating problems, no treatise on energy conservation would be complete without a comment on the potential impact of sources such as solar energy, tidal power and the 'hydrogen economy'. After all, it is probable that we will have to rely increasingly on these alternative sources in the future, and a number of these energy sources are discussed briefly in Chapter 10.

For completeness a number of Appendices are included, covering energy audits, organisations offering advice on energy conservation and manufacturers of equipment in this field. A number of useful conversion factors are listed and an extensive bibliography is included. The Appendices are particularly comprehensive, and it is hoped that the user will find that the data presented will save much time-consuming effort in identifying sources of information and product data.

Current Primary Energy Resources

The major sources of energy currently used by industry, either directly or indirectly via the national electricity supply authorities, all have a finite life. Shortages of natural gas are already having an impact in parts of the United States, and while it is the price of oil rather than its longevity which has spurred much of the work proceeding on the development of alternative energy sources, oil shortages are forecast for later this century. Reserves of uranium and other fuels for nuclear reactors, once considered to be the panacea with which one could overcome all problems associated with fossil fuel resources, at the same time providing low cost electricity, are also subjected to depletion. While the breeder reactor systems should prolong the usefulness of nuclear power generation in the foreseeable future, the current 'environmental lobby' does little to assist such developments.

Of the fossil fuels which may prove to have the most potential in the long term, coal, perhaps surprisingly, offers the most scope. Coal gasification and liquefaction can produce fuels equivalent to oil and natural gas, and the reserves are vast. The extraction of coal must be improved, and the run-down of coal industries in many of the industrial nations, which has to some extent halted, must now be reversed. It is encouraging to note that large sums of money are being spent on developments associated with the extraction, processing and use of coal.

This Chapter discusses existing reserves of fossil and nuclear fuels, and reviews some of the new processes which can lead to wider applications for coal. Alternative sources of oil, such as oil shale, may have long term potential, and data is given on the effect their exploitation could have on the future supply situation.

1.1 Coal

Many regard coal as an energy resource which served its purpose as the foundation of the industrial revolution, but began an inevitable decline in popularity as a primary source of energy in the mid 20th Century. Certainly, if we examine the consumption of coal in the United Kingdom in the years 1969-1975, a continuing trend in the reduction of coal utilisation can clearly be seen, in spite of a general increase in the total consumption of energy from all resources, although the 1974 and 1975 figures reflect the effect of oil price rises. This is shown in Table 1.1.

Compared to oil and gas, discussed later in this Chapter the downward trend in coal consumption is unique, and is not only a feature of the United Kingdom economy.

The reasons for this have no bearing on the relative abundance of these primary resources. Estimates of total reserves of these resources vary considerably, but even allowing for a high degree of error, it can be seen from the following figures, giving recoverable world reserves in oil barrels or equivalent, that coal reserves far exceed those of other fossil fuels (ref. 1.2):

Coal and lignite	20×10^{12} barrels
Oil	2×10^{12} barrels
Natural gas	1.3×10^{12} barrels
Tar sands and shale	4.2×10^{12} barrels

The main disadvantages of coal as a fuel for industrial processes are associated with its inconvenient form when mined using current methods and when subsequently marketed, as opposed to liquid and gaseous fuels. Coal can however be converted into forms which can be readily used in plants using oil or gas; coal liquefaction and gasification plants are the subject of much activity in Europe and the United States, and these and other developments are discussed later.

TABLE 1.1   UK Inland Energy Consumption

Year	Total*	Coal*
1969	318.5	161.1
1970	329.6	154.4
1971	325.9	138.7
1972	331.3	120.9
1973	346.1	131.3
1974	331.0	115.9
1975	320.6	120.4

* Figures are million tons of coal or coal equivalent, not seasonally adjusted (ref. 1.1)

Another reason for the decline in popularity of coal has been a growing lack of competitiveness with alternative fuels. This is largely because of a dirth of effective investment in new extraction techniques, the coal industry, certainly as far as the United Kingdom is concerned, being highly labour-intensive. Legislation designed to protect the environment from the worst effects of fuel combustion has led to a demand for smokeless fuels, and coke and the higher quality coals have proved uncompetitive alongside oil and natural gas from this point of view. The most noticeable evidence of this is in the dramatic switch from coal for domestic heating.

1.1.1 **Types of coal and its properties**   The main fuel constituents of coal are carbon and hydrogen, but coal will also contain oxygen, sulphur and nitrogen, and a variety of mineral matter which is normally described as ash.

The calorific value of coal depends on its quality; the proportion of volatiles in relation to the carbon content, and on the size of the coal particles. The range over which the calorific value extends is large. Commercially available fuels having gross calorific values of over 32 560 kJ/kg are rich in carbon and contain little gaseous matter. At the lower end of the scale, small coal particles having a high proportion of volatile matter can have gross calorific values as low as 23 200 kJ/kg.

Volatile content is one of the two most important characteristics of coal (ref. 1.3). In the United Kingdom the classification of coals, or their 'rank', is on the basis of the volatile content. High rank coals (which have a low rank number), such as anthracite, have relatively little volatile matter within them. Bituminous or low rank coals may have in excess of 40 per cent volatile content. The second important property of coal is its ability to cake. Caking is the tendency for the particles of coal to fuse into a single mass on heating, accompanied by expansion of the coal.

As well as being identified by rank, coal may be classified in grades, relating to its particle size. It may also be in a treated (cleaned) or untreated state.

All of the above characteristics have a bearing on the combustion of the coal. A high volatile content eases the ignition of the coal, but low rank coal has the disadvantage of high smoke output, and all the volatile matter must be burnt to avoid excessive pollution. Smokeless fuels are those having a high carbon content. Caking can obviously affect the passage of combustion air through the bulk of the coal, and hence the burning rate. If the coal has poor caking qualities, grit emission and the loss of unburnt coal particles through the grate may require attention.

Coal which has been cleaned generally results in little ash residue when burnt. In some cases ash is of assistance in protecting grate bars from reaching too high a temperature. The cleaning process can also affect performance. Washing of very fine coal can create handling difficulties, but bonding is assisted by the water, preventing fine coal falling initially through the grate bars.

The airflows used for combustion may be described as follows: Primary air is that passed through the grate into the firebed; secondary air is introduced over the firebed, and, where used, tertiary air is that introduced after the secondary air. Higher carbon content coals such as anthracite do not need much secondary air for satisfactory combustion, but low rank coals use considerable quantities of secondary air in order to ensure low smoke emission.

1.1.2 Prospects for more effective coal utilisation

The existence of very large reserves of coal, and the fact that a considerable proportion of these reserves are present in the industrialised countries, is stimulating a new interest in coal utilisation. In particular, significant amounts of research and development funding are being directed at the investigation of techniques for converting coal into gaseous and liquid forms to enable it to compete on an equal footing with the convenient forms of fuel currently identified by oil and natural gas.

In order to implement the proposed increase in production rate which is forecast in only one of the industrialised nations, the United States, enormous amounts of funding are required to be allocated to coal extracting technology and conversion. Fig. 1.1 (ref. 1.4) shows the magnitude of this task in terms of the current and predicted production rates over the next few hundred years. (Compared with this, US production of oil and natural gas both may achieve their maximum potential during the current decade!)

The Commission of the European Communities have also proposed increasing coal utilisation at the expense of imported oil (ref. 1.5), and as well as improving production technology, studies on improving combustion, coke production, and the conversion of coal into hydrocarbons are being urged. In the United Kingdom the National Coal Board (NCB) have proposed a programme of development on a number of coal conversion processes. Pilot plants are projected to investigate fluidised bed combustion (see Chapter 9), coal liquefaction by both liquid solvents and supercritical gases, gasification with oxygen to yield low calorific value fuel gas. Other projects cover pyrolysis (the production of liquid and gaseous fuels by the thermal decomposition of coal in an inert atmosphere), and coke manufacture (ref. 1.6). The NCB has a close working agreement with the Department of the Interior in the USA which provides for an extensive exchange of research information, and both countries will therefore benefit from successful development projects in the above areas.

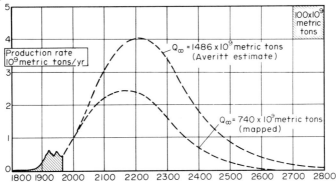

Fig. 1.1 Predicted US coal production   (Courtesy E.S. Rubin, ref. 1.4)

1.1.3 Coal gasification

Coal gasification processes were developed in the 19th Century to produce fuel gas, but were largely superceded by natural gas when this became readily available. If air and steam are driven through a bed of coal heated to about half its normal combustion temperature, 1000°C, these decompose to produce a combustible mixture of carbon monoxide, hydrogen and nitrogen, having a comparatively low calorific value. The heat content of this gas is less than 20 per cent of that of natural gas, making it uneconomical to transport over long distances, (ref. 1.7).

Present coal gasification plants are unable to cope with large amounts of coal, such as the quantities required to produce gas for multi-megawatt electricity generating stations. They are also inefficient and tend to produce considerable quantities of undesirable byproducts such as tar. In the United States a system is being developed which, it is believed, will overcome most of these problems. The technique, known as the 'fast fluidised bed', is illustrated in Fig. 1.2. The fast bed uses high gas flows and provides a cyclone effect for the return of solid particles to the bottom of the bed at a high rate. The gas flow is sufficiently high to permit good mixing in the bed, leading to a more uniform temperature throughout it. Production of wasteful byproducts is minimised by running at a high temperature. The quantity of steam used in this system is much less than that in older gasification plant, and as it is converted almost entirely into carbon monoxide and hydrogen, the efficiency is also increased.

It is envisaged that this type of installation could be integrated with a combined-cycle electricity generating plant, using both steam and gas turbines. The research team working on this project believe that such a complex could have a conversion efficiency of 50 per cent compared to the 35 per cent available using current conventional steam turbine plant.

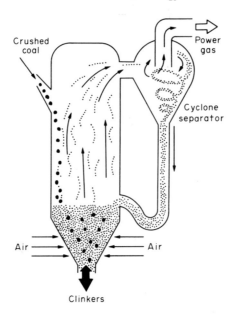

Fig. 1.2 Fast fluidised bed coal gasification process
(Courtesy Mosaic, the magazine of the United States
National Science Foundation)

Other coal gasification research is directed at manufacturing a range of low, medium and high calorific value gases, the production of hydrogen and methane, and the yielding of benzene.

1.1.4 Coal liquefaction The principle aim of coal liquefaction is to produce a low ash, low sulphur fuel capable of meeting all the requirements of Clean Air legislation. While gasification itself can be used to produce liquid products from coal, an alternative method is to attack the coal with solvents (ref. 1.8). In the United States the use of liquid solvents is being actively pursued. In one process coal is mixed with a liquid solvent, itself derived from the coal, then heated and passed with additional hydrogen to a high pressure reactor. Hydrogen and hydrogen sulphide are then separated, the mixture filtered and the solvent distilled for re-use. The final product is recovered either as a liquid or a solid. While this and several other solvent extraction techniques under development require the use of expensive hydrogen, the NCB in the United Kingdom is working on a system which does not require hydrogen.

Another technique leading to liquid fuels based on coal which has been pioneered by the NCB is gaseous extraction. At a temperature between 350 - 400°C and at high pressure, a substantial proportion of coal can be extracted by gases having a critical point within this range. The extract passes directly into the vapour phase, avoiding separation problems which can occur with liquid extraction techniques. Extracts obtained in this manner are lighter, more mobile liquids than those mined using liquid extraction, and are also liquid at room temperature.

Current Primary Energy Resources

In general, there appears to be considerable potential in the use of coal as a base for liquid and gaseous fuels, and the close collaboration between Europe and the United States in this area is encouraging.

1.2 Gas

The production of gas from coal is a method which has all but disappeared, although, as has been shown early in this Chapter, coal gasification is still the subject of continuing development because the world reserves of coal are considerably greater than those of oil and natural gas, on which we heavily rely at present.

It has been estimated that the proven world natural gas reserves were equivalent to about 300×10^9 barrels of oil (ref. 1.9), and that the ultimately recoverable quantity amounted to just in excess of 1000×10^9 barrels of oil. This total represents only 50 per cent of the world oil reserves estimated on the same basis, and other factors show that there is some cause for concern at the rate of depletion of natural gas resources (ref. 1.10). The average rate of depletion of natural gas is 40 per cent of that of oil, with certain regional exceptions. In the United States the gas consumption rate is about 70 per cent of that of oil, and the US reserves/production ratio (this being the ratio of the proven recoverable reserves at the end of a year to the annual production for that year), is currently about 10:1, this being a rather low figure when compared with the fact that even with a growth in coal consumption of 3.5 per cent per annum, the reserves/production ratio would drop to 10:1 only after another 55 years. The figure for coal is based on a comparatively low proportion of the total reserves being economically recoverable, and if necessity demands that this proportion is increased, the 'period of grace' will also be extended.

The consumption of natural gas, expressed as a percentage of total energy consumption, is given in Table 1.2 for the industrialised nations of the Western World.

TABLE 1.2   National Demand for Natural Gas,
as a Percentage of Total Energy Consumption, 1972, (ref. 1.10)

The Netherlands	38%
United States of America	33%
Canada	26%
United Kingdom	13%
Italy	9%
Germany	9%
France	8%
Japan	1%
Scandinavia	<1%

(In most of these countries oil is the most popular energy source from the point of view of consumption). In the United States natural gas is considerably cheaper than in the United Kingdom, and is available to utilities (local electricity generating authorities) at only about two thirds of the cost paid by industry.

In the UK it is estimated that the demand for natural gas will achieve the maximum value in the 1980's, with a drop in demand created by shortfalls in production occurring in the 1990's (ref. 1.11). Peak demand will be of the order of 80 mtce*, probably equivalent to approximately 20 per cent of the total energy requirement of the UK in the 1980's. With one or two minor exceptions, all of the gas supplied by 1980 in the UK will be natural gas, which has a calorific value two times that of the manufactured town gas which formed the basis of supplies until the 1960's, (ref. 1.12). This may come directly from wells such as those in the North Sea, with possible supplementation by gas of similar characteristics manufactured from coal.

Rooke (ref. 1.12) touches upon the delicate matter of methods for ensuring that reserves of natural gas are not too rapidly depleted, particularly in view of the current price advantages over competitive fuels, a fact which is even more evident in the United States, as shown in Table 1.3.

TABLE 1.3   Typical Energy Prices
US cents/MJ (approx.)   June-July 1975 (ref. 1.10)

	UK	US
OIL		
Motor spirit (incl. tax)	1080	445
Domestic heating oil	323	210
Industrial	250	213
Utilities	200	205
GAS		
Domestic	275	152
Industrial	100 (200*)	100
Utilities	–	66
COAL		
Domestic	200	–
Industrial	170	90
Utilities	120	81
ELECTRICITY		
Domestic	1300 (590**)	1200†
Industrial	900	800

* New   ** Off peak   † Eastern States

Natural gas is an attractive source of energy for a number of reasons. It is clean, and when burnt does not produce obnoxious products of combustion. Natural gas can be supplied to a location with relative ease, and intermediate storage is not necessary. It is indigenous to the United Kingdom, as well as to a number of other industrialised countries, and although recent exploitation of oil in the North Sea will (hopefully) assist in obtaining a balanced fuel utilisation in the UK and other European countries having interests in the oil wells, the pricing structure is likely, ultimately, to be a major factor in determining relative consumption. The increasing ability of governments to control this and other factors associated with energy supplies

---

*million tons of coal equivalent

1.2.1 The future of natural gas Much of the natural gas for the United Kingdom is currently extracted from wells drilled on shallow parts of the continental shelf, and future exploration in deeper waters, possibly as deep as 1000 m, will have to be carried out hand in hand with modifications to the production procedure used at present. Rooke foresees the need to recompress the gas from distant offshore wells en route, using an intermediate platform containing gas turbine-driven centrifugal compressors. Liquefaction of the gas at the primary well for transport via tanker to on-shore distribution centres has also been suggested.

While predictions of the effect that deep water drilling may have on the cost of natural gas are difficult, Rooke suggests that, neglecting inflation, improved technology may permit deep sea gas to be produced at a cost about 75 per cent of that of gas obtained from 150 m deep water wells.

Other developments which may become characteristic of the natural gas scene in the 1980's include the widespread use of liquid natural gas (lng) storage facilities. By storing the gas in liquid form at low temperature, large quantities occupy a comparatively small volume, and storage vessels may be conveniently located for meeting the requirements of peak demand or temporary failures in the normal distribution system. Storage of natural gas under pressure is currently being investigated, and the use of caverns formed in deep salt beds for this purpose is promising. The refilling of depleted wells comparatively close to the shore with gas obtained from deep-water wells may also become a reality.

Coal gasification has already been discussed as a means for providing substitute natural gas. Oil may also be used as a basis for sng, using a method known as the catalytic rich gas (crg) process. These plants, which produce carbon dioxide and methane by the reaction between pure light petroleum fractions and steam at $450°C$ over a catalyst, are cheap to erect, and in the United Kingdom this production method is being carried out in converted town gas manufacturing plant. The United States are using the crg process to make up for shortfalls in natural gas supplies.

A technique which makes use of heavier and crude oils for sng production is the fluid bed hydrogenator process. By carrying out the reaction in a fluidised bed of coke particles, a number of gases, including methane, can be produced from the oil, the bed itself retaining the solid carbon residue. Although more expensive than crg, the range of feedstock which the process can accept makes it a very versatile alternative.

The relative importance of substitute natural gas production processes, as well as the use of existing and new natural gas reserves, will be dictated to a large extent by government policy. The European Economic Community indicates (ref. 1.13) that natural gas supply and utilisation problems should be approached with the same thinking as the oil problem. With regard to the optimum economic use of natural gas, the Community calls for some regulation of use over and above that dictated by normal market forces. It is recommended that the use of natural gas in new thermal power stations should be made subject to arrangements requiring prior authorisation in order to keep fuel for applications where it is used to better advantage. Likewise, its present consumption in such stations should be reduced over a reasonable period.

Of more significance to industry is the belief that contracts should be introduced for large industrial consumers which can be interrupted and a "harmonised policy of prices and price scales at Community level of a sufficient degree of 'transparency'", should serve to ensure that natural gas utilisation conforms to the broader aims of any Community energy policy, and in part it appears that this policy is directed at encouraging the production of substitute natural gas.

1.3 Oil

The European Economic Community have proposed that their policy with regard to oil as a primary energy resource be based on the following factors, (ref. 1.13):

(i) Oil will continue to be basic element in the supply pattern, as demand will continue to grow until replacement energy sources are fully developed. In 1985 it is estimated that oil will represent 41 per cent of the Community's total energy supplies, (61 per cent in 1973), thus retaining its position as the largest energy source.

(ii) The oil producing countries are progressively increasing their control over oil production. This can be seen in the Middle East, where a number of the United States or multi-national oil companies are seeing their facilities being taken over by the countries in which they are situated. (In the United Kingdom the government stake taken in off-shore wells is also indicative of this trend).

(iii) New oil deposits, particularly those in the North Sea, will decrease reliance on traditional extra-Community suppliers.

(iv) The trend in (ii) above is affecting the role played by the multi-national oil companies, although the Community see them as taking a continuing part within a changed framework.

(v) Oil requirements of other consumer countries will have a growing influence on the world market.

The policy recommended by the Community as a result of these factors includes the need to develop secure resources and contingencies for dealing with shortages and supply difficulties.

The importance of oil as an energy source for the European Community may be illustrated with reference to Table 1.4, derived from the same source as Table 1.2, which lists the oil consumption of the major EEC countries as a percentage of the total fuel demand, (prior to the oil crisis).

With regard to world consumption of oil related to known reserves, the picture changed abruptly in 1973 as a result of the oil crisis, and whereas predictions made prior to this event were forecasting a 6 per cent per annum growth rate, by the beginning of 1975 the growth rate had been considerably modified, and the trends illustrated in Fig. 1.3 show the effect of the

Current Primary Energy Resources

TABLE 1.4   National Demand for Oil, 1972

Italy	79%
France	70%
The Netherlands	58%
Germany	58%
United Kingdom	52%

the reduction in demand for oil brought about by the increased cost (ref. 1.11). The reserves/production ratio, initially predicted to drop to 15:1 in the early 1980's, is now unlikely to decrease to this value until 1990-2000, and the further drop in oil utilisation in countries such as the United Kingdom, brought about in part by government encouragement to economise, may delay this date still further.

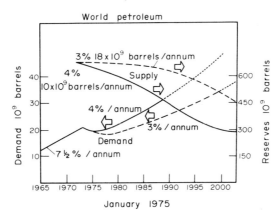

Fig. 1.3   World petroleum demand and reserves, September, 1973

Estimates of the total quantity of recoverable oil reserves in the world have been regularly carried out for many years. In 1949 these estimates varied between 1000×10^9 and 1500×10^9 barrels. By the mid-1960's, this figure had risen to 2000×10^9 barrels plus, and in 1973 estimates remained around this value (refs. 1.10, 1.14). (To put this figure in context, cumulative world oil consumption to 1973 approached 300×10^9 barrels).

While the oil scene has, over the past three years, been dominated by political actions in the Middle East, and, at least for the United Kingdom, the exploitation of reserves in the North Sea, there appears to be a growing belief that dramatic unilateral action which can seriously affect the market for the fuel is undesirable, and we can hope for a period of calm and reasonably steady prices, albeit rather high ones.

1.3.1 Oil shale and tar sands   Liquefaction of coal has already been briefly described as a means of producing a replacement for some oil fractions. However, the reserves of natural oil are not restricted to the conventional oil wells, be they land based or under the sea. Oil extracted from shale (and to a lesser extent from tar sands) is regarded by many as a fossil fuel resource having a long term potential comparable with that of coal, (refs. 1.15, 1.16). The total US reserves of oil extractable from shale are approximately fifteen times as great as the estimated reserves of crude oil which could be extracted from wells, and the total world reserves, expressed as a ratio of oil/shale oil, are about 1:50 (ref. 1.17).

Recovery of the shale by strip-mining techniques similar to that used by the coal industry is possible, but the technology needed to extract the oil from the shale in commercial quantities requires development. A tonne of shale may contain between 45 and 300 litres of oil, and obviously vast quantities of rock will be a visible by-product of the process.

Tar sand treatment has been developed to a commercial level in one State in the United States, but the reserves of oil in this form are comparatively small.

1.3.2 Liquified petroleum gas (LPG)

Over the past decade the sales of liquified petroleum gas in the United Kingdom have doubled, total sales amounting to 1 500 000 tonnes in 1975. Its use in the United States is much more widespread, and France is the largest European consumer. However, in the United Kingdom the effect of landing crude North Sea oil is likely to have a very significant affect on the market, Grangemouth alone receiving about 1 million tonnes per year.

LPG originates from three major sources: from crude oil being brought to the surface, from natural gas when associated with crude oil, and as a by-product in refinery distillation, cracking and reforming plant. It is more commonly known in its two constituent forms, propane and butane, the former being available in much greater volumes. The popularity of LPG is only hindered by problems associated with its storage. Current storage equipment costs (British Petroleum figures) are of the order of £120 per tonne of gas for propane, and half this figure for butane. Bulk price (May 1976) for LPG is £70 per tonne or 14p per therm, comparable to natural gas, but oil suppliers foresee LPG available at £50 per tonne from 1977 onwards, making it a very competitive energy source.

With regard to applications, LPG has for many years been a basic feedstock in the United States plastics industry, and its use in this field is predicted in the United Kingdom. If bulk storage is used (small gas containers, i.e. bottled butane, costs 30 - 50p per therm), LPG can become an important outlet for the oil industry. Suggested applications include dryers, radiant space heating (now available in the form of portable flameless combustion catalytic heaters), heating duties in foundries, and in transport. The gas industry currently uses large quantities in new gas manufacturing processes, and this is likely to continue.

Summarising, oil consumption is now being increasingly influenced by its price, and the traditional oil producing countries are noticing a considerable reduction in demand, which will be aided in the future by the growing use of North Sea oil in the United Kingdom, and in the longer term, the impact of new energy sources. It may well be recognised in the years to come that the oil price rises in 1973 provided the major impetus for energy conservation, and was therefore of considerable benefit to all of us.

1.4 Nuclear Fuel

The generation of electricity using nuclear (fission) energy is becoming increasingly popular, particularly in the United States, and predicted growth rates for nuclear power stations are in some cases very impressive. For example, Fig. 1.4 shows the projected electrical power production in the United States in terms of the primary fuel resource used for this purpose. It is estimated that by 1990 over half of the United States electricity generated will be supplied by nuclear power stations (ref. 1.18).

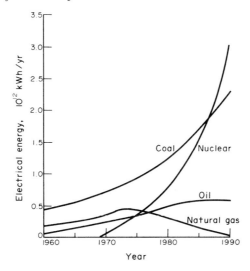

Fig. 1.4 Projected electricity production in the United States in terms of fuel category

It is often not appreciated that the fuel used in nuclear reactors is, like coal, oil and gas, a natural resource which can be depleted quite quickly if rashly used. While some data suggests that reserves of low cost uranium will be seriously reduced by the mid-1980's (ref. 1.19), the projected building programme for nuclear power stations in the United States alone would at first sight appear to neglect the implications of this fact, as shown in Fig. 1.4. There are some factors which suggest that the reserves of uranium could be extended, at additional mining cost, and a considerable reliance is put on the breeder reactor system, which generates fissionable plutonium from the uranium fuel. In addition, thorium, which can be used in high temperature gas-cooled reactors, has been largely unexploited to date. Thorium may also be used in gas-cooled and molten salt-cooled reactors to produce uranium.

Unlike other fuels, uranium resources are categorized in terms of the estimated cost of removal, and it is widely accepted that material available at a price of less than \$20/kg U_3O_8 in concentrate form classifies as ore. Other price catagories used to express reserves are \$20 - \$30/kg U_3O_8 and \$30 - \$60/kg U_3O_8. Thorium estimates are based on material available at a price of \$20/kg ThO_2 in concentrate form (ref. 1.18).

The estimated world reserves of uranium costing less than \$20/kg (excluding Communist countries) are shown in Table 1.5.

TABLE 1.5 Estimated World Reserves of Uranium
Price < $20/kg (ref. 1.20)

	Reserves (10^3 tonnes)	Estimated Additional Resources (10^3 tonnes)
Australia	71	78.5
Canada	185	190
France	36.6	24.3
Gabon	20	5
Niger	40	20
South Africa	202	8
United States	259	538
Others	53	52
TOTALS	866	916

From this table it can be seen that Canada, South Africa and the United States possess the highest amounts of low cost reserves. In South Africa the low cost ore is available only as a by-product of gold mining. In order to be able to relate these reserves to utilisation, it is necessary to have data on the uranium production to date. 1959 was a peak year, with a total production of over 34 000 tonnes, and in 1972 production approached 20 000 tonnes. The estimated total production capability by 1980 of the countries listed in Table 1.5 is 60 000 tonnes per annum.

Resources in the price range $20 - $30/kg U_3O_8 amount to 680 000 tonnes, somewhat less than that available at a lower cost. Reserves having a higher cost are available, but would probably only be used as a last resort.

Bowie (ref. 1.19) concludes that there will be adequate uranium to ensure that there will be no shortages during the 1970's. The problem in the 1980's may be a lack of refining capacity rather than a shortage of the basic raw material, and the current over-supply which is keeping cost down is not assisting the exploration of new reserves which will be required in the late 1980's and early 1990's. It is considered that uranium prospecting holds the key to the future, at least until the year 2000, if nuclear power is to become the major source of electricity. The implementation of breeder reactor construction programmes on a large scale may not be as urgent, but it is envisaged that an increasing number of units of this type should come on-line in the 1990's if the future of nuclear power beyond that date is to be assured.

1.5 Industrial Implications of Energy Resources

We are all aware of the exhortations to save energy, be it used by industry or in the home. The implications of a rise in the cost of a primary energy resource, the most dramatic case being that of oil, have been considerable, meaning a price rise in many industrial products and, to most of us, being emphasised by the increase in the price of petrol.

In the case of oil, this rise is not attributable to market forces moving as a result of a shortage. As yet, we have not been subjected to changes in the price of basic energy forms for this reason, and if sufficient international goodwill, combined with prompt action and investment in new energy resources, is shown to exist, this need never occur.

Obviously government action and policy at both national and international level, is likely to have an increasing affect on the price and availability of existing and future energy resources. In the United States vast sums are being allocated to the better utilisation of gas, oil, and in particular coal. In addition nuclear and solar energy (see Chapter 10) are receiving special attention. In the United Kingdom, we are now overcoming our initial inertia and a number of promising projects associated with new energy sources are being supported. While the current emphasis on exploitation of large coal reserves in the United States has not yet spurred sufficient action here, this will come in time, and our technology in the areas of coal liquefaction and gasification is second to none.

In the Introduction it was emphasised that the energy manager in industry would be most concerned with efficiency within his own plant, rather than the global energy picture: however, it is believed that the data presented so far in this Chapter, together with that in Chapter 10, will serve to emphasise that energy is not inexhaustible, and while the energy manager is not expected to be a philanthropist, this data may put the energy conservation campaign in perspective.

REFERENCES

1.1   Energy Trends. Publ. Economics and Statistics Division, Department of Energy, UK, Feb. 1976.

1.2   Swiss, M. Energy and the Future. Chartered Mechanical Engineer pp 99 - 104, Sept. 1974.

1.3   Senior, J. Coal Utilisation in Industry. Heating and Air Conditioning Journal, July 1974.

1.4   Rubin, E.S. Research and Development Needs for Enhancing US Coal Utilisation. Paper 749160, 9th Intersoc. Energy Conversion Engineering Conference, San Fransisco, August 1974.

1.5   Energy for Europe: Research and Development. Bulletin of the European Communities, Supplement 5/74, Brussels, April, 1974.

1.6   Coal Industry Examination, Final Report. Department of Energy, UK, 1974.

1.7   Coal gasification. Mosaic (Publ. US National Science Foundation), 5, 2 (1974).

1.8   Grainger, L. Future Trends in Utilisation of Coal Energy Conversion. Energy Digest, pp 3-5, Jan/Feb. 1974

1.9   Coppack, C.P. Energy in the 1980's. Natural Gas. The Royal Society London (1973)

1.10 Hawthorne, W.R. Energy: A Renewed Challenge to Engineers. Proc. Inst. Mech. Eng. 189, 52/75, Dec. 1975.

1.11   Varley, E.G. Minutes of Evidence to the Select Committee on Science and Technology, (Energy Resources Sub-Committee) HMSO, London, 12 March, 1975.

1.12   Rooke, D.E. Future Trends in Gas Production and Transmission. Phil. Trans. of The Royal Society, London, A276, 547-558 (1974).

1.13   Towards a New Energy Policy Strategy for the European Community. Bulletin of the European Communities, Supplement 4/74, Brussels (1974)

1.14   Drake, E. Oil Reserves and Production. Phil. Trans. of The Royal Society, London, A276, 453-462 (1974)

1.15   Hottel, H.C. and Howard, J.B. New Energy Technology - some facts and Assessments. Mass. Inst. Tech. (1971).

1.16   Dalal, V. Environment, Energy and the Need for New Technology. Energy Conversion, 13, 85-94 (1973).

1.17   Minerals Yearbook, US Bureau of Mines, Publ. U.S. Dept. of the Interior, (1969).

1.18   Westinghouse Electric Corporation, in Advanced Power Cycles. Hearings before the Committee on Interior and Insular Affairs, US Senate, 92-21 pp 243-256, 1971.

1.19   Bowie, S.H.U. Natural Sources of Nuclear Fuel. Phil. Trans. of The Royal Society, London, A276, 495-505 (1974).

1.20   Anon. Uranium Resources, Production and Demand. NEA/IAEA Compilation, OECD Paris (1973).

Optimum Use of Prime Movers for Power Generation and Other Applications

As well as being provided by national authorities, in the United Kingdom the responsibility for this being held by the Central Electricity Generating Board (CEGB), electrical power can be generated by industry itself. At present the proportion of electricity which is provided by industry is comparatively small, and is more often associated with large plant, such as oil refineries, petrochemical works, and aluminium smelters, where requirements for electricity are often in excess of 100 MW. However systems are available which, if correctly applied, can prove economical when plant demand may be as low as a few hundred kilowatts.

There are a number of systems available for driving electricity generators. These include steam and gas turbines, diesel engines and reciprocating gas engines (some of which are converted diesel engines). These machines may also be used instead of electric motors for driving compressors and large pumps, and this chapter will also briefly discuss these applications. It may be preferable to use a steam turbine in conjunction with a gas turbine, known as a combined cycle, in which the gas turbine exhaust gas is cooled by supplying its heat for raising steam used in the second turbine.

In many instances, the arguments in favour of in-plant electricity generation rest solely on the fact that the waste heat can be effectively used, and that the heat and power generated can be utilised over long periods. Obviously in refineries and other plant where processes are in operation 24 hours per day, 7 days per week, this criterion can be readily met. In other processes where more than one, or at best two, shifts are rarely worked, the economics of the situation require more careful analysis. (Plant installed solely for security, i.e. to fulfil a standby duty, is not considered here).

With reference to use of the waste heat in these installations, a considerable variety of equipment is available for heat recovery in engine water jackets, oil coolers (of more interest on the largest engines), and exhausts, and these are discussed in detail in Chapter 7. Manufactureres of such equipment are listed in Appendix 4.

There are obviously a number of ways in which the efficiency of electricity generation could be improved, and in this area vast sums are being devoted to research and development in the United States, USSR and the United Kingdom. For completeness, brief descriptions of some of these systems are given in Chapter 10, but in general they are of little immediate significance to industry.

2.1 Electricity Generation by National Authorities

The steam cycle, as used in conventional electricity generating plant in all industrialised countries, is shown in Fig. 2.1. Heat from the combustion gases, obtained by burning a fossil fuel, is absorbed by the water in the boiler. The water evaporates and is superheated, and this steam is then expanded through a turbine, where it is converted into work to drive the generator. Having passed through the turbine, the steam is condensed using cooling water. The condensate is then brought up to the initial pressure in the boiler feed pump, heated, and returned to the boiler.

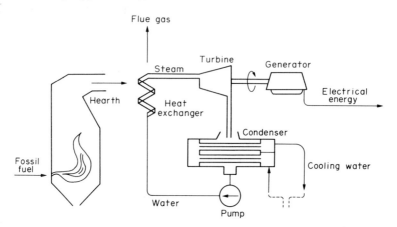

Fig. 2.1 The steam cycle

Preheating of the condensate prior to return to the boiler is carried out using steam tapped from the turbine. This is important because the efficiency of generation is a function of the temperature at which heat is supplied to the turbine. In large turbines, reheat is applied to the steam after it has passed various stages in the turbine by returning it to the boiler, thus improving efficiency still further.

The efficiency of electricity generation in a turbogenerator plant of this type is at best approximately 36 per cent, and most of the energy in the primary fuel is dissipated in the form of heat in cooling towers or as warm water discharged into rivers and estuaries. (In the United Kingdom over 20 per cent of primary fuel consumption is lost in this manner). Present status and future improvements in electricity generation are concisely presented in the study of energy conservation carried out in the United Kingdom by the Central Policy Review Staff, (ref. 2.1).

> "Both the more efficient generation of electricity and more effective use of the rejected heat are important subjects for energy conservation ... In considering the scope for improved efficiency a number of factors peculiar to the electricity generation industry have to be borne in mind. Firstly, the opportunities for improved efficiency are essentially short or medium term in the case of base-load fossil-fueled stations. A long-term programme of research and development is unlikely to be worthwhile if, by the time it comes to fruition, such stations are being progressively phased out ...
>
> No possibilities have been found for achieving immediate improvements in the efficiency of electricity generation. The late commissioning of several 500 and 660 MW generating sets ordered by the Central Electricity Generating Board (CEGB) since the mid-1960's and the still unsatisfactory performance of some of those already in operation remain a cause for concern .... This long-standing problem has helped

to hold down the CEGB's thermal efficiency by inhibiting the retirement of old plant, thus necessitating the excessive utilisation of plant of low or medium efficiency. The thermal efficiency of the CEGB's 500 and 660 MW stations ranges between 33 per cent and 35 per cent at full output compared with the current average of 29 per cent.... If the backlog of commissioning delays could be eliminated and more satisfactory availability obtained from the large modern plant, it is estimated that overall system efficiency could well rise by three percentage points...."

In the short and medium term it is highly unlikely that there will be any significant increase in electricity generation efficiency by the CEGB, and this of course also applies to large generating plant in other countries. As pointed out in the Central Policy Review Staff report, the cost of fuel still does not represent the true cost of investment in capital plant and production costs, and the implications of future rises to compensate for this are likely to be more serious for the national electricity generating bodies, particularly in the short to medium term. It is now no longer believed that nuclear power stations will be the panacea as far as low cost electricity is concerned, although the building programme in countries such as the United States (see Chapter 1) should help to stabilise electricity prices in the long term.

Industry is therefore unlikely to benefit from future investment in large power stations by national authorities in the medium term. Generation efficiency may increase by a few points each decade, but investment in research and development, and the problems associated with much of the large new plant needed to realise these economies offset to some extent the advantages of higher efficiency.

As mentioned above, the current thermal efficiency of the best large electricity generating stations is approximately 35 per cent. The major proportion of the balance of the energy lost in conversion to electricity is in the form of waste heat from the condensers. If the waste heat could be effectively harnessed, thermal efficiencies of 75 to 85 per cent would be possible, and the various ways of using power station waste heat have been the subject of much discussion. In the United Kingdom, there seems to be considerable reluctance in considering serious application of power station waste heat. The siting of greenhouses near power stations, where the thermal discharge can be used to promote growth, is being investigated on a limited scale, and turbine waste heat at one power station is upgraded using heat pumps for on-site conditioning of buildings. One notable exception has arisen as a result of work carried out by Kolbusz at the Electricity Council Research Centre, Capenhurst, on heat pumps, (ref. 2.2). As shown in Fig. 2.2, Kolbusz claims that by using a back pressure turbine (1) to drive the heat pump compressor (2), waste heat from condensing power stations can be usefully recovered. The waste heat, extracted in the condenser (3), which also serves as the heat pump evaporator, is used by the steam-driven heat pump as the first stage (4) in heating water. The second stage of heating (5) is carried out by the condenser of the back pressure turbine. Kolbusz showed that a coefficient of performance (COP) of 6.6 could be achieved, and with the combined operation, each tonne of coal burned in the back pressure system would make available 2 tonnes of coal equivalent of heat to a district heating scheme. (Items (6) and (7) are the boiler and feedpump respectively).

The advantages of such a system are numerous. The plant could be added to existing turbogenerator installations without interfering with normal operation,

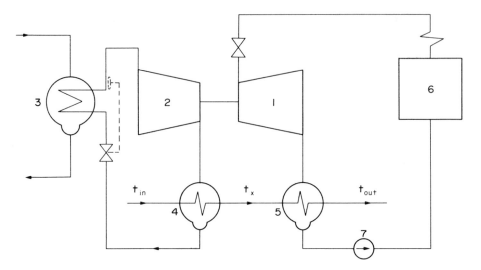

Fig. 2.2 Flow diagram of combined heat pump and back pressure turbine district heating scheme

and the efficiency of electrical power generation would not necessarily be adversely affected. The system could also be linked to a nuclear reactor, the reactor heat being used to raise steam directly for the back pressure turbine. At present it is envisaged that the heat source would be at 28°C and this would be upgraded to provide hot water at 70°C. No doubt by changing condenser conditions, even higher grade heat could be obtained, which could be used by industry for other process applications.

In the United States, the emphasis on power station waste heat has been directed largely at improving plant growth, although a number of other interesting projects have been studied. The Tennessee Valley Authority, (TVA), pioneers in this work, have reported on (ref. 2.3):

(i) Greenhouse horticulture

(ii) Soil or rooting media heating

(iii) Spray irrigation/frost protection

(iv) Organic waste treatment/algae production

(v) Aquaculture/mariculture

(vi) Poultry and swine house conditioning

(vii) Opening of shipping lanes

While this work is of considerable significance, yields of some crops being doubled using warm water irrigation, the benefits to industry of power station waste heat utilisation in most cases seem as distant as ever, and the concept of 'in plant' electricity generation with the user taking advantage of his own source of waste heat is likely to continue increasing in popularity for a large number of years.

Optimum Use of Prime Movers

2.2 Generation of power by industry

Considerable encouragement is given to smaller electricity generating installations with heat recovery by the Central Policy Review Staff:

"The really significant savings to be made arise in the fields of combined systems producing both heat and electricity ... Combined systems can make use of rejected heat for district heating and for industrial processes calling for low-temperature process heat. It is undesirable for environmental economic and safety reasons to site large nuclear and fossil-fueled power stations close to centres of industry and population and this has made it difficult and, in the past, uneconomic to use the rejected heat for industrial, commercial and domestic purposes. But significant scope could still exist for future plant of smaller size to be used in this way. (This implies that such plant would be administered by the CEGB or local authorities. However the arguments also apply to industrial-owned plant - Author). For the medium-sized plant, 'back-pressure' steam turbines with so-called 'waste-heat' recovery facilities offer attractive possibilities, while for small plant, diesel (and gas - Author) generators with waste heat recovery also offer potential energy savings. A 50 MW back-pressure steam turbine would be capable of providing 176 therms of electricity and 750 therms of heat for each 1100 therms of energy contained in the fossil fuel. This represents an overall efficiency of about 84 per cent, compared with 61.3 per cent overall for a central power station and a separate boiler to provide the heat. Similarly, a 1 MW diesel generator with waste heat recovery could provide heat and electricity in approximately equal proportions at an efficiency of 68 per cent compared with only 51 per cent for a central power station and separate boiler for the heat. The different efficiencies for the central power station plus separate boiler reflect the different ratios of heat to electricity in the two examples ...."

The number of factors which must be taken into account when a company is considering generating its own power are numerous, and careful technical and economic assessments must be made. Some of the major stumbling-blocks which can be encountered with the change to 'in-plant' generation are pointed out in an Electricity Council publication, 'Electricity Generation', (ref. 2.4), and are:

(i)   Initial costs: These can escalate considerably, and to the cost of the capital equipment must be added consultants' fees, civil engineering work, commissioning costs etc.

(ii)   Running costs: In addition to fuel, such items as cooling water may prove expensive; ancillary services, rates, insurance, and management time must also be taken into account. In some cases overhaul and maintenance costs may be higher than anticipated, but this depends largely on the type of system adopted.

(iii)   Reliability: Except for the effects of industrial action, national electricity authorities can supply industry with power with only a very slight chance that supplies will be interrupted. In the case of industrial units, this high reliability may not be achieved, and stand-by capacity will be necessary to meet unforeseen breakdowns. This ties up capital. (Of course, for electric power a switch could be made to the national grid in emergencies, but an additional on-site source for heat must be provided).

(iv) Changes in demand: A 'total energy' system has an optimum operating condition, which determines the ratio between heat and electricity produced. Changes in demand may lead to lower operating efficiencies, and while during the normal course of events these may be insignificant, a change in the process may alter the balance completely.

There are a number of general rules which can indicate the acceptability of total energy plant in an industrial situation. If the 'energy cost' or 'energy content' of the manufacturing process is significant, say more than 10 per cent of the total cost, it will probably be worth considering total energy plant. Other favourable factors include the likelihood of a high plant utilisation factor, the local availability of a suitable fuel at acceptable prices, and the ability to argue a case for added security of on-site energy production.

A number of other factors, such as servicing, space used, cost of capital and budgeting, must also be examined. On balance, it appears that on-site generation is desirable in most of the cases where both heat and electricity are required in large quantities, and the equipment now on the market meets the rigorous industrial requirements. This is a growth sector and new developments should greatly increase the popularity of total energy schemes in the future. The techniques available are discussed below.

2.3 Steam Turbines

The steam turbine provides the basis for the large power stations owned by national authorities and utilities, and its performance has already been discussed in some detail. Condensing type turbines are used for total energy systems, and they tend to use a lower steam rate per kW hr than non-condensing turbines. Steam for process and other applications can be extracted from the turbine at some point between the inlet and exhaust, depending upon the pressure required.

One of the drawbacks of the steam turbine for total energy plant in industry is the fact that ASME (in the United States) and other codes restrict the operating pressures for unattended equipment, and consequently in a small plant the steam turbine could well prove uneconomical due to the fact that it would require permanent manning during operation.

Also, while the steam turbine is essentially simple in construction, with a long life and high reliability if correctly used, supporting equipment is needed to ensure that the steam quality supplied to the turbine is sufficiently high. The need to maintain a high boiler feedwater quality, and to incorporate superheaters and the like, adds to maintenance problems as well as increasing capital cost. Further discussion on this topic is given in reference 2.5.

The generator is an integral part of any power system, be it solely used for electricity generation, or as a combined heat and electricity plant, and a wide range of standard generators are available. Advice on generator selection will commonly be made by the manufacturer of the prime mover; indeed in many cases involving steam and gas turbines and reciprocating engines, the generator may be included in the 'package' and it is therefore not considered necessary to discuss it in detail here.

An example of the use of steam turbines in an industrial process, resulting in more efficient plant operation, is given in data published by the Standard Oil Company (ref. 2.6).

In this particular plant, a surplus of steam was available, and along with this was the necessity for continuous operation of a number of steam turbine and motor drives. The critical nature of air cooler drivers had led to consideration of utilizing small (20 kW) steam turbines instead of the usual electric motors not an economical size for a steam turbine. In addition, other critical pumps were being considered for steam turbine drives partly to use up the surplus steam.

Energy requirements for this combination are shown on Fig. 2.3. Neither the steam nor electrical cycle were very efficient, the overall combined efficiency being 30 per cent.

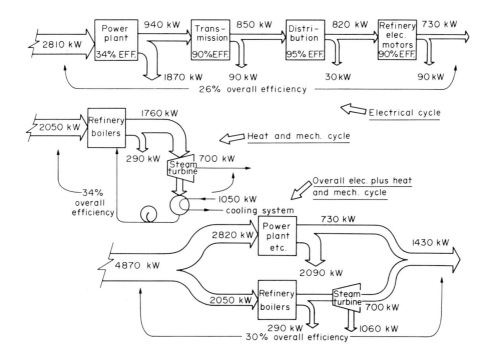

Fig. 2.3 Proposed system using small steam turbines at Standard Oil Company plant

The next step was to consider providing a high level of electric power reliability while still utilizing the available steam. As shown in Fig. 2.4, the alternative selected was to install a single large condensing steam turbine driving an electric generator, which in turn provided electrical energy for all of the critical drivers. The overall cycle efficiency increased to 48 per cent, a 60 per cent increase over the original plan. Estimated fuel savings were 8900 barrels per year of equivalent fuel oil.

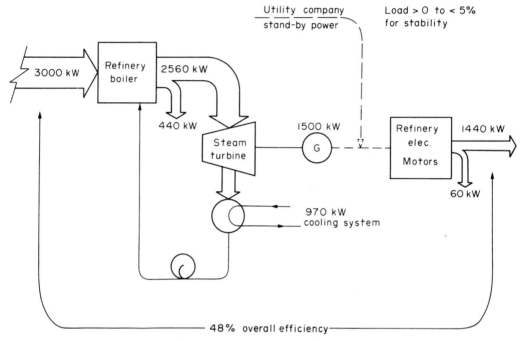

Fig. 2.4 Adopted layout at Standard Oil Company, giving high reliability and good efficiency

In order to provide the high level of electric power reliability, the plant generator is operated in parallel with the serving utility. Loss of the generator for any reason causes critical loads to be assumed by the utility system (or national grid) without interruption.

Using the utility system as the backup for the plant generator offered several advantages. Aside from better utilisation of energy and an improved steam balance, the arrangement is quite logical. Normally, the plants have a steam turbine emergency generator to provide backup for the utility electrical service. Standard Oil's experience with small infrequently operated in-plant generators has shown that the stand-by source is not always reliable. There have been a number of cases where the stand-by system failed to operate just at the time when it was most needed. On the other hand, utility systems are typically well over 99 per cent available. Thus, the move to reverse the stand-by arrangement adds to the reliability. There are, of course, the complications of interconnected local plant and utility electrical systems, coupled with stand-by power costs. In this case, these problems were resolved satisfactorily.

Even though the cycle efficiency is not outstanding, the alternative still offered an attractive investment as compared to the original plan.

Figure 2.5 summarizes how the added investment can be paid out, depending on the cost of power and fuel.

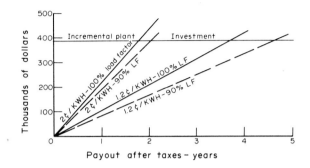

Cost of purchased power:	Load factor	1.2c/kW hr	2c/kW hr
Incremental plant investment $		380 000	380 000
Purchased power saving, $/yr	0.9	140 000	234 000
	1.0	156 000	260 000
Added fuel cost (for elec. generator), $/yr @ $1.40/kJ	0.9	350 000	35 000
	1.0	390 000	39 000
Added annual maintenance & operation costs, $/yr	0.9	12 000	12 000
	1.0	13 000	13 000
Net annual savings, $/yr	0.9	83 000	187 000
	1.0	104 000	208 000
Payback period before taxes, years	0.9	4.6	2.0
	1.0	3.7	1.8

Fig. 2.5 Economics of the system adopted by the Standard Oil Company

2.4 Rankine Vapour Cycle Turbines Using Freons

Water is by far the most widely used fluid in vapour cycle turbines, and the search for alternative fluids which could operate in small turbines at different conditions of temperature and pressure has been going on for a number of decades.

As early as 1912, nine alternative fluids had been tried in turbines (ref. 2.7). To date mercury has met with some success in turbines, particularly in binary systems where both mercury and steam cycles are used, the mercury condenser acting as the steam boiler. Reference 2.8 gives an excellent review of the requirements of vapour cycle working fluids, covering a wide range of operating pressures and temperatures. In the context of energy conservation, however, it is the turbine running on fluorocarbons such as Freon 11 or Freon 113 which is of interest.

Several factors have helped to generate interest in these fluids for turbines. Much more is known about the thermodynamic properties of these synthetic compounds, also, improvements in technology, particularly in the aerospace industries, have made highly reliable sealed systems possible. In parallel with this, the thermodynamics and fluid mechanics of turbines are sufficiently understood to allow designs for other fluids to be carried out. More recently, the desirability to use even low grade heat as effectively as possible to conserve energy means that new fluids having lower boiling points than water may be applied.

A Freon turbogenerator system is marketed by Ishikawajima-Harima Heavy Industries (IHI) of Japan, with outputs ranging up to 3.8 MW. Primarily designed to utilise process waste heat, the IHI 'Fron' Turbine System, shown in Fig. 2.6 is available as a packaged power unit for generating electricity or driving other machinery (omitting the generator).

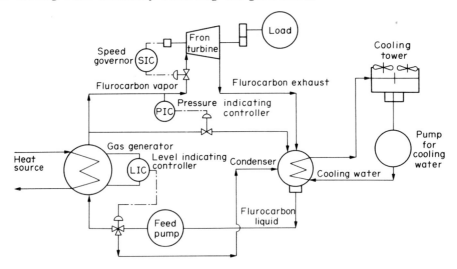

Fig. 2.6 Flow diagram of 'Fron' turbine system

Liquid fluorocarbon pressurised by the feed pump passes into the gas generator, where it receives heat from the heat source and vaporises under constant pressure. The high pressure vapour is then fed into the turbine, undergoing adiabatic expansion. Subsequently the exhaust is condensed and recycled through the feed pump.

A wide variety of heat sources have been suggested for this system, including low temperature waste steam, hot waste water and other liquid effluent, solar energy and geothermal energy. The 'Fron' system can also be incorporated in a combined cycle plant (discussed with respect to steam and gas turbines later in this chapter) to utilise waste heat from the steam turbine exhaust.

2.5 Gas Turbines

The principal advantages of the gas turbine are its compact size, being an internal combustion engine it needs no boiler and condenser, and the fact that the pressures achieved are much lower than those of the steam turbine cycle, (the air compressor typically achieves only about 10 bar).

A gas turbine cycle is shown in Fig. 2.7. The fuel, which may be one of a

Optimum Use of Prime Movers

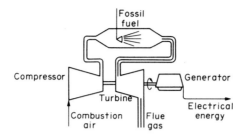

Fig. 2.7 The gas turbine cycle

variety of gases or light oils, is burnt in a pressurised combustion chamber in the presence of air supplied by the compressor. This combustion gas is then expanded to atmospheric pressure through the turbine, which drives the compressor and also the electricity generator. Most gas turbines, particularly those used for industrial duties, are of the single shaft type, with the air compressor and turbine mounted on a common shaft. However, it is possible to obtain split shaft units which use two turbine stages, one driving the compressor and the other providing shaft output for generator drive (or for other rotating equipment).

It is possible to examine the factors which affect the amount of shaft power that a gas turbine can produce with reference to the temperature-entropy diagram in Fig. 2.8, and the pressure-volume diagram in Fig. 2.9.

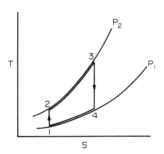

Fig. 2.8 Temperature-entropy diagram - gas turbine cycle

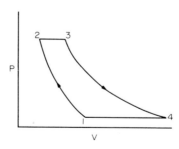

Fig. 2.9 Pressure-volume diagram - gas turbine cycle

The compressor work is : $W_{12} = Cp(T_2 - T_1)$

The turbine work is : $W_{34} = Cp(T_3 - T_4)$

The heat supplied in the combustion chamber is given by: $Q_{23} = Cp(T_3 - T_4)$

where T_1 = air inlet temperature to compressor
T_2 = air outlet temperature from compressor
T_3 = temperature of gas leaving combustion chamber
T_4 = turbine exhaust temperature
Cp = specific heat

The cycle efficiency may therefore be written as the ratio of the net work done (i.e. the useful work from the turbine) to the quantity of heat supplied.

i.e.
$$\eta = \frac{Cp(T_3 - T_4) - Cp(T_2 - T_1)}{Cp(T_3 - T_2)}$$

Alternatively the cycle efficiency of the gas turbine can be expressed in terms of the pressure ratio, P_2/P_1, denoted by r_p:

For isentropic compression and expansion,

$$T_2 = T_1 r_p^{(\gamma-1)/\gamma}$$
$$T_3 = T_4 r_p^{(\gamma-1)/\gamma}$$

where γ is the ratio of specific heats.

$$\therefore \quad \eta = 1 - \left(\frac{1}{r_p}\right)^{(\gamma-1)/\gamma}$$

The work ratio, r_w, is the ratio of the net work output to the total turbine work, and is given by

$$r_w = \frac{Cp(T_3 - T_4) - Cp(T_2 - T_1)}{Cp(T_3 - T_4)}$$

$$= 1 - \frac{T_1}{T_3} r_p^{(\gamma-1)/\gamma}$$

Thus while the ideal cycle efficiency (which neglects to take into account the isentropic efficiencies of the compressor and turbine) is a function of the pressure ratio, r_p, the work ratio depends upon temperatures T_1 and T_3. In a gas turbine, for maximum work output it is therefore desirable to have the air inlet temperature as low as possible, and the combustion chamber gas temperature as high as permitted by gas turbine materials etc. At present turbine inlet temperatures are limited to 850 - 950°C, giving overall thermal efficiencies approaching 30 per cent, with some industrial gas turbines

Optimum Use of Prime Movers

operating with efficiencies below 20 per cent.

The raising of the turbine inlet temperature could lead to thermal efficiencies of over 40 per cent, eliminating the major disadvantage of the gas turbine when compared with steam turbine plant.

The efficiencies of the turbine and compressor are typically in the range 80 to 90 per cent, and this can also have a significant affect on plant efficiency.

Apart from increasing turbine inlet temperatures, which can only be carried out using blade cooling and/or new materials (this is the subject of many development programmes, temperatures of 1300 to 1400°C being the aim), the use of inter-coolers, reheaters and regenerators can assist.

The application of an intercooler necessitates a two-stage compressor, as shown in Fig. 2.10. With reference to the temperature-entropy diagram, it can be seen that the compressor work, $Cp(T_2 - T_b) + Cp(T_a - T_1)$, will be less than that without intercooling, $Cp(T_2' - T_1)$.

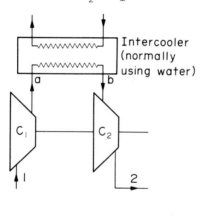

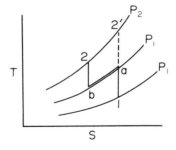

Fig. 2.10 A gas turbine system incorporating an intercooler between stages

Reheating involves at least two turbine stages, with a combustion chamber added between each stage. Again it can be shown that the turbine work output, $Cp(T_3 - T_c) + Cp(T_d - T_4)$, is greater with reheat.

A third technique for improving the efficiency of the gas turbine, and one which will depend on the users' requirements for shaft power and heat, is to use exhaust heat to raise T_2, the combustion chamber inlet temperature. Incorporation of all the above modifications results in the cycle shown in Fig. 2.11.

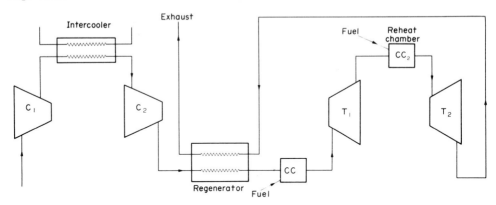

Fig. 2.11 A gas turbine incorporating intercoolers, reheat and regenerators

Some users of industrial gas turbines have opted for combined cycle units, in which the exhaust from the gas turbine is used to raise steam which then drives a steam turbine. Combined systems can operate at efficiencies in excess of 40 per cent, and are discussed in Section 2.6.

2.5.1 Industrial gas turbines

Because of its low thermal efficiency when used solely to provide power for driving generators etc., and the fact that it does not in general offer cost advantages over steam turbines and other prime movers such as diesel engines, industrial use of the gas turbine as a generator drive for continuous duty should only be envisaged when the exhaust heat can serve a worthwhile function. Thus the industrial gas turbine should form part of a 'total energy' package, particularly where fuel costs are high. (In this context the term 'industrial gas turbine' includes modified aero engines, although, as pointed out by manufacturers of industrial units designed from the outset as such, maintenance may be needed more regularly on aero engine-derived units. The Ruston gas turbine, designed solely for industrial use, will run for 20 000 to 30 000 hours between blade inspections and 80 000 to 100 000 hours between overhauls).

In the area of industrial energy applications, gas turbines power compressors, pumps, fans and electric generators. Their exhaust heat is used for steam generation, process energy, drying, space heating and air conditioning. The shaft power output of gas turbines ranges from 50 kW to 100 MW, with up to four times as much heat being available, depending upon the particular thermal efficiency of the machine.

The applications in industry may also be depicted as in the diagram supplied by Ruston Gas Turbines (Fig. 2.12). Ruston highlight two main areas where their 'total energy' units have been successfully used, and the comments appertaining to these applications are concisely described in their brochure (ref. 2.5). (It must be pointed out that total energy sets based on gas turbines form only a small proportion of the applications satisfied by Ruston Gas Turbines. Between 80 and 90 per cent of production has been on behalf of the oil and gas industry for use in pumping stations etc.).

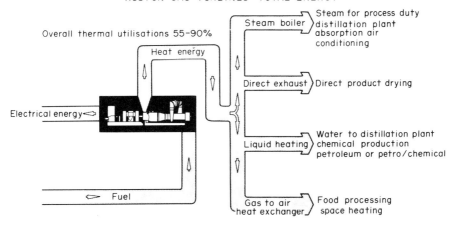

Fig. 2.12 Typical total energy applications using a Ruston Type TA industrial gas turbine as the single power and heat source

A second major manufacturer of gas turbines for total energy installations is Centrax Ltd. Details of the operating costs of these units, and the payback periods, are presented in Appendix 2 as part of an example on the economic appraisal of energy conservation equipment. However, more general points related to the use of industrial gas turbines of this type are included below.

It is of course necessary to have a requirement for both steam (heat) and electricity in a plant before considering a total energy set, and in the case of a gas turbine the low thermal efficiency as a prime mover, possibly as little as 17 per cent, suggests that the heat/power ratio should be of the order of 4:1 or higher. Thus in a gas turbine installation the heat generated is of considerable importance in making an economic case for installation. Quantifying heat and power outputs, a Centrax CS600-2 unit in its basic form requires a fuel input equivalent to 3100 kW to produce 500 kW of electrical power. Taking into account small losses, the waste heat in the turbine exhaust is 2500 kW, which could be directly employed in, for example, a drying process. This would raise the overall efficiency to 65 per cent.

Alternatively, considerable quantities of steam may be raised. It is common practice to supplement the exhaust heat capacity of the gas turbine, making use of the oxygen present, by supplementary firing. Applied to the Centrax unit, a supplementary fired waste heat boiler could produce approximately 3.5 kg per second of steam, increasing the overall efficiency to almost 80 per cent. (This assumes a boiler efficiency of 80 per cent). The heat input via auxiliary firing in this case is considerable, accounting for 75 per cent of the total heat input to the boiler. However it does ensure optimum use of the waste heat produced by the gas turbine, although the total heat/power ratio is

16, possibly rather high for most industrial applications. If the gas turbine set is only providing a portion of the total electricity requirements of the plant, the balance can be more easily accommodated, however.

The economics of a system such as the above appear very favourable, Centrax estimating payback periods of between 12 300 and 12 400 hours, even with allowances for maintenance costs. Leasing arrangements, discussed in Appendix 2, ease the burden even further.

Ruston see the gas turbine set in industry as alleviating some of the problems associated with stand-by generators. Stand-by sets are most commonly based on diesel or gas reciprocating engines having a lower capital cost when compared with some industrial gas turbines. As utilisation of stand-by generators is low, maintenance is also inexpensive and long life reliability is not an important consideration. A gas turbine, however, is able to fulfil both continuous and stand-by duties. Even in industries where the power requirements are high, the relationship of the generation capacity needed under emergency conditions to the heat capacity under normal operating conditions could well fall within the range of power/heat ratios available from a total energy system based on the gas turbine. Thus gas turbine units sized for the emergency power requirement can be operated to supply a proportion of the base load electricity needs while also supplying the heat requirements, with or without auxiliary firing. In an emergency situation, the gas turbines would of course, maintain their thermal duty.

The economics of total energy gas turbine operation may also be examined with reference to experience at the Standard Oil Company, (ref. 2.6). This particular plant possessed a large furnace. When electricity purchased from utilities and national authorities was comparatively cheap, all plant electric motors were supplied with power from outside the refinery, and although the furnace efficiency was high, the combined conversion efficiency of furnace and power station → electric motor is low, as shown in Fig. 2.13.

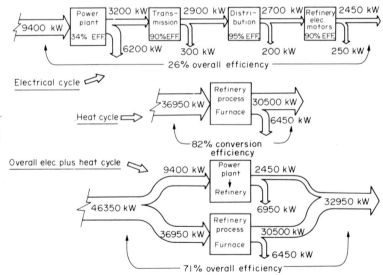

Fig. 2.13 Standard Oil Company electrical and steam cycles before installation of gas turbine plant

A more attractive scheme, selected for installation in a Standard oil plant in 1975, is shown in Fig. 2.14. In spite of the fact that the furnace was already installed, a gas turbine unit was conveniently fitted local to the furnace, ensuring only the minimum amount of ducting between the two units. This is attributed to the compactness of the gas turbine, and its low auxiliary system requirements. The gas turbine was selected so that its exhaust gas stream met the combustion air requirements of the furnace.

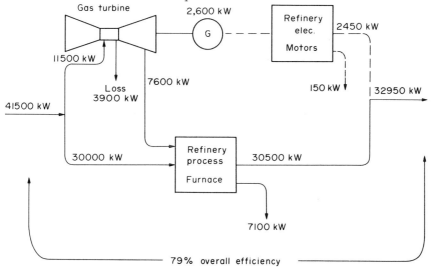

Fig. 2.14 On-site generation of electricity using a gas turbine, waste heat being applied to furnace preheating

The relatively large generator eliminated the need for other steam turbine emergency generators that would otherwise have been required. In this case, the gas turbine generator supplies critical motors normally connected to the refinery emergency power system. The utility supplied electric power thus becomes the 'emergency' source, i.e., critical motors are transferred to the utility system during short periods when the gas turbine generator is inoperative. After adding the gas turbine generator, the overall cycle efficiency increased from the 71 per cent shown in Fig. 2.13 to 79 per cent shown in Fig. 2.14.

Economics applying in this example are shown in Fig. 2.15. The savings involved are modest compared to a large refinery system. On the other hand, there are a number of possibilities in any typical refinery, so in aggregate they can add up to significant savings.

This concept can be utilized to provide a considerable share of the refinery's electric power if several units are operated in parallel. One normally thinks in terms of centralized power generation, but with modern speed control governors, fast-acting voltage regulators and reliable protective relaying, multiple units located strategically in various process plants are a real possibility.

A second assessment of the technical and economic factors in the selection of a gas turbine total energy package, this time based on a Centrax 500 kW generator set, is presented in Appendix 2.

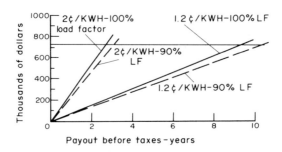

	Load Factor	1.2c/kW hr	2c/kW hr
Cost of purchased power:			
Incremental plant investment $		724 000	724 000
Purchased power savings, $/yr	0.9	246 000	411 600
	1.0	271 000	452 400
Added fuel cost (for elec. generator) $/yr @ $1.40/kJ	0.9	164 000	164 000
	1.0	180 200	180 200
Added annual maintenance & operation costs $/yr	0.9	12 000	12 000
	1.0	13 000	13 000
Net annual savings, $/yr	0.9	70 000	235 600
	1.0	78 000	259 200
Payout before taxes, years	0.9	10.3	3.1
	1.0	9.3	2.8

Fig. 2.15 Economics of a gas turbine generator with waste heat used for furnace preheating

2.6 Combined Cycles (Gas Turbine Plus Steam Turbine)

If the requirements of an industrial plant demand considerable generation of electricity in preference to very large quantities of waste heat, the use of a gas turbine alone to drive the generator would be uneconomical. However, because of the high ($\simeq 600\,^\circ\text{C}$) exhaust temperatures of the gas turbine, its waste heat can be used to raise steam, which in turn can drive a second turbine. This combined cycle, involving gas and steam turbo-generator plant, is illustrated in Fig. 2.16.

Use of this system can raise the thermal efficiency to 42 to 44 per cent, and it retains several advantages over the isolated steam turbine installation. It is possible to start up and shut down this combination rapidly, and if a number of gas turbines are used in conjunction with a single steam turbine for generating large powers, high efficiencies can be maintained even if the load is sometimes as low as 75 per cent of peak value.

The potential of the combined cycle will be even greater when gas turbine inlet temperatures can be raised well above $1000\,^\circ\text{C}$. It has been predicted that efficiencies in excess of 50 per cent are achievable when high inlet temperature units are combined with steam plant.

The type of steam turbine used in conjunction with the gas turbine depends upon the duty. If electricity generation is the most important consideration, a condensing steam turbine will be used. If in addition there is a need for

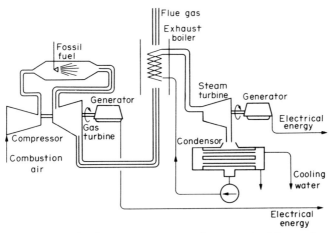

Fig. 2.16 Combined power cycle, using gas turbine exhaust to raise steam for driving a second turbo-generator

process heat in the form of steam, a back-pressure turbine, in which the steam exhausts into the condenser at a relatively high pressure, is used. Alternatively, an extraction (or pass-out) steam turbine can meet this second requirement; in this system steam for process use is extracted between turbine stages at the required condition.

Several operational units using the combined cycle are described in the literature (refs. 2.6, 2.9, Appendix 5), and the first such installation in Europe was commissioned in Germany in 1973, (ref. 2.9). With a peak thermal efficiency of 39.8 per cent using coke oven gas as fuel, and 42 per cent using natural gas, this system has a net output of approximately 35 MW, the emphasis in this instance being on power generation. Apart from the increased thermal efficiency compared with an isolated steam turbine, other advantages claimed are:

- (i) Faster start-up capability, full load being achieved in 30 minutes.

- (ii) Lower cooling water consumption, only about 33 per cent of that of a corresponding conventional steam turbine plant.

- (iii) Lower atmospheric pollution (NO_x)

- (iv) Lower cost of investment per kilowatt installed. In this case the installed cost has been quoted as just less than $100 per kilowatt.

Obviously it will be possible to 'ring the changes' when selecting a combined cycle system, depending upon the power/heat requirements. At present operational experience in Europe is limited, but there are a number of equipment manufacturers producing both prime movers, and in many cases the combined cycle will prove more attractive than isolated steam or gas turbines.

2.7 Diesel and Gas Engines (Reciprocating)

As with gas turbine and steam turbine generators, it is much more desirable to be able to use the waste heat generated in reciprocating prime movers than to exhaust it directly to the air or coolant water. A similar argument applies to the use of these engines to drive air compressors, heat pumps etc., where they are considered as replacements for electric motors.

Reciprocating engines have several advantages over the gas turbine when a substantial mechanical output is required. In general their shaft efficiency is greater than that of the turbine, ranging from 30 per cent to 35 per cent, and their cost is less (although the trend is towards an equalisation of cost between reciprocating and turbine units). On the debit side, the ancillary systems required for reciprocating engines tend to be more complex particularly in view of the water cooling required, and oil consumption can be considerable. Units have a low power to weight ratio, and exhaust gases can be a source of considerable atmospheric pollution if maintenance is neglected. Maintenance is one of the features which, certainly in the past, has held back wider use of reciprocating engines in industry, in particular where very high utilisation is required, and this is discussed in some detail later in this section.

There are two basic types of reciprocating engine, dual fuel and spark ignition units. Dual fuel in this context applies to an engine which runs on the diesel cycle, using a small amount of diesel fuel as the igniter when using natural gas as the main fuel. This results in simultaneous combustion of atomised liquid fuel and gas, and the engine has similar economy and emission characteristics to the pure natural gas engine. In instances where a company may have interrupted natural gas supplies, this type of engine can be switched immediately to 100 per cent diesel fuel, without swtiching off the power plant. (In some literature on these engines, 'dual fuel' implies that the unit is capable of operating using a variety of gaseous fuels, but the use of the terminology in this context is not usual). The diesel fuel requirement to initiate combustion in a dual fuel engine is approximately 5 per cent to 10 per cent of full load diesel requirement.

Spark ignition engines have in the past been restricted to use with petrol and diesel fuels, particularly in the United Kingdom. This is because the price of town gas, and its high hydrogen content, restricted the performance of spark ignition engines. Only very low compression ratios were obtainable, but with natural gas the compression ratio can become high before detonation occurs. As a result, large spark ignition gas engines are mainly manufactured in continental Europe and the United States, most United Kingdom-made types being dual fuel.

In general, the use of natural gas as the fuel is preferred in industrial applications. It burns cleanly, hence reducing maintenance of the engine as no carbon is deposited within it. No oil dilution occurs and hence the life of the lubricant is considerably extended. As the main constituent of natural gas is methane which has a high 'knock' resistance, compression ratios of up to 12:1 can now be obtained in spark ignition engines, leading to high efficiencies. It is easy to make the fuel-air mixture entering the engine homogeneous, so that cold starting is possible without enrichment. Also, the improved fuel-air mixture composition results in more economic running and lower levels of exhaust gas toxicity, (ref. 2.10).

In order to increase the output of these engines, it is common to use turbo-charging. Ths shaft power delivered by a turbo-charged high-compression (10.5:1) engine may be as much as 35 per cent greater than a naturally aspirated low compression (7.5:1) unit having the same displacement. However because of the possible need for larger bearings and crankshafts, and more exacting cooling jacket design, the initial cost per kilowatt for a turbo-charged engine may approach that of a naturally aspirated engine rated at the same output.

TABLE 2.1 Gas Fuelled Prime Movers (ref. 2.13)

Gas Turbines		
Centrax	Industrial	500 - 700 kW
Ruston	Industrial	1.1 - 3.7 MW
John Brown	Industrial	3 - 54 MW
GEC-English Electric	Industrial	6 - 60 MW
Rolls Royce	Industrial	13 - 20 MW
	Gas Generator	13 - 25 MW
Waukesha		330 kW
Kongsberg		1.2 - 1.4 MW
Solar		60 - 2600 kW
Avco Lycoming		750 - 3000 kW
Sulzer		3 - 20 MW
Reciprocating Engines		
W.H. Allen	Dual Fuel	240 - 2300 kW
English Electric	Dual Fuel	600 - 2600 kW
Mirrlees	Dual Fuel	750 - 5000 kW
Ruston	Dual Fuel	1000 - 3500 kW
GEC Diesels	Spark Ignition	40 - 200 kW
Cummings		140 - 300 kW
Caterpillar		100 - 700 kW
Waukesha		100 - 1400 kW
White-Superior		100 - 1550 kW

Table 2.1 shows typical power outputs of reciprocating gas engines, (Appendix 4 gives addresses of manufacturers), and Fig. 2.17, (ref. 2.11) illustrates the effect of running at part load on the fuel economy of a natural gas engine.

For the lower power requirements, where the engine may be used for driving a small heat pump compressor, for example, the work being carried out at the British Gas Corporation Midlands Research Station is very significant, (ref. 2.13). A development programme has been running for several years directed at investigating the potential of gas engines based on conversions of mass-produced automotive petrol and diesel engines.

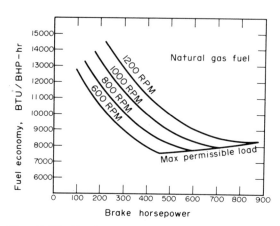

Fig. 2.17 Fuel economy of a natural gas engine running at part load

As a manufacturer of highly specialised natural gas engines for industrial purposes will stress, the conventional automotive engine is designed for a comparatively short life, is less reliable and hence more costly to maintain. However, a converted Ford diesel engine which can be used as a compressor drive, fuelled by natural gas, may have a capital cost less than 35 per cent of that of the equivalent purpose-built gas engine. The performance of some of these automotive engines when converted to natural gas is shown in Table 2.2 and more detailed performance characteristics of a converted Ford engine are given in Fig. 2.18. Conversion of the engines necessitate a new carburettor, gas carburettors being readily available from United States manufacturers such as Impco.

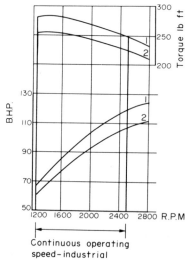

Fig. 2.18 Performance characteristics of an automotive engine converted to natural gas (Courtesy Power Torque Eng. Ltd.)

TABLE 2.2 Data on Automotive Type Engines When Fuelled
With Natural Gas (ref. 2.13)

Engine	Capacity cc	Cylinders	Maximum kW on natural gas at:		
			1500 rev/min	1800 rev/min	max. rated speed
					2400 rev/min
(a) Perkins 3.152	2500	3	23	26	29
Perkins 4.203	3340	4	30	36	40
Perkins 4.236	3860	4	37	42	49
					2100 rev/min
Ford 2503E	2590	3	20	25	28
Ford 2511E	3294	3	28	32	35
Ford 2513E	4196	4	35	40	47
					3000 rev/min
Rolls Royce B61	4880	6	46	57	89
Rolls Royce B81	6560	8	61	74	119
(b) Ford 2261E	1100	4	7.5	9	16
Ford 2264E	1600	4	11	14	28
Ford 2602E	2000	V4	17	22	37
Ford 2614E	3000	V6	25	29	52
					2250 rev/min
(c) Ford 2711E S.I.4	4150	4	33	40	49
Ford 2715E S.I.6	6220	6	51	62	74

(a) Typically used in heavy duty applications e.g. tractors, fork lift trucks, stand-by generators etc.
(b) Car engines, seldom used for stationary duties.
(c) Converted diesel engines, marketed initially for LPG operation.

Electronic governor systems may be used for constant, stepwise or variable speed control, and offer closer governing than the conventional mechanical type. These advantages must be weighed against considerably higher costs, however, and an electronic system may only be required in a small proportion of applications.

As with the large gas engines, heat recovery from the water coolant and the exhaust is possible, although on the smallest units it is worth examining the capital cost involved in, for example, an exhaust gas heat exchanger, as the quantities of heat involved can be quite small.

2.7.1 Heat recovery from reciprocating engines

The quantities of heat which can be recovered from natural gas reciprocating engines are shown in Table 2.3, and compared with gas turbines (which have significantly lower efficiencies) in Table 2.4. This data is presented in a different manner in Table 2.5, itemising the quantities extracted in the cooling jacket and engine exhaust, (ref. 2.12).

A system based on a diesel generating set, using both exhaust and coolant water heat, installed at the Petbow factory in the United Kingdom, is illustrated in Fig. 2.19, (ref. 2.14). The diesel engines are 500 kVA Rolls Royce units, one of the three installed always being kept on stand-by.

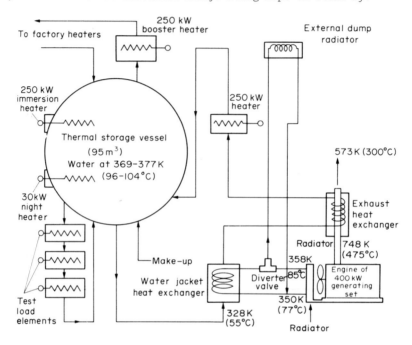

Fig. 2.19 Petbow diesel generator/waste heat storage combination, installed in 1975
(The Engineer, 23 Oct. 1975)

The system for recovering heat centres round a 95 m^3 (21 000 gal) thermal storage vessel. Here water is kept near boiling point by extracting heat in heat exchangers through which the engine radiator cooling water and exhaust gases are passed. Water from the storage vessel is then pumped to the factory areas to special heating units in each bay at roof level. Hot air is circulated round the bays by fans blowing cold air over hot pipes. The temperature in each bay is controlled individually by thermostats. Refinements include a 30 kW immersion heater to keep the water in the storage vessel hot overnight, 250 kW booster immersion heaters to provide quick start-up early in the morning, and an external radiator to dump heat during exceptionally warm spells.

The company aims to use its system to gather data on how total energy works

Optimum Use of Prime Movers

TABLE 2.3 Typical Reciprocating Gas Engine Inputs and Recoverable Outputs

Shaft Load (%)	75 kW; C 12:1; 1800 rpm NA; P. 11 mm Hg				175 kW; C 10:1; 1200 rpm TA; P. 100 kN/m²				225 kW; C 10:1; 1200 rpm TA; P. 100 kN/m²				450 kW; C 10:1; 1200 rpm TA; P. 100 kN/m²			
	Fuel rate (kW)	Outputs			Fuel rate (kW)	Outputs			Fuel rate (kW)	Outputs			Fuel rate (kW)	Outputs		
		S (kW)	H (kW)	S/H		S (kW)	H (kW)	S/H		S (kW)	H (kW)	S/H		S (kW)	H (kW)	S/H
100	293	75.0	148	1/2.0	638	175	247	1/1.4	805	225	311	1/1.4	1610	450	624	1/1.4
90	272	67.5	134	1/2.0	595	158	222	1/1.4	735	203	276	1/1.4	1475	405	556	1/1.4
80	253	60.0	121	1/2.0	545	140	198	1/1.4	688	180	250	1/1.4	1370	360	495	1/1.4
70	233	52.5	109	1/2.1	504	122	177	1/1.4	624	158	218	1/1.4	1240	315	433	1/1.4
60	214	45.0	97	1/2.2	457	105	154	1/1.5	562	135	189	1/1.4	1110	270	375	1/1.4
50	194	37.5	87	1/2.3	408	88	132	1/1.5	500	113	162	1/1.4	980	225	319	1/1.4

Notes: C – compression ratio
        NA – naturally aspirated
        TA – turbo charged
        S – shaft output
        H – heat output
        P – gas pressure

TABLE 2.4 Typical Gas Turbine Inputs and Recoverable Outputs

Shaft Load (%)	150 kW; C 3.4:1; 30 000 rpm; P. 690 kN/m²				200 kW; C 4.1:1; 35 000 rpm; P. 690 kN/m²				300 kW; C 6:1; 43 500 rpm; P. 1030 kN/m²				900 kW; C 4:1; 6000 rpm; P. 690 kN/m²			
	Fuel rate (kW)	Outputs			Fuel rate (kW)	Outputs			Fuel rate (kW)	Outputs			Fuel rate (kW)	Outputs		
		S (kW)	H (kW)	S/H		S (kW)	H (kW)	S/H		S (kW)	H (kW)	S/H		S (kW)	H (kW)	S/H
100	1690	150	820	1/5.5	2170	200	1080	1/5.4	1850	300	760	1/2.5	6630	900	3290	1/3.7
90	1615	135	815	1/6.0	2040	180	985	1/5.5	1670	270	700	1/2.6	6220	810	3040	1/3.8
80	1545	120	805	1/6.7	1890	160	895	1/5.6	1600	240	615	1/2.6	5800	720	2790	1/3.9
70	1475	105	790	1/7.5	1760	140	800	1/5.7	1490	210	585	1/2.8	5420	630	2550	1/4.1
60	1405	90	760	1/8.5	1640	120	710	1/5.9	1330	180	530	1/2.9	5040	540	2310	1/4.4
50	1330	75	700	1/9.3	1530	100	620	1/6.2	1240	150	470	1/3.1	4630	450	2080	1/4.7

Notes: C — compression ratio
        S — shaft output
        H — heat output
        P — gas pressure

TABLE 2.5 Typical Heat Recovery Rates of Various Types of Engines

(kcal/kW hr)
(based on 100 kN/m^2 steam pressure and 38°C ambient)

Type of engine	Fuel input at rated load	Heat recoverable at rated load & bmep		
		Jacket water		Exhaust unit
		Air-cooled manifold	Water-cooled manifold	
Two-cycle				
Mechanical supercharged gas	2800	590	750	410
Naturally aspirated gas	4100	1100	1300	510
Blower charged diesel	2800	560	660	370
Four-cycle				
Naturally aspirated gas	2900	650	790	420
Naturally aspirated diesel	2900	650	790	420
Turbo-charged diesel	2500	370	460	410
Turbo-charged dual-fuel	2200	320	370	340
Gas turbine				
Simple cycle	6200			4300
Regenerative cycle	4400			2500

in practice. A full set of instruments will permit continuous monitoring of data such as fuel and power consumption, and air temperature.

The total cost of the system is reported to be £200 000. The diesel engines will only be used to generate electricity when there is a demand in the plant for the waste heat from the engines, and the use of this energy for space heating suggested that it would be most economical to generate electricity 'in-house' during the day for eight months in the year, night and summer power requirements being obtained from the national grid. The electricity authorities were found to be most helpful in implementing this scheme.

There are many other applications where waste heat can be successfully utilised. As well as generating electricity, reciprocating engines can be used in air conditioning and refrigeration, and gas engine-driven chiller units are marketed on a wide scale, where modulation may be applied easily when variations in load occur. Drives for centrifugal compressors, which may have rotational speeds up to six times those of the gas engine, are another application. It has been found that to obtain the best compromise between equipment first cost (engine, couplings and transmission) and maintenance cost, engine speeds of 900 rev/min are recommended.

One of the most interesting applications of the diesel/gas engine is in driving a heat pump compressor. In this case the ability to use the waste heat to full advantage makes the system much more efficient on energy consumption than an electric system, and this is discussed fully in Chapter 7. Other applications are listed in references 2.12, 2.15 and 2.16.

2.7.2 Maintenance and reliability

Engine maintenance costs are not as easily estimated as fuel consumption, recoverable heat, or initial cost; they are, however, not entirely elusive. The increasing use of guaranteed maintenance and service contracts most common in the United States has eliminated much of the estimating formerly required in feasibility studies. Maintenance contracts vary from complete maintenance and service, including all parts, supplies and labour, to contracts that provide only a guaranteed cost for engine rebuild. For this reason, a maintenance cost figure is meaningless unless well defined. Complete maintenance costs are composed of three basic items:

(a) The miscellaneous maintenance and service cost, including service manual recommendations plus make-up oil (excluding labour to perform this routine duty).

(b) The overhaul maintenance cost, usually expressed in terms of cost per engine operating hour. This item should cover all labour and parts necessary to perform major and minor overhauls at the recommended intervals.

(c) The third item is the labour cost necessary to perform the miscellaneous service for Item (a).

Items (a) and (b) will vary considerably with the severity of service the engine must perform. For on-site power installations the conditions under which the engines operate, the quality of fuel, the routine maintenance the engine receives, are all usually considered to be good. Item (c) will vary with labour costs and the location of the engine plant with respect to the point from which service personnel must be despatched. Because of the many variables, it is difficult to provide realistic figures that would be useful for all applications. Maintenance costs should be based on past experiences in the area being considered.

In many applications of the gas engine, this past experience may be somewhat limited. In the United Kingdom, British Gas have installed several units comprising refrigeration compressors for air conditioning duties driven by Ford, Perkins and Pelapone automotive engines. Their experiences, reproduced below, (ref. 2.13) are very encouraging to potential users of these systems:

(i)   Long Eaton. 280 kW Unit This plant, installed in 1969, was operated for 2 years on town gas before conversion to natural gas. The Pelapone engine drives a Trane compressor, and the system has proved extremely reliable, and has required minimum attention.

(ii) Midlands Gas Research Station. 37 kW Unit A Perkins G4.236 water cooled, naturally aspirated gas engine was installed in 1970 to drive a 160 kW capacity Trane chiller for an air conditioning duty. To meet part load requirements the engine can operate at 1200, 1475 and 1750 rev/min. After 4000 hours of operation over a three year period at an average of 38 per cent full load, it was concluded that the replacement of an electric motor by this arrangement for driving a compressor was technically successful.

(iii) Eastern Gas Region. 88 kW Cooling Capacity Ford 2503 engines linked to Dunham Bush chillers are used at two offices. They

Optimum Use of Prime Movers

incorporate automatic start/stop, loading and unloading of cylinders, and stepwise speed modulation. Standard engine and chiller trips and controls have permitted unmanned operation.

Quantification of maintenance costs, and any other cost associated with the operation of plant, can be misleading because of the time factor involved, inflation making it desirable, where possible, to use the latest data available. Shearer, (ref. 2.17) presents figures based on 1976 costs for reciprocating engines running on gas oils (Class A/B), residual oils (Classes F,G) and natural gas (in dual-fuel systems), and he takes a low average value for the recovered waste heat of 0.45p/kW hr for the exhaust heat and 0.95p/kW hr for total heat recovery (including water jacket heat). The costs are given in Table 2.6 for reciprocating engines in the size range 1 to 2 MW.

TABLE 2.6 Operating Costs on Total Heat Recovery - 1 MW to 2 MW Unit Size of Engine (ref. 2.17)

Fuel Type	Class A/B	Class F	Class G	Dual fuel
Cost p/therm	13.4	11.4	9.6	11.0
Engine Speed rev/min	750	750-600	600-500	750-600
Capital Cost £/kW				
a) Elect/Mech. equipment	110-124	120-170	150-190	120-180
b) Switchgear and cables	12-20	12-20	12-20	12-20
c) Civil work erection cooling system	20		20	20
d) Total £/kW	142-164(153)		182-230(206)	152-220(186)
Fixed annual charges				
Insurance, Interest & Depreciation on $1\frac{1}{2}$ MW unit £/kW installed (average)	1.5304	1.8107	2.0608	1.8607
Operating Cost p/kW hr				
a) Fuel	1.292	1.004	0.850	1.100
b) Lub. Oil	0.050	0.053	0.056	0.050
c) Spares and Labour	0.027	0.033	0.036	0.023
d) Staffing	0.067	0.133	0.200	0.067
e) Total p/kW hr	1.436	1.223	1.142	1.240
Capitalisation p/kW hr				
a) 50 per cent utilisation (4380 hr)	0.349	0.413	0.471	0.425
b) 10 per cent utilisation (876 hr)	1.747	-	-	-
Gross generating costs p/kW hr	3.183	1.636	1.613	1.665
Waste Heat Recovery				
a) Exhaust only 0.45p/kW hr				
b) Total 0.95p/kW hr	0.950	0.950	0.950	0.950
Net Generating Costs p/kW hr *	2.233	0.686	0.663	0.715

*Using total recovery figures for heat value

2.8 Installation and Control Systems

Installation of many of the prime movers used for electricity generation and total energy schemes has been considerably simplified during the last few years by the adoption by many manufacturers of the 'packaged' system. The customer is required to provide all services for the 'package', but in many instances installation can proceed without significant interruption to the normal plant processes.

This comparatively straightforward routine is probably most common in gas turbine generation and total energy systems. When packaged units have to be designed for applications such as offshore oil rigs and production platforms, factors such as small size, low weight, and minimum installation and servicing are necessary requirements, and this is applied to many of the land-based systems.

The most common prime mover for total energy plant is the reciprocating engine, running on liquid or gaseous fuels. In general, the installation requirements for ensuring long life of these units are more demanding than with rotating prime movers. This is primarily due to the need to introduce cooling water into internal coolant passages in the engine, and possible complexities in the heat recovery systems used (see also Chapter 7). The installation therefore necessitates the use of water circuits, pumps and valves which can be internally cleared before use. Cleanliness of the coolant is also important, it must be non-corrosive and free from material which could be deposited on hot engine surfaces or form a sludge in areas of relatively slow fluid velocity.

Other factors to keep in mind when using water cooled engines are as follows:

(i) Ebullient cooling, because it involves evaporation in the engine, requires special precautions as far as water treatment is concerned.

(ii) In a multi-engine installation, separate cooling systems to each machine should be used to avoid total shut-down should one coolant system component fail.

(iii) Care should be taken to ensure provision for condensate removal, this forming in exhaust passages when the engine cools.

(iv) Water level control of separate engine cooling systems that produce steam must cater for the possibility of backflow through the system nozzle of units which are not running. Use of a steam check valve is recommended.

There are several publications which give excellent accounts of the many factors to be taken into account when installing this type of plant, and a manual such as the ASHRAE Systems Handbook provides much useful data, (see Appendix 5).

Control systems are commonly specified by the prime mover manufacturer, and, particularly in the case of packaged units, the control panel is an integral part of the 'package'. For example, the Centrax 500 kW gas turbine/alternator unit includes the following control features:

Fuel feed control

Speed governing

Maximum temperature control

Automatic shutdown and visual fault indication

Automatic voltage regulation

Starting controls and instrumentation

Control gear for independent or parallel operation with grid

2.9 Conclusions

Power generation by industry itself really makes sense from the point of view of energy conservation when applied as part of a 'total energy' package. An energy audit will indicate whether the correct 'mix' of electricity and heat is required in any particular plant, and as indicated by the Tables in Section 2.7, the selection of prime mover type allows considerable latitude in this ratio. Shearer, (ref. 2.17) gives an indication of the criteria on which prime mover selection may be based: if the heat/power balance is less than 0.6 kg steam per MJ (5 lb steam per kW hr) of electricity used on site, reciprocating engines are in favour. If greater, gas turbines or steam turbines are preferable.

Electricity authorities may argue against in-house power generation, and many of the technical and commercial considerations which have to be taken into account, listed in Section 2.2, at first sight appear to tip the balance in terms of the electricity authorities, however until these authorities are able to provide total energy plant in industrial locations, and this requires a major policy change, the growth of in-house generation is likely to increase.

REFERENCES

2.1   Anon. Energy Conservation. A study by the Central Policy Review Staff, HMSO, London, (1974).

2.2   Kolbusz, P. The use of heat pumping in district heating schemes. Electricity Council Research Centre Report ECRC/M700, Feb. 1974.

2.3   Furlong, W.K. and King, L.D. Beneficial uses of waste heat - status of research and development in the USA. Proc. Joint USA/USSR Conf. on Heat Rejection Systems, Washington, D.C., June 17 - 19, 1974.

2.4   Anon. Electricity Generation. Electricity Council Publication EC 3124, London, Oct. 1973.

2.5   Stocks, W.J.R. Total energy. Publicity Data of Ruston Gas Turbines Ltd., P.O. Box 1, Lincoln, England.

2.6   Levers, W.H. The electrical engineer's challenge in energy conservation. IEEE Trans. on Industrial Applications, 1A-11, 4, (1975)

2.7   Ennis, W.D. Vapours for Heat Engines. D. Van Nostrand, New York, (1912)

2.8   Wood, B.D. Applications of Thermodynamics. Addison-Wesley, Reading, Massachusetts, (1969)

2.9   Mitchell, R.W.S. and Gasparovic, N. Total energy. Steam and Heating Engineer, pp 34 - 39, July 1973.

2.10 Karim, G.A. and Souza, M.V. Combustion of methane with reference to its utilisation in power systems. J. Inst. of Fuel. 45, 335 - 339 (1972)

2.11 D'Amour, R.A. Natural gas engines as prime movers for air conditioning. SAE Paper 876A, Society of Automotive Engineers, New York, (1974)

2.12 Anon. Gas Engineers Handbook. The Industrial Press, New York, (1965)

2.13 Butler, P. A showpiece to save money by promoting waste heat recovery. The Engineer, 239, 23 Oct. 1975.

2.14 Horsler, A.G. et al. Natural gas fuelled prime movers in the United Kingdom. British Gas Corporation Research Report MRS E 235, Feb. 1974.

2.15 Jones, R.A. Caterpillar on total energy. Diesel Engineers and Users Assn., Publ. 345, London (1972)

2.16 Diamant, R.M.E. Total Energy. Pergamon Press, Oxford, (1965)

2.17 Shearer, A. Selection of prime movers for on-site energy generation. Power Generation Industrial, No. 3, May 26, 1976.

The Energy Intensive Industries - 1

The manufacturing industries in the United Kingdom account for approximately 40 per cent of the total energy consumed there. In the Netherlands the figure lies between 35 and 40 per cent and in the United States the proportion of energy consumed by industry is also similar to that in the United Kingdom.

Figures 3.1 and 3.2 show energy consumption on a heat supplied basis for

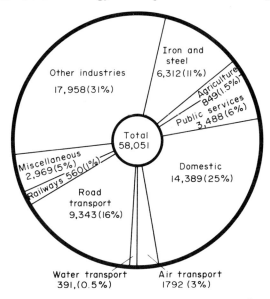

Fig. 3.1 Energy consumption by final users in 1972, on a 'heat supplied' basis. (Million therms) (HMSO Publication Figure)

all users in the United Kingdom, and the breakdown in terms of industries other than iron and steel, (ref. 3.1). Table 3.1 gives the primary energy consumption by various industrial sectors in the Netherlands for 1972, (ref. 3.2). Of the total energy used by the metal industry, 115×10^{15} J are accounted for by the iron and steel industry.

From the data in these figures and Table 3.1, it can be seen that the metal industries, particularly iron and steel, the chemical industry, and oil refining account for a large proportion of energy consumption in the industrialised countries. As a result, these industries have attracted most attention and financial support as far as the development of techniques for energy conservation is concerned, and this chapter discusses these industries in this context.

TABLE 3.1 1972 Primary Energy Consumption in Various Industrial Sectors in the Netherlands

Industry	Power Used (10^{15} J)	Percentage of Industrial Consumption	Percentage of National Consumption
Chemical Industry	272	34	12
Oil refining	177	22	8
Metal Industry	162	20	7
Food and allied products	69	9	3
Building materials	50	6	2
Paper Industry	30	4	1
Textile Industry	14	2	1
Others	21	3	1
TOTAL	795	100	35

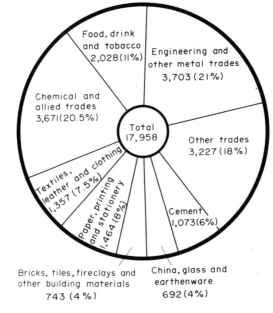

Fig. 3.2 Energy consumption of industries other than iron and steel in 1972, on a 'heat supplied' basis (Million therms). (HMSO Publication Figure)

3.1 Iron and Steel Production

As can be seen from examination of the figures at the beginning of this chapter, the production of iron and steel is the most significant user of industrial energy in the United Kingdom (and also in the United States). Iron and steel energy consumption also ranks high compared to most other

The Energy Intensive Industries - 1

industries in a considerable number of other industrialised countries. While at present the energy consumption of the iron and steel industries throughout the world is very high, (in 1970, for example, the world steel industry used the equivalent of 750 million tonnes of coal, or 11 per cent of the world's total energy consumption (ref. 3.3), of which 60 per cent was used in the form of coking coal), the future steel requirements are likely to be much greater.

Allen G. Gray, Technical Director of the American Society of Metals, writing in 'Metal Progress' (ref. 3.4), emphasises the important role that steel will play in most of the plant needed to develop and exploit existing and future sources of energy. In the United States alone, the projected growth in construction of nuclear power plant will require the use of approximately 2 million tonnes of high quality stainless steel between now and the year 2000. Steel equipment for the United States coal industry, the potential of which has been discussed in Chapter 1, and the exploitation of geothermal energy, will also involve vast quantities of material. Currently about 160 000 km of oil and gas pipelines are projected throughout the world, and as the search for fuel reserves moves further away from industrialised areas, transportation of gas, oil and coal, however it be accomplished, will make great demands on steel production.

3.1.1 Present energy use in steel making

The breakdown of energy consumption in the steelmaking processes in the United Kingdom is presented in a NEDO report on energy (ref. 3.5). As shown in Table 3.2, the past few years have shown an increase in the efficiency of energy use in this industry, brought about by several technological improvements and by the use of more high grade imported ore than in previous years. The major savings have been accounted for in the blast furnace operation, but more recently the introduction of basic oxygen steelmaking (BOS) and continuous casting of steel have contributed towards higher efficiency.

TABLE 3.2 Trends in Energy for Steelmaking (UK)

	1955	1960	1965	1972
Crude steel production, 10^6 tonnes	20.1	24.7	27.4	25.3
Net energy supplied, 10^6 GJ	710	768	759	623
Gross energy including electricity overheads, 10^6 GJ(a)	765	846	863	718
Specific net energy for crude steel, GJ/tonne(b)	35.3	31.1	27.7	24.6
Specific energy for crude steel, including electricity overheads, GJ/tonne(a)	38.1	34.2	31.5	28.4
Average ratio finished steel to crude steel	na	0.731	0.727	0.734
Specific energy for finished steel, including electricity overheads, GJ/tonne(a)	na	46.8	43.3	38.7

(a) Electricity overheads are calculated on the basis of 28 per cent conversion efficiency

(b) Specific net energy in the low production years of 1962 and 1966 was 31.9 and 28.4 respectively, these may be compared with those above for the peak production years of 1960 and 1965. Source: ISI Statistics

Table 3.2 shows that the production of steel in 1972 was lower than that in 1965 in the United Kingdom. This may be accounted for in part by more efficient use of steel, in particular high strength lightweight steels, in many applications.

The NEDO report also highlights the way in which the type of fuel used in steelmaking has changed in the past two decades. Table 3.3 shows the dramatic decline in the use of raw coal, countered by increased requirements for fuel oil and gas. It may well be that this trend will have to be reversed, unless the use of nuclear plant for supplying heat for the steel industry is advocated (see later).

TABLE 3.3 Fuel Substitution in Steelmaking (UK)

	% of total energy used by the industry			
	1955	1960	1965	1972
Coking coal and products	61	63	57	52
Non-coking coal	23	12	6	0
Fuel oil etc.	7	13	22	25
Electricity plus overheads[a]	7	10	13	16
Gas (town and natural)	2	2	2	7
TOTAL	100	100	100	100

[a] Calculated at 28 per cent conversion efficiency, coal is main fuel for providing electricity

Source: ISI Statistics

A method of indicating the various energy consuming processes in the iron and steel industry, used in a report compiled by the University of California, is shown in Fig. 3.3. The four major stages in production as far as energy use is concerned are mining and ore preparation, manufacture of pig iron, manufacture of steel, and the steel finishing processes.

The manufacture of pig iron is preceded by conversion of the coal to coke, with a coke oven thermal conversion efficiency of the order of 85 per cent, and the preparation of sinter, a mixture of coke, iron ore and limestone, which is roasted for injection into the blast furnace.

A modern blast furnace is a very large steel structure having as its main member a refractory-lined steel cylinder, typically 30 m high with a hearth 10 m in diameter at its base. Ore, sinter, coke and limestone are fed in at the top of the blast furnace, and air is supplied via the base. Furnaces operate at a temperature of about 1100°C, and the products are molten iron and slag. About 35 per cent of the energy used to produce bulk carbon steel is consumed in the blast furnace. The pig iron produced by the blast furnace is then fed into steel furnaces, of which there are three main types, as indicated in Fig. 3.3. Open hearth furnaces can produce typically 250 tonnes of steel in 5 to 8 hours, but this production rate, even with the improvements brought about by uses of oxygen and fast firing rates, is inferior to that of basic oxygen converters which are taking over in many plants. A typical charge of 350 tonnes can be removed as steel within about 40 minutes of

The Energy Intensive Industries - 1

commencement of operation of this type of furnace. As can be seen from this figure the energy consumption per tonne of steel is only about 30 per cent of that required in an open hearth furnace. The only disadvantage of BOS as compared to open hearth furnaces is the reduced amount of solid scrap steel with which it can cope.

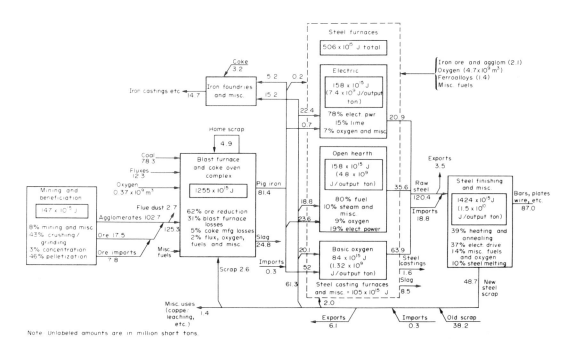

Fig. 3.3 US energy use in the iron and steel industry, 1971

The third type of steel furnace is the electric arc unit. Originally used primarily for the production of special steels, units of 300 tonnes capacity are now available which can process a 100 per cent charge of solid scrap. As far as energy consumption is concerned, electric arc furnaces are more expensive to operate than other types, but there are advantages in the use of electricity in similar small plants, and these are discussed in subsequent chapters.

The ingots formed by cooling the molten steel from the furnace in moulds are then subjected to a variety of rolling and finishing processes, most of which are highly energy-intensive. Electricity is used in a large number of these processes, not least in providing the drive for rollers producing plate, rod and strip. Up to 15 per cent of the energy used in these stages can be saved by adopting continuous casting techniques which are widely used in the non-ferrous metals manufacturing industry. This procedure eliminates the need to produce steel ingots, as the molten steel can be fed directly into open-ended water-cooled moulds to produce lengths of steel ready for rolling and other finishing processes.

3.1.2 Energy conservation in the iron and steel industry

There are a large number of ways in which energy consumption can be reduced in the iron and steel industry, many of which have already been implemented in the United Kingdom, United States, Japan and elsewhere. The use of higher grade ore, when obtainable, is one obvious way of improving efficiency, but this section will concentrate on improvements which can be brought about using engineering technology.

In Japan, which has to rely heavily on imported coking coal, (ref. 3.3), reduction of the coke rate in the blast furnace has been achieved by several modifications. These include:

(a) Improvement of blast furnace burden properties, and increase in sinter plus pellet ratio

(b) Increase in blast temperature and pressure

(c) Improvement in coke quality

(d) Injection of auxiliary fuel from tuyeres

(e) Improvements in charging and blast furnace control technology.

As a result the coke rates in Japanese blast furnaces average 340 - 350 kg/tonne, and this compares with 580 kg/tonne in the United States, where it is hoped to reduce this to 470 kg/tonne by 1980, and 500 kg/tonne in the United Kingdom; (in 1952 the figure for the United Kingdom was 1000 kg/tonne). The low coke utilisation in Japanese furnaces is offset by higher fuel rates (430 kg/tonne compared with 28 kg/tonne in the United Kingdom). The incentive for saving coke in the United States is not as great as elsewhere because of its comparatively low cost.

Similar techniques in the Netherlands have led to a reduction in coke rate from 850 kg/tonne to 430 kg/tonne over a 25 year period, (ref. 3.6), only 50 per cent of the energy used in steelmaking now originating from coal.

In the United Kingdom blast furnace operation has also benefited from economies of scale, where single units can produce up to 8000 tonnes of pig iron per day, and by improvements in refractory lining, increasing time between maintenance. It is in plants such as this that control technology can be important, and the use of computer control in conjunction with more accurate instrumentation is one step towards ensuring optimum fuel feed, firing temperatures, etc.

One of the most interesting methods for conserving energy in the blast furnace is the use of turbines to recover power from the blast furnace top-pressure. Substantial reductions in pressure in the furnace must be made to achieve line pressure, and this is normally done using a let-down valve or scrubber. However by releasing the pressure through a turbine, substantial amounts of power can be generated. Such a system installed on a blast furnace in Japan in 1974 generates 6.5 MW, amounting to an energy recovery of 80 to 85 per cent. Pressure fluctuations in the furnace can be accommodated using a bypass system, ensuring a constant load through the turbine. Predicted power outputs vary from 5 MW for an 8 m diameter hearth to 20 MW for a 14 m diameter furnace. While this system appears very attractive based on Japanese experience, Dutch studies suggest that a payback period of seven years would result from the use of turbines, and it is concluded that steel industry finance could be used more beneficially in other areas (ref. 3.6). (Further information on this system may be obtained from Chemical Construction Corporation, listed in Appendix 4).

It is of interest to note the emphasis in the above discussion on the movement away from the use of coke and coal in blast furnaces and steelmaking in general. Reservations concerning this policy have been expressed in West Germany (ref. 3.7), and because of possible future shortages of natural gas and the rising price of oil, it is argued that there is a new economic importance in creating steelworks operating purely on coal and coal-derived products, namely coke and coke oven gas. It is believed that a necessary condition for this is that blast furnace gas must be used for the firing of the coke ovens. In view of the coal reserves in the United States and Europe, these arguments are likely to gain weight in the future. The National Steel Corporation of America has reviewed new coal technology with this trend in mind, and the use of formcoke processes, enabling the range of coking coal to be greatly extended, is one encouraging line of development (ref. 3.8).

Both the BOS and electric arc furnaces for steelmaking are lower in capital cost than the open hearth system, and the operating cost of the BOS type is likely to lead to its very widespread use. In the United Kingdom it is also felt that the electric arc furnace will increase in importance by using pre-melted scrap on a large scale.

In many of these items of plant, as with existing open hearth furnaces, it is in the area of heat recovery that further energy savings could be implemented. A modern steel plant rejects up to 40 per cent of the energy used in the form of medium and high temperature gases, 15 per cent as low temperature steam and hot water, and 10 per cent in radiation loss. If this heat, in particular the high grade proportion, can be recovered and re-used in the plant (or outside), the savings would be substantial.

The Corporate Engineering Laboratories of the British Steel Corporation (BSC) have been working for a number of years on ceramic high temperature recuperators and regenerators for such applications (refs. 3.9, 3.10). The ceramic recuperator, which can preheat combustion air to 650°C, is illustrated in Fig. 3.4. A prototype system commenced operation in November, 1973, and its performance is considerably better than metallic recuperators, particularly as far as restrictions on operating temperature are concerned. Leakage problems are also minimised by the use of flexible ceramic seals.

BSC are also investigating rotating regenerators capable of giving preheat temperatures of up to 1000°C (ref. 3.11). Further data on these and other heat recovery systems are given in Chapter 7 and Appendix 4.

By recovering the heat in waste gases from BOS steel furnaces, and using the heat to preheat scrap metal prior to being fed into the furnace, the usage of scrap can be increased, (ref. 3.12). A pilot plant investigation undertaken by the United States Bureau of Mines, showed that by passing

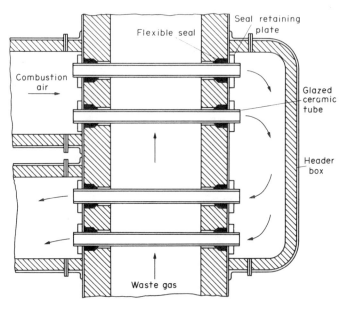

Fig. 3.4 British Steel Corporation ceramic recuperator

off-gases generated during the oxygen blowing of a molten pig iron and scrap charge through a static bed of shredded scrap, scrap usage could be increased by 43 per cent. The thermal energy recovered by this preheating method could account for up to 44 per cent of the energy necessary to melt the scrap. The system adopted is shown in Fig. 3.5. Scrap quantities used varied between

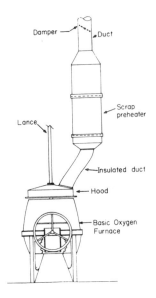

Fig. 3.5 Scrap preheating system developed in the United States
(Courtesy US National Bureau of Mines)

20 to 40 per cent of the total charge of the furnace (200 kg). Similar work has also been carried out on electric furnaces, (ref. 3.13), and proposals have also been discussed in the United Kingdom, (ref. 3.14).

The continuous casting of steel, mentioned earlier in this chapter, is particularly important, but the very high capital cost involved in replacing existing casting techniques will limit the rate at which any changeover can occur, particularly in the United Kingdom. Here there may well be future opportunities for use of the large amounts of low grade heat which will result from the cooling in the moulds. Unlike current 'batch' ingot cooling, the continuous availability of low grade heat may encourage more serious consideration of its re-use, possibly following upgrading.

In the long term, the use of nuclear energy, either converted to electricity or supplied to the steelworks directly in the form of heat, is likely to greatly influence steel production. In order to be fully effective, the iron ore would have to be reduced using gas, rather than being melted in a blast furnace. Difficulties also exist in balancing the demand for the heat and electricity output of the reactor complex. Several studies are being undertaken based mainly on the High Temperature Reactor (HTR), which uses helium as the reactor coolant, but any applications are unlikely before 1990 and benefits are therefore of a very long term nature. One possible site layout is shown in Fig. 3.6, based on the HTR system.

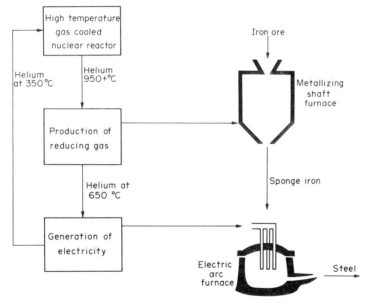

Fig. 3.6 Proposal for the use of nuclear heat in steelmaking
(Courtesy Institute of Physics)

3.2 Aluminium

The manufacture of primary aluminium is carried out in three stages:

(a)   Mining of the bauxite

(b)   Refining of the bauxite to extract alumina

(c)   Reduction of the alumina to obtain primary aluminium.

Further stages involve rolling and extrusion to obtain aluminium sheet, sections and tube. The mining process involves only a very small proportion of the energy used in the above stages. The Bayer process, used to obtain the alumina from the bauxite, uses a considerable quantity of fuel. This plant first digests the bauxite ore in caustic soda, producing aluminium hydrate which is then converted to anhydrous alumina by calcination at a temperature of $1200^{\circ}C$. The calcination stage requires approximately 22.3 GJ per tonne of alumina. (Four tonnes of bauxite yield two tonnes of alumina, which then can be reduced to give one tonne of aluminium). The alumina is reduced using the Hall-Heroult process, which is based on electrolysis. As shown in Table 3.4, this stage involves use of large quantities of electricity. The efficiency of electricity generation therefore has a considerable bearing on the total energy utilisation. In the Table, a conversion efficiency of 28 per cent is used, but higher efficiencies are likely where aluminium producers generate their own power, (ref. 3.5).

TABLE 3.4 Energy Used in Aluminium Production

Plant and product	GJ per tonne of product at each stage				
	Electricity		Thermal energy, gas, LPG or fuel oil	Total energy	Cumulative total per tonne Al.
	Net energy	Overheads*			
Mining, bauxite	0.02	0.05	1.0	1.1	4.4
Bayer plant, alumina	1.1	2.8	18.4	22.3	49.0
Anode preparation:			2.4	2.4	51.4
Carbon anodes			18.5(carbon)	18.5	69.9
Reduction plant, ingots, blocks, etc.	64.8	166.6	13.7	245.1	315.0
Alternative finishing plants					
Rolling mill A, plate, sheet, strip	5.5	14.1	16.8	36.4	
Rolling mill B, plate, sheet, strip	7.2	18.7	31.4	57.3	
Extrusion plant A, sections	4.7	12.1	17.7	34.5	
Extrusion plant B, sections	8.3	21.3	9.0	38.6	
Extrusion plant C, sections, tube	5.9	15.1	21.6	42.6	

*Electricity overheads have been calculated from a conversion efficiency of 28 per cent, though for some aluminium plants a higher figure might be obtained (see text)

The Hall-Heroult electrolytic process has been considerably improved over the last twenty years, the electrical energy required to produce one tonne of aluminium having been reduced by about 35 per cent during this period. An approximate energy balance sheet for the electrolytic reduction process is given in Table 3.5, (ref. 3.4).

TABLE 3.5 Energy Balance Sheet for the Hall-Heroult Process

Energy supplies (GJ/tonne)		Energy consumed (GJ/tonne)	
Electrical energy	54.0	Reaction heat (Al_2O_3)	32.4
Reaction heat	13.0	Warming up plant	3.3
Aluminium reoxidation	1.8	Heat losses (radiation etc)	33.1
	68.8		68.8

The large heat losses evident, amounting to about half of the total energy consumption, are primarily due to the large inter-polar distance (5 to 6 cm) between the electrodes. Utilisation of this heat in other areas of the plant may be possible, but the investment needed may be prohibitive.

Most reviews of the aluminium industry (refs. 3.5, 3.6) agree that the efficiencies achieved to date in the Hall-Heroult process represent the realisation of most of the possible energy savings, and research is now concentrated on other improvements, such as the manufacture of non-consumable anodes.

Two new processes are under development for the manufacture of aluminium. One of these, the Toth process, illustrated in Fig. 3.7, is based on a chemical route rather than an electrolytic one (ref. 3.15).

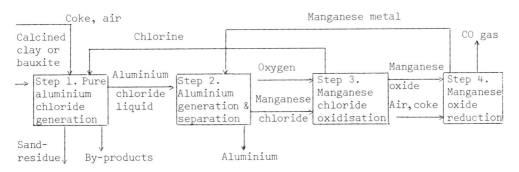

Reaction: Step 1 $Al_2O_3 + 3C + 3Cl_2 = 2AlCl_3 + 3CO$
Step 2 $2AlCl_3 + 3Mn = 2Al + 3MnCl_2$
Step 3 $2MnCl_2 + O_2 = 2MnO + 2Cl_2$
Step 4 $MnO + C = Mn + CO$

Net: $Al_2O_3 + 1\tfrac{1}{2}O_2 + 6C = 2Al + 6CO$

Fig. 3.7 The Toth chemical route to aluminium

As far as energy utilisation is concerned, the principle difference between the Toth process and the Bayer/Hall Heroult system is that the former relies on coal (or coke) for its energy, rather than electricity, thus dramatically changing the emphasis. Theoretically, 55 GJ will be required to yield 1 tonne of aluminium, and in the Toth process there appears to be much more scope for energy conservation within the plant using various recovery techniques.

Several alternative electrolytic processes are also being investigated, the most important being the Alcoa system, which obtains aluminium from the electrolysis of aluminium chloride. It is claimed to offer savings in energy of approximately 30 per cent over current plant. Operating temperatures are substantially lower, and the carbon anode is not consumed during electrolysis. Fuel savings would amount to 60 GJ per tonne.

Scrap produced within an aluminium plant amounts to about 40 per cent of the input. Most is recycled, but the energy content is quite high, and a reduction in the scrap output could save significant amounts of energy. The major producers of aluminium (United States, Japan, West Germany and the United Kingdom) recycle aluminium scrap from outside the industry in varying proportions. 33 per cent of the aluminium produced in the United Kingdom is obtained from scrap, but in the United States, the major aluminium producer, the figure is only 20 per cent.

At present there appears to be little scope for further energy conservation in aluminium plant, except possibly in the 'Total Energy' concept where manufacturers are generating their own electricity. However, if the Toth process is adopted for future use, there may be opportunities for heat recovery and utilisation similar to those existing in the steel industry.

3.3 The Chemical Industry

Unfortunately the chemical industry is so diverse, and the number of production processes so numerous, that a satisfactory treatment of the industry in total is not possible in a book of this type. However, by selecting a number of processes, and describing the associated plant and energy conservation methods used or proposed, it is hoped that the reader will be able to identify other areas in his specific field where similar arguments could apply.

The manufacture of one inorganic and two organic chemicals will be discussed, and these have previously been the subject of published treatments. The chemicals are ammonia (ref. 3.6), methanol, (ref. 3.16) and ethylene (ref. 3.17). Methanol and ethylene both feature strongly in the manufacture of polyester fibre; ethylene provides the basis for polyethylene and is also an important constituent in the production of synthetic rubber. Ammonia is used in the production of fertilisers.

Before discussing these processes, however, it may be beneficial to put into perspective the total amounts of energy consumed by the chemical industry. This is shown in terms of the type of energy used for the United Kingdom chemical industry in Table 3.6.

After the iron and steel industry, metals and engineering, the chemical industry is the highest user of energy in the United Kingdom. As shown previously, the chemical industry in the Netherlands accounts for 12 per cent of that country's energy consumption, using over one third of the total industrial energy requirement. Thus the industry is a prime target for energy conservation, and the chemical manufacturers have probably led the field in applying techniques to this end for many years.

TABLE 3.6 Energy Consumption in the UK Chemicals Industry

	10^6GJ		
	1960	1965	1972
Solid fuel	180	139	27
Town gas and COG	4	4	4
Natural gas	–	–	125
Petroleum	40	77	148
Electricity (net)[a]	27	30	43
Electricity overheads[b]	69	77	111
Total energy	320	327	458
Petrochemical feedstock (thermal content)[c]	na	104	200
Gross energy	na	431	658
Total chemical output[d]	56	75	108

Source: UK energy statistics; UK chemical industry statistics

[a] Additional electricity was generated by the industry
[b] Electricity overheads based on 28 per cent conversion efficiency
[c] Thermal content as if all naphtha
[d] 1970 output is 100.

As can be seen in Table 3.6, the trend in the United Kingdom chemical industry has been away from solid fuels, these being replaced by natural gas and petroleum. The heavy reliance on natural gas and oil products at a time when both these resources were available at low cost will obviously lead to a re-appraisal of energy conservation techniques. The 1973 rise in oil prices will contribute to this, but in addition many major chemical manufacturers will be renegotiating long term contracts for the supply of natural gas, and it is probable that these may lead to large price increases to these companies. We may see a growing use of process gases as fuel, and, if some of the developments in coal liquefaction and gasification show major promise, a return to this type of fuel (see Chapter 1).

3.3.1 **The manufacture of ammonia**   Sjoerdsma and Over (ref. 3.6) have written an excellent concise assessment of this process, and their work is used as the basis for the following discussion. Since their report was published, additional possibilities for heat recovery using heat pumps in industrial processes have appeared, and these are briefly mentioned here. (The heat pump is fully discussed in Chapter 7).

Ammonia is manufactured from methane in a number of separate reaction stages, as shown in Fig. 3.8, (ref. 3.19).

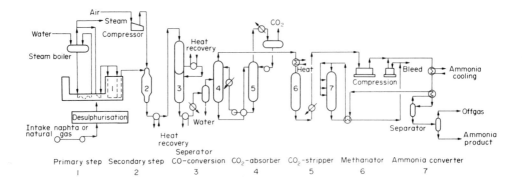

Fig. 3.8 Schematic diagram of a large ammonia plant

Reformer Section

In the reformer section steam is used to convert methane according to the reaction:

$$CH_4 + H_2O \rightleftharpoons CO + 3H_2$$

After the greater part of this reaction has taken place in the primary reformer stage a quantity of air is added to the reaction gas to produce a well-balanced $N_2 - H_2$ mixture. The resulting reactions are:

$$CH_4 + 2O_2 + 8N_2 \rightarrow CO_2 + 2H_2O + 8N_2$$

This combustion reaction raises the temperature to around 1000°C, reducing the methane content to 0.2 per cent. The result is a gaseous mixture of CO, CO_2, H_2, H_2O and N_2.

Conversion

The gas mixture is cooled in the conversion section to approximately 350°C. When brought into contact with the appropriate catalysts the reaction

$$CO + H_2O \rightleftharpoons CO_2 + H_2$$

occurs in the direction of the right-hand term in the equation.

CO_2 removal

The CO_2 is removed from the reaction mixture by absorption in a suitable solvent. The final traces of CO and CO_2 are converted into CH_4 by methanisation.

Synthesis

The $N_2 - H_2$ mixture is then compressed to approximately 250 bar, heated to 450°C and converted into NH_3. The reaction is:

$$N_2 + 3H_2 \rightleftharpoons 2NH_3$$

Since the gas mixture contains traces of CH_4 and argon part of the synthesis gas has to be blown off (between 5 and 8 per cent) in order to minimise the build-up of pollutants.

The thermal effect of ammonia synthesis from methane

The thermal effect of the four synthesis reactions is:

$$CH_4 + H_2O \rightarrow CO + 3H_2 - 206 \times 10^3 \text{J/mol C}$$

$$CO + H_2O \rightarrow CO_2 + H_2 + 41 \times 10^3 \text{J/mol C}$$

$$N_2 + 3H_2 \rightarrow 2NH_3 + 46 \times 10^3 \text{J/mol NH}_3$$

$$CH_4 + 2O_2 + 8N_2 \rightarrow CO_2 + 2H_2O + 8N_2 + 804 \times 10^3 \text{J/mol C}$$

The total amount of heat released by synthesis based on methane, steam and air is equivalent to $+34 \times 10^3$ J/mol NH_3. Since 28×10^3 J/mol NH_3 is required for the formation of the theoretical steam requirement there is a reaction heat surplus after correction for the heat requirement for steam generation of 6×10^3 J/mol NH_3.

An ammonia plant capable of producing 100 tonnes of ammonia per day requires a considerable quantity of compressor power to achieve pressures necessary for the reformer and synthesis processes, totalling approximately 30 MW. In practice much of the heat and energy requirements are provided by waste heat boilers. They utilise heat from the reformer gas and the synthesis process. Steam is generated at 100 to 120 bar and is expanded to 30 bar through turbines for use in the reformer. The steam also meets a number of other plant duties. As a result the overall efficiency of an ammonia plant can be as high as 83 per cent, the energy requirement being approximately 35×10^9 GJ per tonne of ammonia produced using methane as the raw material.

In theory it should therefore be possible to make further energy savings of about 15 per cent. In practice, however, there are some losses which are irrecoverable, including condenser losses of 2.1 GJ per tonne, and these reduce the practical amount which could be saved to about 6 per cent. Savings in energy consumption have been achieved already in the Netherlands by using large rotary compressors powered by steam turbines. Gas turbines, which have an efficiency of only 20 to 25 per cent, can also be used for compressor drive, provided that the waste heat from the gas turbine exhaust can be effectively used to preheat air for the reformer burners. This can reduce consumption to 33 GJ per tonne of ammonia.

If it is possible to find uses for the low grade heat which to date has been wasted, further economies would be achieved. It would probably necessitate the integration of the ammonia plant with other processes, commercial complexes, etc. It may be in a situation like this that the heat pump could be considered. If the heat could be upgraded sufficiently to produce additional low pressure steam using a heat pump compressor driven by a gas engine or gas turbine, further economies could result. It would of course be necessary to make full use of the waste heat generated by the engine/turbine drive, but very high overall efficiencies could be obtained. For reasons discussed in Chapter 7, it would be impracticable to use heat pumps driven by electric motors for a duty such as this, from an energy conservation point of view.

Sjoerdsma and Over identify one other feature of the ammonia production process which, if changed, could lead to quite dramatic energy savings. This is associated with the synthesis process. If a catalyst active at 300°C instead of 450°C, as at present, were available, the operating pressure could be reduced and the resulting elimination of the synthesis gas compressor could save 13 per cent of primary energy consumption in the plant.

3.3.2 Low pressure methanol plant

Imperial Chemical Industries Ltd., (ICI) produce methanol over a copper based catalyst via the following reactions (ref. 3.19):

$$CO_2 + 3H_2 \rightarrow CH_3OH + H_2O$$

$$CO + 2H_2 \rightarrow CH_3OH$$

The plant required for this process is illustrated in Fig. 3.9, (ref. 3.16). The primary feedstock, natural gas, is preheated in the reformer box and sulphur is removed. It is then mixed with medium-pressure steam, distributed to the tubes of the down-fired furnace box for heat to progress the endothermic reforming reaction. Flue gas from the furnace is used for preheating the natural gas feedstock, and also for superheating steam. The steam raised in the reformer is used to drive the compressor and synthesis loop circulator, and flue gas is also used to preheat furnace combustion air.

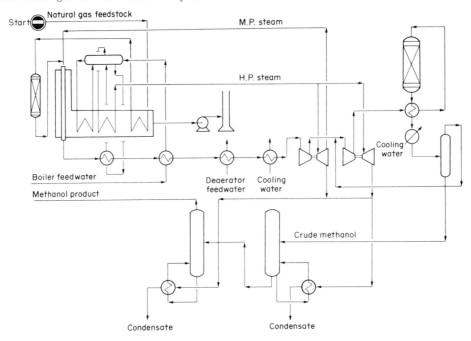

Fig. 3.9 Classical ICI methanol plant
(Courtesy Gulf Publishing Co.)

Hot gas from the reformer then passes through heat exchangers for raising steam, preheating boiler feedwater and preheating deaerator water. Cooling water removes the rest of the heat in the reformed gas, without further recovery.

The Energy Intensive Industries - 1

The gas is then compressed to $1.03 \times 10^4 kN/m^2$ prior to injection into the synthesis loop, with fresh make-up gas, all heated to the reaction temperature by exhaust effluent from the converter. In the reaction described by the above equations, the hot effluent gas is cooled by the feed gas and also by water, resulting in the condensation of crude methanol. Remaining gases are purged, the purge being used for fuel, and the clean gas compressed for recycling via the converter.

The distillation of the methanol is carried out using turbine low pressure exhaust steam ($482 kN/m^2$).

A typical plant produces approximately 1100 tonnes per day using 3.39×10^4 MJ per tonne.

Davy Powergas Ltd. have proposed a number of plant modifications involving relatively inexpensive equipment which could lead to a reduction in energy requirement of approximately 10 per cent (ref. 3.16). Four modifications have been proposed, all of which illustrate how even a highly efficient chemical process can, with careful study, be improved even further using equipment which is readily available. These are detailed below.

(i) <u>Improvements to distillation stage</u>

By analysis of data obtained from 'conventional' low pressure methanol plant, it was found that heat loads in the distillation stage could be reduced without affecting the quality of the final product. Because of the reduction in reboiler steam use, power for the compressor and circulator turbines had to be augmented by $8.3 \times 10^3 kN/m^2$ steam, but the net energy consumption was reduced to 3.36×10^4 MJ per tonne.

(ii) <u>Reboiler heat supply (distillation)</u>

It was found that there was sufficient heat available in the reformed gas to eliminate the need for any steam heating in the distillate plant. This led to a reduction in the cooling water requirement, while retaining the ability to use the balance of the heat in the reformed gas to maintain preheating of feedwater and water passing to the deaerator. As a result less fuel was required for steam raising and the plant consumption was reduced to 32.5×10^9 J per tonne, a 5 per cent saving. The modified plant is shown in Fig. 3.10, (ref. 3.16).

(iii) <u>Use of waste heat for raising medium pressure steam</u>

In the above plants, the heat rejected to the cooling water after the synthesis process has been considerable, and by using some of this heat to raise medium pressure steam ($2930 kN/m^2$) for use in the reformer furnace, a reduction of 9 per cent in total fuel consumption can be achieved.

The modifications needed are shown in Fig. 3.11, involving addition of a number of adiabatic reactor vessels with steam raising boilers at the exit of each vessel.

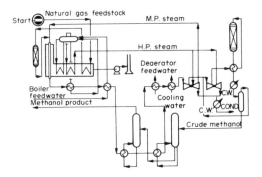

Fig. 3.10 Changes needed in methanol plant to use reformed gas heat to supply the distillation reboiler
(Courtesy Gulf Publishing Co.)

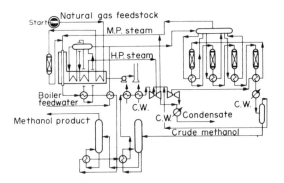

Fig. 3.11 The raising of medium pressure steam as an alternative to heat rejection to water coolant
(Courtesy Gulf Publishing Co.)

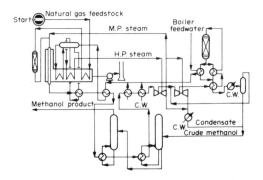

Fig. 3.12 Reaction heat used to raise the temperature of boiler feedwater
(Courtesy Gulf Publishing Co.)

The Energy Intensive Industries - 1

(iv) <u>Supply of synthesis heat to boiler feedwater</u>

Davy Powergas believe that a more effective way of using synthesis heat than in (iii) above, would be to heat the boiler feedwater. As well as requiring less additional plant, as shown in Fig. 3.12, the fuel required is reduced even further, amounting to a saving of up to 11 per cent. The distillate stage heat can still be supplied by the reformed gas line, and a reduction in condenser duty (for the loop gases and crude methanol) of 30 per cent is obtained.

It is also hypothesised that the 1.03×10^4 kN/m^2 purged gas from the synthesis stage could also be used to provide about 3.7 MW of electricity, more than sufficient to supply the plant electrical power demands, if expanded through a turbine linked to a generator.

The above analysis does not present the total picture as far as the implications of energy conservation are concerned. In the documentation of cost savings, no account is taken of the economies accrueing to the reduced requirement for cooling water, (the consumption of the basic plant is 1.76×10^5 litres per tonne of methanol, whereas proposal (iv) consumes only 1.32×10^5 litres per tonne). If the water is discharged to waste prior to being cooled, this saving can be very significant. (In the United Kingdom water can cost up to 11p per 1000 litres). Alternatively, if the water is cooled and recycled, the reduction in energy consumption in cooling towers and pumps, and the lower capital cost resulting from this load reduction, can be beneficial. The greatest savings are obtained if <u>mains</u> water cooling is replaced by recirculation.

In addition, the four schemes proposed above all involve increased capital expenditure, to a greater or lesser degree. Because of the demands on capital in industry, and high interest rates on loans, an assessment should be made of the payback period (i.e. the number of years of operation required to recoup the capital expenditure by savings in energy costs). Further details of this are given in Appendix 2.

3.3.3 The manufacture of ethylene As with the production of methanol, described above, several options may be open to the engineer considering energy conservation techniques in the manufacture of ethylene, particularly by using some of the numerous methods available for waste heat recovery.

The analysis carried out by Mol (ref. 3.17) is of particular interest because it examines two options available when considering heat recovery in multiple installations, each being nearly identical. In this context the particular items of equipment used for heat reclamation are incidental; the emphasis is placed on the relative merits of recovering the energy before it leaves each individual installation, (Integrated Waste Heat Recovery, IWHR), or recovering part of the waste heat in heat exchange plant located further downstream, in for example, a common exhaust duct leading to a single flue serving the multiple installations, (Central Waste Heat Recovery, CWHR). These two systems are shown in Figs. 3.13 and 3.14 as applied to cracking furnaces for the production of ethylene.

The amount of high grade heat released during pyrolysis in a large (1000 tonnes per day) ethylene plant is of the order of 120 MW, hence it is desirable

Industrial Energy

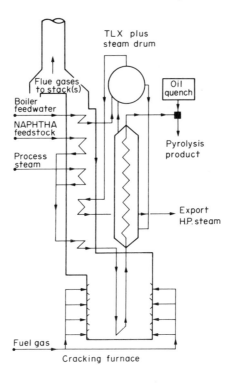

Fig. 3.13 Cracking furnace with integrated waste heat recovery system (IWHR). (Courtesy Gulf Publishing Co.)

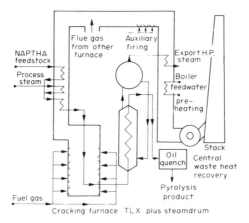

Fig. 3.14 Cracking furnace with central waste heat recovery system (CWHR). (Courtesy Gulf Publishing Co.)

to recover as much of this energy as possible for re-use. Fig. 3.13 shows the IWHR system where hot gases exhausting from the cracking furnaces at 1100 to 1200°C are immediately passed over heat exchangers for preheating boiler feedwater, preheating the feedstock, and generating steam for the process. In addition some superheating of steam for use outside the pyrolysis process is carried out.

In the CWHR system, illustrated in Fig. 3.14, heat exchangers are retained in the exhaust duct from each furnace, but these are for duties involving process steam generation and naptha preheating only. All the boiler feedwater preheating and additional superheating of steam are implemented with the aid of heat exchangers located in a common flue a considerable distance downstream of each furnace. By this stage, the flue gases have been cooled to 500 to 600°C, and the combination of lower flue gas temperature and the higher pressure drops will necessitate induced draught stacks, whereas in the IWHR system natural draught will probably suffice.

As well as thermal efficiency, the selection of the system must take into account the investment needed. Mol suggests that IWHR is desirable where a plant contains eight or less cracking furnaces; for larger plant CWHR appears more attractive, although the total cost of the heat recovery equipment remains only a very small proportion of plant expenditure in both cases.

From a reliability viewpoint, obviously a failure in any of the high pressure lines will necessitate shutdown of only that particular cracker if an IWHR system is incorporated. Using CWHR, total production in the complex may be halted while repairs to the steam lines are undertaken. However, in situations where tall stacks are mandatory because of environmental pressures, the CWHR is preferred as only one stack is required.

In this example, a number of additional factors which must be taken into account when considering energy conservation measures have been identified, showing that improvements in plant thermal efficiency are only one of a number of trade-offs which must be done when designing plant or retrofitting energy conservation equipment.

3.4 Oil Refineries

The world oil refinery capacity (excluding the Soviet Bloc and China) amounts to 65×10^6 barrels per day, this capacity being divided between over 700 oil refineries, more than half of which are in the United States and Western Europe. The output of the oil refineries includes motor and aircraft fuel, heating oil, diesel engine fuel, fuel oil, liquified petroleum gas, lubricating oil and chemical plant feedstuff. (Crude oil is refined to naptha, the basis for acetylene, methanol, ammonia, and many other diverse chemicals). In the United States, which has 22 per cent of the total world refining capacity, 13.4 per cent of the energy requirement of industry was consumed by oil refining, ranking it third to the primary metals and chemical industries on this basis, (refs. 3.20, 3.21).

It is again to Sjoerdsma and Over (ref. 3.6) that one must turn to obtain an up-to-date assessment of energy conservation potential, this time in the oil refining industry. In the Netherlands the oil industry is the second largest consumer of energy, although the total refinery capability is only about 65 per cent of that in the United Kingdom. Being one of the initiators of the 'multi-national' concept in industry, oil companies are much more likely to employ common energy conservation techniques in plant, regardless of their geographical location, than most other industries. Thus nothing is lost in applying the experience of Dutch refiners in other localities.

At least 25 per cent of the operating cost of an oil refinery is attributable to energy consumption. Because of this, designers of process plant for oil refining have always been energy conscious, and most of the comparatively straightforward methods available for conserving and recovering energy are already used in refineries. The primary users of energy in all oil refineries are the distillation columns, strippers and splitters, which separate the crude oil into the numerous end products, ranging from propane to heavy fuel oil. Of the energy used in these processes, about 50 per cent is required in the primary fractional distillation column, shown in Fig. 3.15, (ref. 3.22), this energy being used to heat the crude oil and to raise steam used in the column. A further 35 per cent is consumed in subsequent conversion plant, and the remaining 15 per cent is used in the final treating and finishing of products

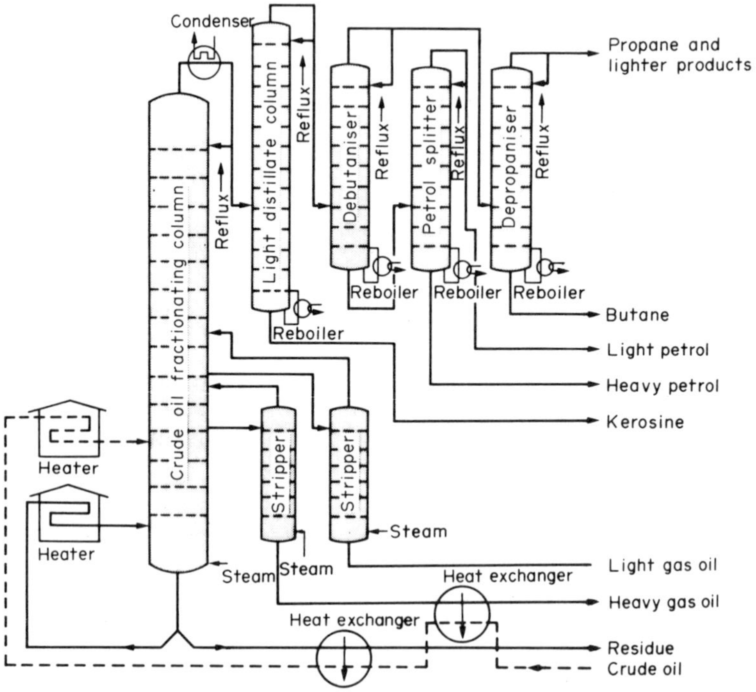

Fig. 3.15 Layout of a crude oil distillation plant

In addition to the implications of increased fuel prices on refinery operations, the Dutch report emphasises the fact that environmental considerations, in particular the need to minimise product sulphur content and the effect of legislation regarding the use of low-lead petrol, will increase fuel consumption at the refinery. Sjoerdsma and Over considered a number of options available to refinery operators which could contribute towards increased fuel efficiency:

(i)   Better control of operations using computer techniques

(ii)   Improved waste heat recovery

(iii)   Improved furnace efficiencies

The Energy Intensive Industries - 1

(iv)   Increased distillation plant efficiency by using additional stages

(v)   Upgrading of heat using heat pumps

(vi)   'Total energy' schemes

(vii) The use of low grade waste heat for space heating.

(This checklist would be applicable to a large number of energy-intensive industries).

Of the above, only item (iv) can be considered as being unique to an oil refinery or petro-chemical complex. The use of computerised control systems in plant is directed at ensuring high utilisation with optimum fuel efficiency, and offers many advantages in the monitoring of individual component performances. This method is adopted in most, if not all, new large plants, and the features of control systems as they affect energy utilisation are discussed in Chapter 5.

As can be seen from Fig. 3.15, considerable use of waste heat recovery equipment is made in existing oil refinery plant. The potential for further investment in heat recovery equipment depends upon the relationship between capital equipment costs and the price being paid for energy. The Dutch study indicates that if sensible economic restraints are imposed on the purchase of additional heat recovery equipment, a saving of 5 per cent on energy costs could probably be achieved. This excludes improvements in furnace efficiency which could be implemented by using more air preheaters, particularly on the lower efficiency furnaces. The use of heat pumps in the context of heat recovery in refineries is discussed later.

The design of distillation plant is a major topic in most chemical engineering textbooks, and is not discussed in detail here. Heat input to a distillation column is required for the separation process, and also for heating of the feed, which may be carried out before injection into the column. By increasing the number of trays in the column, the heat input could be reduced. This would necessitate equating the energy saved to the additional cost of new trays.

Of more general application is the use of heat pumps and 'total energy' schemes. Sjoerdsma, in discussing the application of heat pump compressors with electric drive in oil refineries, correctly states that when taking into account the low efficiency of electricity generation, in a global energy context this type of heat pump loses some of its attractiveness. However electricity is not the sole energy source for compressor drives, as his subsequent section on total energy plant reveals. As applied to a refinery having a capacity of 10×10^6 tonnes per annum, by using all the fuel for the refinery processes to power a turbine, 200 MW of electricity could be generated in addition to providing sufficient exhaust gas at $600^\circ C$, to supply the process heat requirements. One proposal involves a proportion of the turbine mechanical output being used to drive an air compressor, this supplying combustion air at 10 to 12 bar for the turbine. The 200 MW obtainable is the balance of the available turbine shaft output. It would also be feasible to use shaft power to drive a Freon compressor, providing a heat pump duty without the high cost in primary fuel resources which may be incurred using an electric driven heat pump. In this way Sjoerdsma's objections to the use of the heat pump in refineries (and other industrial installations) may be overcome.

The use of a heat pump based on a gas turbine drive may obviate the need to recoup waste heat by integrating the refinery with space heating (vii). It has been estimated that the low grade heat discharged by a refinery, if applied to local district heating schemes, could lead to overall savings of up to 20 per cent in fuel consumption, but obviously the expenditure need to implement such a project would be high, and the potential must be regarded as medium- to long term.

REFERENCES

3.1 Anon. Energy Conservation. A Study by the Central Policy Review Staff, HMSO, London (1974)

3.2 Anon. Technology of efficient energy conservation. Report of a NATO Science Committee Conference, France. (NATO Scientific Affairs Div.) Oct. 1973.

3.3 Toyoda, S. Energy in the steel industry. Metals Engineering Quarterly, May 1975.

3.4 Gray, A.G. Steels for energy survival. Metal Progress, October 1975.

3.5 Anon. Energy conservation in the UK - achievements, aims and options. National Economic Development Office. HMSO, London 1974

3.6 Sjoerdsma, A.C. and Over, A.J. Energy conservation - ways and means. Future Shape of Technology Publications, No. 19, The Netherlands, (1974)

3.7 Friedl, R. Saving energy in steelmaking. Metals Engineering Quarterly May 1975.

3.8 Smith, E.J. Present and future of coal in steelmaking. Metals Engineering Quarterly, May 1975

3.9 McChesney, H.R. Recovery of heat from metal processing furnaces. Paper 7, Waste Heat Recovery Conference, Inst. Plant Engineers, London, 25 - 26 Sept. 1974.

3.10 Jones, A. Energy economy within B.S.C. British Steel. pp 12 - 16, Spring 1974

3.11 Winkworth, D.A. and Blundy, R.F. Trends and new developments in high temperature air preheater equipment. Iron and Steel Institute Slab Reheating Conference Proc., pp 117 - 126, June 1972.

3.12 Drost, J.J. et al. Thermal energy recovery by Basic Oxygen Furnace offgas preheating of scrap. Bureau of Mines Report RI 7929, US Department of the Interior, (1974)

3.13 Chatterjee, A. Economics of preheated scrap usage in the LD process. Iron and Steel Internat., pp 325 - 331, August 1973.

3.14 Laws, W.R. Prospects for scrap preheating for the Basic Oxygen Furnace. Steel Times, pp 679 - 682, Sept. 1972.

3.15 Fidler, J. Aluminium at half the price - if Toth's idea will scale up. The Engineer, 237, 6125/6, 38 - 40 (1973).

3.16 Pettman, M.J. and Humphreys, G.C. Improve designs to save energy. Hydrocarbon Processing, 54, 1, 77 - 81 (1975).

3.17 Mol, A. Which heat recovery system? Hydrocarbon Processing, 52, 7, 109 - 112, (1973).

3.18 Quartulli, O.J. and Wagener, D. Technologie der Ammoniaksynthese in Vergangenheit. Gegenwart und Zukunft. Erdöl und Kohle, 26, 4, 192 - 198, (1973).

3.19 Bolton, D.H. and Hanson, D. Economics of low pressure in methanol plants. Chemical Engineering, pp 154 - 158, Sept. 1969.

3.20 Reistad, G.M. Available energy conversion and utilisation in the United States. Trans. ASME, Journal of Engineering for Power, pp 429 - 434, July 1975.

3.21 Anon. Oil and Australia, 1974 - The figures behind the facts. Petroleum Information Bureau (Australia), (1975).

3.22 Anon. The Petroleum Handbook, Shell International Petroleum Co. Ltd. London, 1966.

The Energy Intensive Industries - 2

This chapter, in common with Chapter 3, concentrates on a number of industries where energy costs have a significant effect on efficient operation and final cost of the product. However, in the examples discussed below, which include the food, glass, paper and textile industries, the companies are often without the resources of large research laboratories or government funding, and tend to be highly geared to mass production. This last fact in itself can lead to difficulties in developing energy conservation equipment because of the reluctance to run a production line at less than optimum capacity during any experimental trials. As a result the industries concerned are more likely to use equipment developed in other industries or by their appropriate Research Associations, and the relationship between the industries described, in that several common processes or instruments exist, will become evident later. This is most important, because one company can learn a great deal about energy conservation from other firms, although they are producing totally different goods.

4.1 The Pulp and Paper Industry

Energy consumption statistics for the pump and paper industry tend to be contradictory to some extent, and it is difficult to identify the energy content of each major product of the industry. This is partly due to the fact that some countries classify the pulp and paper industry as a part of a greater group of industries encompassing plastics and tinplate - the packaging industry. Paper manufacture, printing and stationery form a major proportion of the output of the industry, however it may be classified, and the manufacture of board also accounts for a significant part of the energy consumed. The manufacture of tissues and paper towels is also becoming a major activity in the pulp and paper industry.

The energy consumed in the United Kingdom by the paper, printing and stationery industry accounted in 1972 for about 5.7 per cent of the total industrial energy used. The trend over the past decade was showing a slight increase in energy use by this industry as a proportion of the total, (in 1960 the consumption was 5.4 per cent). In the Table below, no account is taken of the energy consumed by the industry as a result of internal waste recycling.

Although the specific energy content of the material cannot be assessed using Table 4.1, various estimates have been made. Over and Sjoerdsma (ref. 4.2) state a value of 18.8 GJ per tonne for paper produced in the Netherlands, while in the United States higher values are quoted, typically 30 to 32 GJ per tonne. A NATO study (ref. 4.3) produced a figure midway between these two, 25 GJ per tonne. The NATO study also attempted to show how the cost of the energy needed to make a particular product was related to the total value of the product, although the latter is difficult to quantify. However, on a comparative basis, this ratio was high for paper (0.3), only surpassed by the values for aluminium (0.4) and cement (0.5).

The Energy Intensive Industries - 2

TABLE 4.1 Energy Used in the UK for Paper Manufacture, Printing and Stationery (ref. 4.1)

	10^6 GJ		
	1960	1965	1972
Solid fuel	94	87	43
Gas	2	2	23
Petroleum	25	50	79
Electricity	6	9	11
Electricity overheads*	16	22	27
Total	143	170	183
% of industry total	5	6	6

*Electricity overheads are calculated assuming a 28 per cent conversion efficiency

The ratio for paper was identical to that given for steel.

It is estimated that the manufacture of paper and board in an integrated plant uses steam and electricity in the ratio 3:1, (ref. 4.1). Typical values for steam and electricity consumption per tonne of product are 12 GJ and 4 GJ respectively. This can of course be identified as an application for a total energy unit, where the process heat requirement is at least two times as great as the electricity demand in the plant. A particular application of a total energy unit, based on a gas turbine, is described later in this section.

It is believed that the trends in energy consumption over the past few years have been influenced by four major factors (ref. 4.1):

(i) An increasing emphasis on the production of high quality goods, which have a higher energy content.

(ii) Progress in effluent treatment, applicable to both water and gaseous waste, which also tends to increase the energy consumption at a plant.

(iii) Increased recycling of waste, contributing towards a reduction in the use of wood resources.

(iv) Modernisation programmes, including better process control and plant integration of the 'total energy' type, also leading to reductions in energy consumption.

4.1.1 The manufacturing process The basic raw material for paper is wood, prepared by removing the bark from the logs, cutting the logs into small pieces and then cooking these pieces under pressure with a sodium or ammonia based liquor. This chemical pulping process dissolves the lignin in the wood, allowing the fibres to separate. The fibres are then packed in bales, or made into sheets, for transport to the paper mill. (In some cases the production of the fibres is carried out in the same plant as the manufacture of the finished paper or board).

A beneficial result of the transformation of the wood into fibres is the ability to use the bark and the spent liquor as fuel. The bark has a calorific value of approximately 8141 kJ/kg, while the waste liquor, with a much higher calorific value, of the order of 16 282 kJ/kg, can also be used for steam raising, once its water content has been sufficiently reduced. Because of the difficulties associated with the behaviour of the waste liquor flue gas, boilers successfully developed for this type of application are understood to be a considerable technical achievement, the importance of such units being illustrated by the fact that they were often the only steam raising unit in a pulp mill (ref. 4.4). If a sodium-based liquor is used, the ash resulting from combustion can be further processed to recover chemicals which are suitable for re-use in the cooking of the wood chips.

The fibres, having been mixed with water to form a 'stock', then commence the continuous process leading to a completed roll of paper. A typical paper manufacturing machine is shown in Fig. 4.1, (ref. 4.5). The stock is fed on to a continuous moving wire screen, where a sheet is formed and a major proportion of the water introduced with the fibre is drained away. Further water removal is effected by applying suction below the mesh screen, followed by a mechanical drying procedure involving passing the paper, now supported on a felt conveyor, through presses. As the paper leaves the final press, the water content of the sheet has been reduced to approximately 60 per cent.

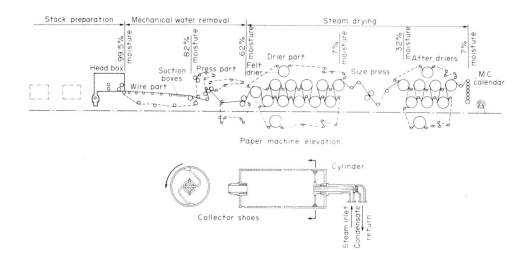

Fig. 4.1 A layout of a typical paper manufacturing plant
(Courtesy Combustion Publishing Company, Inc.)

The paper now undergoes the most energy intensive process in its manufacture, drying until the water content is reduced to about 7 per cent. Investigations are being carried out on a number of other paper drying methods, but the technique normally applied is to evaporate the water from the paper as it passes over drums which are internally heated using steam. The heating of the paper commences fairly slowly, until all the water in the paper is at a sufficiently high temperature to start evaporating into the surrounding air.

Although the paper will pass over several cylinders before this occurs, the exposed surface which has been in contact with the cylinder will lose some water by evaporation as it passes through the space between the cylinders.

The drying process must be carefully controlled. If the cylinder is too hot, particularly during the early stages of drying, rapid evaporation can cause the paper to lift away from the cylinder, possibly creating damage. Similarly, the uniformity of the heat input to the roll, particularly across the cylinder, is necessary for even quality. Excessively rapid drying during early stages can also result in brittle fibres. While the steam input to the cylinder, and its distribution, is critical in determining the uniformity of heating, removal of the condensate from the cylinder must also be rapidly facilitated to prevent build-up on the cylinder periphery.

The drying process can be greatly assisted by the quality of the air in the vicinity of the cylinders, and under normal circumstances provision is made to accelerate the flow of humid air from this area. More recently, however, the application of forced convection heat transfer on the outer side of the paper as it passes around the cylinders has helped to increase the production rate, and the way in which the heated air is applied to the surface is shown in Fig. 4.2, (ref. 4.5). This has enabled the evaporation rate (typically 15 kg/m^2 of cylinder surface per hour) to be trebled in some cases. In general, the hot air is provided by direct fired air heaters.

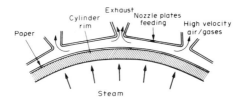

Fig. 4.2 Forced convection heat transfer for drying paper on a drying cylinder. (Courtesy Combustion Publishing Company, Inc.)

A heat balance for a paper machine is presented in Fig. 4.3, (ref. 4.5). There is considerable potential in the application of heat recovery techniques to recover heat from the moist air being taken from the dryer prior to its discharge to atmosphere. One installation in Eastern Europe recovers a total of 3 MW in this way, the heat being used to warm incoming air for roof heating (to prevent condensation), underfelt heating and various other uses. In a plant where the forced convection air heating is used to assist the drying of the outer surface of the paper, preheating of the air being fed to the heater bank would also be beneficial, (see Chapter 7).

4.1.2 Combined power and steam generation A paper machine having a capacity of 10 tons per hour will use approximately 30 MW of heat and 3.5 MW of electricity. While much of the steam required for heating is at a low pressure, it is generally raised at comparatively high pressures, making back pressure power generation attractive. In a paper mill which combines pulp production and paper manufacture, steam at 2400 kN/m^2 is used in the hardboard mill and in the chemical digester, 760 kN/m^2 steam is used in the packaging paper mill, while steam at only 172.5 kN/m^2 is passed through the drying cylinders of the machine manufacturing newsprint paper. Among the major uses of electricity are the log grinders, the digester, and the drives for the cylinders in all the paper drying plant.

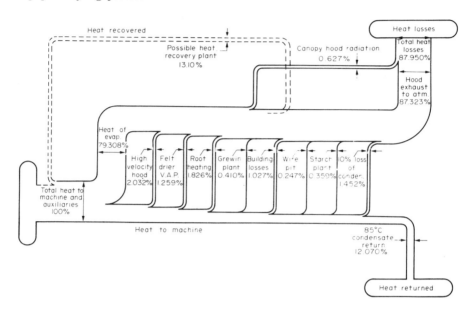

Fig. 4.3 Paper machine heat balance
(Courtesy Combustion Publishing Company, Inc.)

This would indicate that a total energy plant capable of meeting the demand for both steam and electricity would be attractive in a paper mill. However, these figures relate to one specific mill, and it is very difficult to satisfy the ratio of steam:power consumption in all plant of this type. Owing to the differences in the grades of paper produced and the different production techniques used to manufacture these various grades, this ratio can vary considerably. Similarly, raw material composition and product quality can also affect energy usage.

The rate of production also affects this balance. The energy consumption figures shown in Fig. 4.4, (ref. 4.6) relate to the performance of an Escher Wyss tissue machine, which has been in operation since 1974, (the fuel gas used is for heating air used in the cylinder drying process). Demand for steam required per kilogram of product reduces with increasing machine speed, whereas the electricity consumption, calculated on the same basis, is essentially constant. Thus a total energy plant would be required to have characteristics which permit the steam output to be raised, while maintaining a constant electrical load.

A further influence on the relative heat and power requirements of the plant not previously mentioned is the number of machine stoppages. Most of these are likely to occur because of a break in the paper web, and although the stoppages are often of fairly short duration, they are accompanied by a sudden drop in the steam consumption. In spite of this, processes further upstream such as log preparation, and finishing processes downstream of the cylinders, may proceed as normal, leaving the demand for electricity comparatively unaffected.

The Energy Intensive Industries - 2

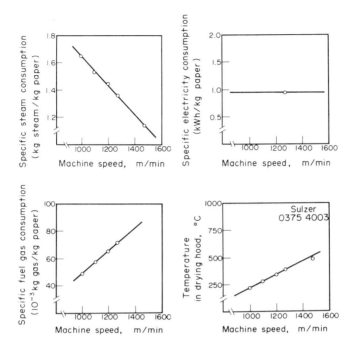

Fig. 4.4 Typical energy requirement structure in the paper industry, based on an Escher Wyss tissue machine
(Courtesy Sulzer Technical Review)

Sulzer argue that the power coefficient S, defined as the ratio between the specific power requirement (MJ per tonne of product) and the specific heat requirement Gcal per tonne of product), may vary from 540 to 3600 MJ/Gcal, depending upon the state of the plant and the product. This variation in power coefficient may be difficult to meet with a conventional back-pressure steam turbine total energy plant, mainly because the electrical output from the available steam rate is too low.

This leads to a surplus of steam and lack of power, which, with the trend towards higher power coefficients, makes the back-pressure turbine alone less attractive. The unit is also less efficient when operating outside its optimum range, as may be required due to the various factors discussed above which can create changes in plant power coefficient.

The solution proposed by Sulzer utilizes a gas turbine/generator set, the waste heat from the gas turbine exhaust being used to raise steam which in turn is fed to a back-pressure turbine. This turbine generates a limited amount of electricity, and the steam, having expanded through the turbine, is then available for process use at a pressure of about 3 bar. The system is illustrated in Fig. 4.5.

Compared with a system relying solely on a back pressure turbine, this arrangement can give a peak value of S 3 times as great, while producing 25 per cent more steam. An energy balance carried out on the gas turbine/

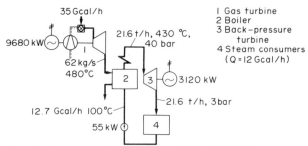

Electrical output at terminals 9680 + 3120 = 12800 kW
Process steam heat 12 Gcal/h

Power/heat value $S = \frac{12800 \text{ kW}}{12 \text{ Gcal/h}} = 1067$ kWh/Gcal

Specific heat consumption for power generation:

$$\frac{35 - \frac{12}{0.88}}{12800} \times 10^6 = 1669 \text{ kcal/kWh}$$

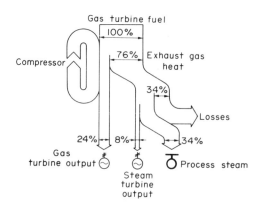

Fig. 4.5 Combined cycle heat and power plant, based on a
a gas turbine/back pressure turbine package
with heat flows
(Courtesy Sulzer Technical Review)

steam turbine plant reveals that 34 per cent of the energy input is lost as exhaust gas heat, and greater efficiencies might be attainable if a proportion of this heat could be used to warm air used for drying above the cylinders. This is already done in tissue plant, where the power coefficients are very high, (ref. 4.7).

4.1.3 Infra-red heating Other drying techniques are being investigated in the paper industry. Tests have been carried out using gas-fired infra-red heating elements covering the width of a paper machine (ref. 4.5). Installed between the mechanical presses and the first drying cylinder, the units have an output of 900 kW. The aim is to heat up the web of paper sufficiently from the outer surface, so that heat applied to the inner surface by the first cylinder is sufficient to create immediate evaporation of water from the web, thus increasing output. Preliminary experiments demonstrated that the application of heat in this way did not damage the web, unlike the use of higher heating rates generated within the cylinder dryers.

The only major disadvantage of infra-red heating is associated with the wide range of wavelengths that it produces. The comparatively narrow part of the spectrum which is absorbed by water, creating evaporation or solely sensible heat, means that the radiation produced is not being used very efficiently.

4.1.4 R.F. and microwave heating
R.F. and microwave heating, described fully in Chapter 9, can be effectively applied to paper drying (and other drying processes) because it takes the heat directly to where it is required, i.e. into the water, without having to overcome the resistances to heat transfer associated with other techniques. Also, because the amount of energy used is proportional to the quantity of water present, no excess heating is used once the required degree of dryness has been achieved, (ref. 4.8).

It has been shown (ref. 4.9) that in the last one-third of the paper making process, where moisture content is particularly important, primary energy consumption is reduced by about 50 per cent using the electrical method. Losses are also reduced from 8 per cent to 4 per cent, and the Electricity Council, having obtained detailed energy balances in this and other similar processes, state that the installation of this heating system can be recovered in two years by savings in energy.

4.1.5 Alternative production methods
It has been suggested, (ref. 4.3) that alternative non-wet manufacturing methods should be given greater development emphasis. The use of heat pumps (see Chapter 7) for pulp drying, and electrostatic laying of dry fibres have been proposed, saving large amounts of energy. These are long term developments, and considerable funding is required for full implementation.

4.1.6 Short-term energy conservation measures
Much of the above discussion deals with methods which, while able to offer substantial savings in energy, will cost considerable amounts of capital to implement, and would best be incorporated in new plant, or when the replacement of old equipment is envisaged.

A recent survey of the pulp and paper industry in one area of the United States, at a time when the industry was in a relatively poor economic position, revealed a large number of ways whereby energy consumption could be reduced, many of which could be put in hand quickly without undue expenditure, (ref. 4.10). The survey was comprehensive, involving interviews with mill management and plant operators, visits, reviews of new technologies in the industry, and the sending of questionnaires to fifty manufacturers (with a 60 per cent response).

The results of the survey, in terms of the types of conservation measure being applied in the industry, and the degree of implementation, are shown in Tables 4.2 to 4.4.

In general, it was found that more attention was given to short-term 'good housekeeping' conservation measures (many of which are more fully described in Chapter 5), steam and water recovery, and recycling. Less interest was shown in techniques requiring capital expenditure. The energy savings

TABLE 4.2 Conservation of Electric Power in the
Paper Industry - Responses from the Survey

Conservation Technique	Already Incorporated	In Process
Increasing initial steam pressure to produce electric power needs through by-product turbine power	24	3
Maximization of power use during off-peak periods	21	7
Installation of capacitors to raise power factor	52	3
Elimination of oversize, duplicative and inefficient pumps and motors	38	31
Conducting preventive maintenance on rotating machinery components	69	14
Installation of more efficient pulping equipment	10	14
Replacing pulpwood by waste paper as raw material	31	-
Limitation of power demand to predetermined maximum	21	7
Converting from fossil fueled to battery powered industrial trucks	21	-

(Status of Implementation Per Cent of Plants Surveyed)

TABLE 4.3 Conservation of Steam in the Paper Industry
- Responses from the Survey

Conservation Technique	Already Incorporated	In Process
Operation of boilers at maximum efficiency (through proper maintenance, computer control)	72	21
Elimination of steam leaks	69	38
Avoidance of producing steam in excess of process needs	66	7
Insulation of steam pipes, fittings and vessels	55	21
Maintenance of steam demand	41	3
Limitation of steam demand to predetermined maximum	7	-
Purchase of process steam from electric utilities	3	-
Reduction of steam boiler pressure	-	-

(Status of Implementation Per Cent of Plants Surveyed)

The Energy Intensive Industries - 2

TABLE 4.4 Conservation of Process Heat Energy
- Responses from the Survey

Conservation Technique	Status of Implementation Per Cent of Plants Surveyed	
	Already Incorporated	In Process
Installation of mechanical water extractors prior to dryers (presses, vacuum)	62	3
Collection and return of condensate in order to heat boiler feed water	79	28
Reuse of hot washing water	45	3
Reduction in quality of water used	69	31
Avoidance of overdrying	59	17
Recovery of waste heat from boiler flue gases	34	3
Recovery of waste heat from air over paper machines	17	10
Expansion of use of cold water cleaning	17	3
Cleaning of dryers to maximize heat transfer efficiency	52	-

resulting from the various measures undertaken ranged between 0.1 per cent and 20 per cent, most individual measures saving between 0.5 per cent and 2.0 per cent.

4.2 The Glass Manufacturing Industry

In the United Kingdom there is no separate listing in the energy statistics for the production of glass, but figures available suggest that over 90 per cent of the energy used in the china, earthenware and glass industry is consumed in the section producing glass containers, flat glass, and other glass products. The total energy consumption in the industry is shown in Table 4.5.

TABLE 4.5 Energy Consumption in the China, Earthenware and Glass Industry

	10^6 GJ		
	1960	1965	1972
Solid fuel	30	14	2
Gas	12	12	18
Petroleum	23	39	46
Electricity	4	5	7
Electricity overheads*	10	13	18
Total	79	83	91
% of Industry total	3.0	2.8	2.8

*Electricity overheads are calculated on the basis of 28% conversion efficiency
Source: UK energy statistics

The displacement of solid fuel as an energy source is most marked, with natural gas and oil taking over the role vacated by coal and coke.

The production process, and the relative energy consumption during each of the steps in the manufacture of glass, may be described with reference to the consumption diagram in Fig. 4.6, (ref. 4.11). The basic raw materials, silica, limestone and sodium carbonate, with in some instances crushed waste glass, are mixed and heated until the metal oxides combine in the solid state in the temperature range 600 to 1200°C. The resulting compounds melt in the furnace to form liquid glass, which is then passed to the forming process. At this stage the glass may be moulded or blown to produce containers, drawn to produce tube, or formed into glass sheet using float tanks and other techniques. During the forming process the cooling of the glass may be aided by forced air convection or other methods, which involve some energy utilisation.

In most of the forming processes, stresses occur in the glass during the rapid solidification. These stresses are subsequently removed by passing the glass product through an annealing lehr, where it is reheated and cooled under conditions of close control.

Where large quantities of glassware are manufactured, the production is a continuous process, with plant operating 24 hours per day with possibly one plant shut-down period per year. The bulk of glass produced is manufactured in continuously operating plant.

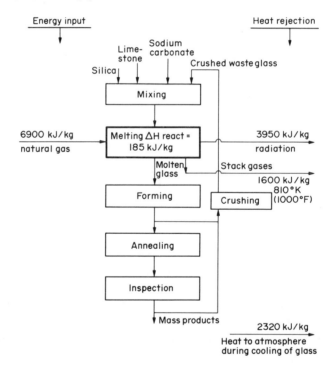

Fig. 4.6 An energy flow diagram for the glass industry
(Courtesy US Environmental Protection Agency, Office of Research & Development, Industrial Environmental Research Laboratory, Research Triangle Park, North Carolina 27711.)

The energy required for finished glass products is approximately 15 to 20 GJ per tonne for bulk production, with a considerably higher energy requirement for technical and hand-made glassware, (approaching 90 GJ per tonne). Estimates vary as to the amount of energy consumed in the furnaces in a glass plant, but generally between 70 and 80 per cent of the total energy used in the glass manufacturing process is taken up in producing the molten glass.

4.2.1 Glass Melting Furnaces

Because of the large proportion of the total energy consumption attributable to the glass melting process, the furnace has received most attention within the industry from the point of view of energy conservation. There are two forms of glass melting furnace, these being classified in terms of the heat recovery system employed, namely regenerators or recuperators.

In furnaces of the regenerative type, the melting tanks are usually rectangular and are divided into two compartments, a large melting section and a smaller refining compartment. A crown above the walls of the tank provides a space for combustion, and in some units part of the hot gas is recirculated to preheat the products entering the tank. Regenerators installed between the tanks and the exhaust flue recover some of the exhaust heat for combustion air preheating. The temperatures in the melting tank range up to 1450°C, and exhaust gas temperatures downstream of the regenerator are of the order of 400°C. On a gas-fired furnace, heat losses up the stack are about 1600 kJ/kg, while convection and radiation losses from the furnace itself can be as high as 3950 kJ/kg.

With regard to convection and radiation, good maintenance procedures and a well insulated structure can help to minimise losses. It has been suggested in the United States (ref. 4.11) that the application of submerged combustion (described in Chapter 5) may also be beneficial.

While the regenerators on these furnaces recover a substantial proportion of the heat in the exhaust, and studies are being undertaken by the British Glass Industry Research Association concerning the measurement of efficiencies and the relationship between model and full scale performance (ref. 4.12), there is still a considerable amount of useful energy retained in the flue. It has been shown (ref. 4.13) that a waste heat boiler installed in the flue can recover substantial quantities of heat. A unit fitted to a continuous rolling furnace with an output of 250 tonnes per day and a gas consumption of 4300 m^3 per hour has produced 5 tonnes of steam per hour. This boiler utilises exhaust gases at 320 to 340°C, and is installed at the Salavat Technical Glass Factory in the USSR. Provision is made for bypassing the boiler, and it was necessary to install a flue pump having a capacity of 140 000 m^3 per hour to provide the draught. However, operating experience has shown that the installation of the boiler did not adversely affect operation of the glass furnace; in fact the artifical draught assisted in stabilising glass tank operation.

If the output of the waste heat boiler is of little use in the plant, a secondary metal regenerator or recuperator may be installed downstream of the main heat recovery unit.

The development of recuperative glass melting furnaces has, until recently, been hindered by lack of suitable materials. While melting temperatures were gradually being increased, no acceptable refractories were available to form the recuperator tubes. However this is no longer a problem and ceramic recuperators with good sealing are now available (see also Chapters 3 and 7). Teisen Furnaces Ltd., of the United Kingdom, have applied these recuperators to an increasing number of large melting furnaces, locating the recuperator below the glass tank, thus saving floor space. Additional development work on improving port design, and optimising the automatic control systems on the furnace has, claim the manufacturers, resulted in an economical furnace with consistent product quality.

In a comparison of running costs of regenerative and recuperative glass furnaces, it was shown that while fuel consumption in a regenerative furnace increased at a rate of about 1 per cent per month, resulting in a 50 per cent higher consumption after a four year life, this was not the case with the recuperative system. The latter type showed a fuel consumption increase due to ageing of only 25 to 30 per cent. The waste gas is drawn from the furnace chamber into two vertical ducts, at the bottom of which are slag pits; the gas then enters a mixing chamber before passing through the tubes in the three recuperator chambers before being fed through separate feed ducts to the burners. Individual variation of recuperator capacity is permitted so that optimum flame conditions can be obtained at each port.

The trend towards recycling of scrap glass is increasing, and as well as conserving materials, improvements in melting can result because of the higher thermal conductivity of the scrap glass compared to the raw material mix forming the bulk of the charge.

4.2.2 Combustion efficiency

In glass manufacture, in common with many other industries where furnaces are major energy consumers, the effect of combustion efficiency can have a significant effect on the fuel bill. Improvements in combustion efficiency are generally relatively simple to implement, (see Chapter 5), but in order to do this it is preferable to be able to base changes on analyses of flue gases. BGIRA have developed oxygen analysis equipment which has been used successfully on gas-fired regenerative furnaces. Prior to installation assessments were made of the furnace combustion conditions at regular intervals by measuring the vertical distribution of oxygen concentration in the furnace ports. BGIRA stress the need for exercises such as this before the oxygen analysis equipment can be used to best advantage for furnace combustion control, (ref. 4.12).

Reding and Shepherd, (ref. 4.11), suggest that a potential area for energy conservation is the sensible heat in the nitrogen in the combustion air, which accounts for losses of up to 690 kJ/kg. This loss could be reduced if oxygen enrichment of the combustion air was carried out.

4.2.3 The glass forming process

There is probably little that can be done to conserve energy on any significant scale in the glass forming process. During forming the glass must be cooled, and forced convection cooling using air is commonly employed. Improvements in the cooling of glass containers brought about by extended surfaces on the moulds and other more sophisticated modifications could lead to a lower unit cost in terms of fan power. The amount of heat to be dissipated (2320 kJ/kg) during this process is high, but the geometry of many forming machines is such that recovery of this heat would be impractical.

Where compressed air is used to actuate the moulding machines, normal 'good housekeeping' procedures as described in Chapter 5 should be implemented. Over a decade ago work was reported in the United States on freeze drying of compressor air prior to its entry to the compressor, (ref. 4.14). This was done using heat recovered from the aftercoolers, and it was claimed that compressor power consumption could be reduced by up to 25 per cent. Alternatively a heat exchanger could be used to recover heat extracted during cooling to remove water, returning this heat to the dry air to increase its volume.

4.2.4 The annealing process

Following forming of the glass into the required component shape, stresses are removed by passing the glass through an annealing lehr. The modern recirculating lehrs now available and shown in Fig. 4.7 consume substantial amounts of energy, with fans of up to 45 kW for temperature control and cooling, and burners with combustion rates in excess of 600 kW.

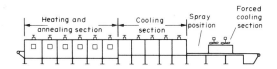

Fig. 4.7 Modern recirculating lehr for glass annealing, with total temperature control
(Courtesy American Ceramic Society)

In a lehr the glass is first heated to a temperature a few degrees above the annealing point, (for soda-lime container glass this is approximately 1000 to 1050°C). It then passes through an annealing zone, where it is gradually cooled to its strain point, the rate of cooling being determined by the need to minimise temperature differentials between the inside and outside surfaces of the glass. After annealing, the glass must be cooled sufficiently for it to be handled, packaged etc. As it leaves the annealing section, the glass contains 80 per cent of its original heat, and effective cooling is necessary to keep lehr length to a minimum and throughput at an acceptably high level.

The Hartford Division of Emhart Corporation in the United States recently reported on a comprehensive study of energy conservation in lehrs and have successfully tested a number of systems incorporating modifications aimed at reducing running costs (ref. 4.15). In one unit, three of the five annealing sections, which had previously incorporated recirculating fans and burners, were replaced by plain sections, as shown in Fig. 4.8. Adjacent sections retaining burners, fans and the temperature control systems were found to be sufficient to maintain the correct annealing temperature profile, with savings

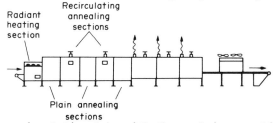

Fig. 4.8 A recirculating lehr with three plain annealing sections
(Courtesy American Ceramic Society)

in gas and electricity consumption. This design has been further modified by replacement of all except the final annealing section (where stabilisation of temperature is most important) with natural convection cooling, as in Fig. 4.9. As well as reducing still further the energy consumption, capital and maintenance costs are also proportionately lower.

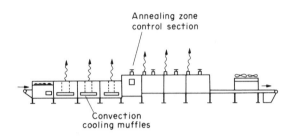

Fig. 4.9 The use of natural convection in a lehr, eliminating most of the fans. (Courtesy American Ceramic Society)

Yet another innovation in lehr design introduced by Emhart was the use of high velocity burners. These were sufficiently effective in ensuring heat circulation that the fans could be eliminated, as the products of combustion were able to entrain air and create a sufficiently high level of turbulence to guarantee uniform temperature control.

Emhart were also able to quantify reductions in energy brought about by returning the lehr conveyor through the inside of the furnace, thus preventing it from cooling down and hence having to be reheated as it picked up the glassware for re-introduction into the lehr. Also, before glass on the conveyor can be reheated uniformly, the conveyor must reach at least the bottle temperature, otherwise it will act as a heat sink. As the base of the bottle is often the thickest part, a preheated conveyor will assist considerably in reducing reheat time.

Based on a lehr with a throughput of 1 kg glass per second, the heat required to raise the conveyor from $38^\circ C$ to annealing temperature is 260 kW. To heat the glass from $480^\circ C$ (the lehr is located close to the moulds) to $565^\circ C$ requires 85 kW. Assuming a 67 per cent efficiency, total natural gas consumption will be 13.8 litres per second. If the conveyor is returned inside the lehr, with only 20 per cent heat losses from the conveyor, the total gas consumption falls to 5.5 litres per second. Thus the saving per annum, based on a gas cost of 10p per therm, amounts to £9070.

Proposals have been put forward by the Electricity Council Research Centre at Capenhurst concerning the use of thermal energy storage in electrically heated annealing lehrs (ref. 4.16). It was considered that lower operating costs could result if a proportion of the energy input could be provided by a storage unit built as an integral part of the inlet end of the lehr. Such a system would only be appropriate where the load was likely to fluctuate during the 24 hour operating period. At times when the electricity load is sufficiently below the 'maximum demand' level, the energy available could be absorbed by the storage unit for use later to avoid incurring additional 'maximum demand' charges. In other words the thermal store is acting as a load levelling device.

An end elevation of the unit, located on one side of the lehr, is shown in Fig. 4.10. A storage capacity of 360 kW hr, with a 4 hour recharge time, would provide a peak output of 100 kW with an average demand capability of 40 kW. Further information on the Electricity Council work on thermal storage is given in Chapter 8.

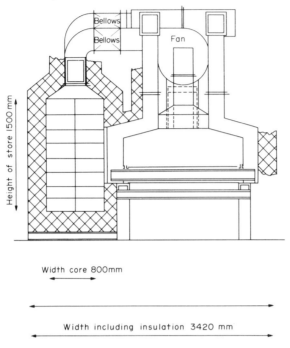

Fig. 4.10 A thermal storage system developed by the Electricity Council for use in lehrs

While the lehr only accounts for a small proportion (typically 5 per cent) of the energy consumed in glass making, these and other 'good housekeeping' measures can result in fuel savings quickly and with little capital expenditure. The layout of the overall plant, with short transit times between furnaces, moulds and annealing furnaces can also assist in this aim.

The most dramatic reductions in energy consumption, in terms of the energy content per unit of production, can be obtained if containers (or other glass products) are reduced in weight. New lightweight glass milk bottles (170 g) have a lower heat capacity, hence can be annealed using less energy. It must also be realised that many times during its life the bottle is likely to be washed or subjected to pasteurisation or sterilisation plant which involves heating and cooling. Bottles with lower masses will also consume less heat in these processes, hence contributing to energy conservation in dairies, breweries etc. Section 4.3 gives information on the food industries, and these topics are discussed in more detail there. (Many of the processes are common to the brewing and soft drinks industry). Dairies also use refrigeration plant, which reject heat, and uses may be found for this heat in other parts of the food preparation process.

4.3 The Food Processing Industry

The preparation and packaging of foodstuff involves many different kinds of processes, a large number of which take the form of either heating or refrigeration. Because food processing in most cases is closely integrated with packaging, be it in milk production, where 'packages' range from glass bottles to plastic and cardboard cartons, or canned foods, it is preferable to include packaging as well as food preparation in this section.

Quantification of the energy consumption in the food industry is difficult because of the many diverse processes involved and, until recently, energy use was regarded as being of minor significance. In the United States total energy utilisation by the food industry was estimated to be 900×10^{15} J in 1971, (ref. 4.11). A more detailed analysis carried out in the Netherlands, where the food industry ranks fourth in listings of the major areas of energy consumption, revealed a consumption of 69×10^{15} J in 1972, (ref. 4.2), this being 9 per cent of the total energy required by Dutch industry. Statistics in the United Kingdom include food in the classification 'food, drink and tobacco', listing consumption in 1972 as 265×10^{15} J, or approximately four times as much energy as that for the Dutch industry, although the bases of the statistics are not identical.

Obviously it would be a major task to study in depth the whole food industry, with its extensive range of processes, and engineers closer to the industry are much better qualified to do this. However, one of the main purposes of this book is to assist the transfer of energy conservation technology between different industries, and there are a number of techniques, including instrumentation, developed primarily for the food industry which could be of considerable interest to users of similar plant in other branches of engineering. The majority of these techniques are closely linked to the packaging of food products in one form or another, hence the emphasis on this aspect made earlier in the introduction.

Two specific areas of the food industry will be considered; the production of canned foods, and the dairy industry (in particular milk processing and bottling). In canning the measurement of temperature and techniques for heating and sterilising the product are of interest. The dairy industry uses a substantial amount of energy in preparing packages for acceptance of the milk, as well as treating the milk itself.

These areas of the industry are noted for their intensive use of water and steam, and the resulting large quantities of effluent, having varying degrees of contamination. It is important to conserve water in food production processes, particularly in the brewery, soft drinks and dairy industries, and this is becoming a commodity as valuable as energy because of increasing costs. To the 'bought-in' cost of water must be added the levies imposed by water authorities for treatment of effluent.

The quantity of liquid waste in any plant may be minimised by good housekeeping. Prevention of unnecessary spillage is an obvious measure and the recirculation of water, be it used for heating, cooling or washing, should where possible be implemented even if some heat transfer or purification treatment has to be carried out prior to re-use.

Because of the significance of this aspect of energy conservation, a section in Appendix 4 is devoted to manufacturers of water treatment plant.

The Energy Intensive Industries - 2   93

4.3.1 Canning of foods   The processes involving energy expenditure in the canning of foods, in this case canned fruit and vegetables, are illustrated in Fig. 4.11, (ref. 4.11). In common with a large number of heat transfer processes, temperature control can be a major aid in ensuring that overheating, with consequent wasting of fuel, is not carried out. Other factors influencing the energy use are associated with the maintenance and insulation of the plant, and also the form of the heating process. In the latter case, for example, it is much more desirable to operate where possible on a continuous flow-line basis than to use batch production methods. While this may not be practicable during cooking, it is more readily achieved when considering sterilisation.

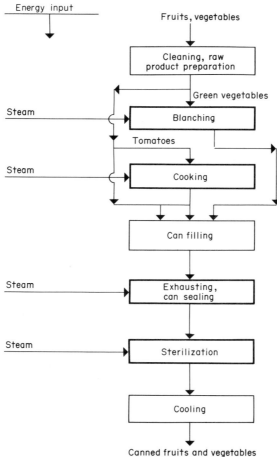

Fig. 4.11 An energy flow diagram for the food canning industry
(Courtesy US Environmental Protection Agency)

The British Food Research Association (see Appendix 3) is actively involved in all of the above areas, and among the suggestions they make concerning energy conservation in this area of the industry are the following:

(i) Use the minimum amount of water addition to a process. This water may have to be evaporated at a later stage in the process, increasing the heat load.

(ii) The type of cooking should be investigated from the point of view of effective energy utilisation. In theory pressure cooking is more efficient than ambient or vacuum cooking.

(iii) In continuous processes more opportunities may exist for heat recovery. During cooling of the product, use recovered heat elsewhere if possible. If the product does not have to be rapidly cooled, use natural convection in areas where the heat can contribute towards space heating.

(iv) Where a batch process is unavoidable, use different pans for heating and cooling of the product.

These and many other energy saving measures may involve investment in new plant. However, as a start, a large number of minor improvements which could, when totalled, save a significant amount of heat, coming under the heading of 'good housekeeping' may be made. The reader is referred to Chapter 5 for information on these improvements, which are often sufficiently general to be of relevance to many different industries and items of plant.

Changing plant or adding insulation are excellent measures in many respects, but inefficiencies in processes may be inherent regardless of these improvements if adequate control of the process is not carried out. In the production of canned foods, (and also in other packaged food production), the monitoring of temperatures has often been neglected, with the result that heat input to processes has been excessive, partly to give a margin of safety. In sterilisation, for example, excessive heating may be applied because of a lack of knowledge of the heat transfer characteristics of the can and contents necessitates a form of 'over-kill' to ensure product quality.

In order to determine the heating required, a knowledge of the thermal history of the product is necessary because the destruction of bacteria is a function of both temperature and time. For example treatment for one minute at $120^{\circ}C$ has approximately the same effect on lethality as treatment for one hundred minutes at $100^{\circ}C$. Because this effect is highly non-linear, accuracy of temperature measurement becomes more important as the treatment temperature increases.

If the temperature, or in some instances the pressure, in the container can be measured during the heating process, the quantity of heat required in a certain time to create sterile conditions can be accurately assessed. One technique for measuring in-can temperatures is radiotelemetry, (ref. 4.17). Using radiotelemetry probes it is possible to measure the temperature inside moving cans remotely, and thus the optimum conditions for heating the cans and maintaining them at the required temperature, determined as indicated in the previous paragraph, can be set up. The technique can also be used to optimise cooling curves for various products, (ref. 4.18). Normally a single can may be instrumented by inserting a thermistor into the centre of the contents. This changes in resistance as the temperature varies, and if the signal is converted into a frequency of about 1 kHz, this can be picked up by a receiver, changes in frequency then indicating changes in temperature, (ref. 4.19).

The Energy Intensive Industries - 2

The cooking of food prior to and after canning is one of the most energy intensive aspects of the food industry - in a cannery it amounts to approximately one half of the total heating requirements. Sterilisation can be carried out in a number of ways other than by conventional steam heating. Ionising radiation can be used, and bacteria can be killed by the application of intense high frequency fields (ref. 4.3). It is possible that this could permit sterilisation without heating the material appreciably, hence leading to important energy savings.

The Food Research Association are studying the use of heat transfer in fluidised beds as a means of improving the processing of canned foods, and this could also offer promising increases in efficiency. Fluidised bed technology is described in detail in Chapter 9.

4.3.2 The dairy industry The dairy industry provides us with an increasingly wide variety of liquid and solid food products, the most important of which is bottled milk. A number of recent papers have considered the overall energy consumption in dairies (ref. 4.20) and in packaging (ref. 4.21) in the United Kingdom, and similar studies have been carried out in the United States (ref. 4.11). Some aspects concerning the use of plate heat exchangers in pasteurisation are also discussed in Chapter 7.

It is intended to concentrate here on one particular aspect of the dairy industry which is receiving an increasing amount of attention from the energy conservation point of view, namely bottle washing. (The types of machine used for washing milk bottles are also used in the brewing and soft drinks industries). Machines operating on similar principles are used in a wide variety of other industries, including the motor component manufacturers, and in dishwashing.

The units are excellent examples of plant in which both heat and water conservation can play a major role, and also where the items being processed have a significant bearing on the heat transfer within the machine. Because of these factors, the washing procedure will be described in some detail.

Operation of bottle washing machines: A bottle washing machine, which can process several tens of thousands of bottles per hour, conveys the bottles through a number of stations, as shown in Fig. 4.12, at each of which the bottles are treated with water or a detergent solution. The form that this

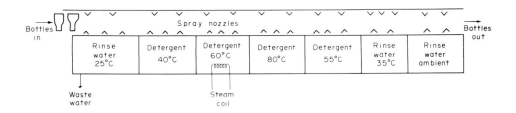

Fig. 4.12 Sectional view of a bottle-washing machine

treatment takes depends upon the type of bottle and its application, and machines capable of jetting the bottles with hot detergent solutions, soaking the bottles (a process which assists label removal),or using a combination of jetting and soaking, are available. In the dairy industry most bottle washers use jetting systems, spraying the inside and outside of the bottles with high pressure jets of water and detergent. Heat is supplied to the water and detergent solutions using steam coils immersed in the tanks below the spray bars.

In order to be effectively cleaned, this involving removal of both visible contamination and bacteriological matter, the bottles have to be raised to comparatively high temperatures, of the order of 70 to $80^{\circ}C$. As the bottles entering the machine are normally at ambient temperature, and for pasteurised milk, are required to be cold when delivered to the filling machine which is situated a short distance downstream of the washer, heating and cooling is carried out in a number of stages. A warm pre-rinse with ordinary water removes gross contamination before the bottles pass into detergent solution sprays. The bottles are subjected to sprays at temperatures gradually increasing towards the peak value, and are subsequently slowly cooled, a final rinse with fresh mains water bringing the bottles back to the required ambient conditions ($10^{\circ}C$). A bottle takes three to four minutes to pass through the machine, and while a dairy may be filling bottles for eight hours per day, breweries and soft drink manufacturers tend to operate for twenty hours per day, a four hour period being reserved for cleaning, maintenance etc.

Considerable quantities of heat, typically in excess of 500 kg/hr of steam, depending upon the capacity of the bottle washing machine, are needed to maintain water and detergent solution temperatures during operation. In addition, water from the rinse used to remove gross contamination flows straight to waste, at rates of up to 10 000 litres per hour and a temperature of 20 to $30^{\circ}C$. Heat transfer within the machine itself is quite complex. Large quantities of heat are carried from one washing section to another by the bottles and associated transport mechanisms, and this heat is given up to the cooling sections and the final rinse water.

Some heat recovery and water conservation is currently employed on these machines. For example, the final cold rinse water is heated, in part by contact with the bottles, for use in the warm rinse, and is then passed through a heat exchanger located in one of the detergent tanks on its way to the inlet to the machine. Here it supplements the pre-rinse before being discharged to waste.

Obviously a substantial amount of the heat input to a machine of this type is lost by radiation and convection, but if one considers that steam costs approximately £2.50 to £3.00 per 1000 kg, and water supplied by local authorities is currently approaching 13p per 1000 litres, recovery of heat and water becomes in principal, attractive. Additional charges are levied in most areas for treatment of process effluent, and water discharged from bottle washing machines may attract a levy of up to 7p per 1000 litres over and above the initial cost of supply.

Heat and water recovery in bottle washing machines: On the assumption that a particular heating and cooling pattern must be followed in bottle washing, due to the need to give detergent solutions an opportunity to be effective, one is left with two alternatives as far as energy conservation is concerned. One may redesign the 'package', in this case a glass bottle, so that less energy is used in treating it. Alternatively one can invest in a heat and/or water recovery system.

As far as the package is concerned, opportunities exist for conservation of energy extending back to the manufacture of the bottle. A reduction in the mass of the bottle, characteristic of the new light-weight milk bottles which weigh approximately one half of their predecessors, saves glass and the 'energy content' of the bottles is reduced in proportion. The significance of a lower bottle weight with respect to the washing is concerned with the amount of heat which is required to raise the temperature of the bottle to values corresponding to that required to carry out a satisfactory treatment. The quantity of heat taken up by the smaller mass of glass will reduce the amount of heat transferred by the bottles from one tank to another in the machine. Hence the heat input to each tank will be reduced.

More immediate and equally significant energy conservation measures can be taken by installing a comprehensive heat and water recovery system. As discussed above, a number of heat exchangers are currently employed on most bottle washing machines for recovery and redistribution of heat given up to the cooler water by bottles moving out of hot sections into cold sprays. Recently bottle washing machines have been identified as a suitable medium on which to apply the heat pump principle. Heat pumps, described fully in Chapter 7, are particularly appropriate in an application such as this, where there are a number of low grade heat sources and high grade heat sinks. The form which the heat pump might take in this application is discussed below.

Electric heat pump system (Milpro NV): The Milpro heat pump system, the subject of a patent which is listed in Appendix 5, includes water treatment facilities, and it is claimed that this unit provides all the heat required for operation of a bottle washing machine. A simplified layout of the heat and water recovery plant is shown in Fig. 4.13. The bottle washing machine follows a similar temperature profile to that of the unit illustrated in Fig. 4.12.

A conventional heat exchanger is used to recover heat from the final detergent tank, this heat being used to raise the temperature of the pre-rinse spray water. The input to the pre-rinse sprays is supplemented from the final rinse tank ($30^{\circ}C$) and also by overflow from the final detergent tank ($45^{\circ}C$). A filtration system, which rejects a proportion of the pre-rinse water, is used to maintain pre-rinse water in a comparatively clean condition.

Once some of the heat has been moved from the water passing out of the final detergent tank for transfer into the pre-rinse water, the final detergent tank water is then passed through an evaporative heat exchanger, serving as one of the heat sources for the heat pump, before being returned at $30^{\circ}C$ to the final warm rinse and final detergent tanks. Prior to passing through the evaporator, a proportion of this flow is taken for filter treatment and cooled to 12 to $15^{\circ}C$ in a second evaporator. This clean water is then used as the final cold rinse water supply.

The heat recovered in these evaporators is taken up in the form of refrigerant vapour, which is then compressed, raising its temperature by the addition of the work of compression. This heat is then rejected at the higher temperature in a condenser to water from the first detergent tank (raised from 55 to $65^{\circ}C$) and also in a second condenser, where water for the high temperature section is heated from $65^{\circ}C$ to $80^{\circ}C$.

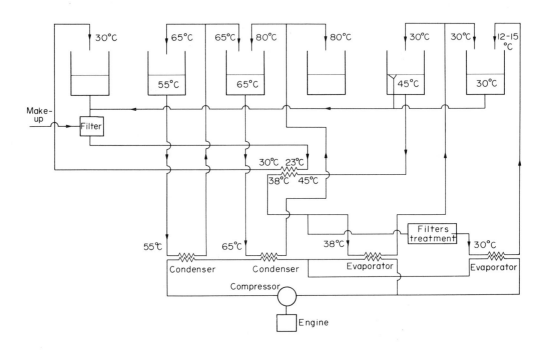

Fig. 4.13 A heat pump and water recovery system developed by Milpro NV for bottle washing machines

Any make-up water required is added to the pre-rinse water circuit. Based on a machine washing 30 000 bottles per hour, consuming, with conventional heating and water usage, 13 600 litres and 600 kW heat energy per hour, Milpro claim that the heat pump and filter system can reduce consumption to 70 kW and 2600 litres per hour. Payback periods of two to three years are claimed for the unit, which can run either on natural gas or electricity.

(There are a large number of techniques for treating water, including chlorination, filtration, electroflotation and reverse osmosis. Manufacturers of treatment plant are listed in Appendix 4 for the benefit of those readers who are interested in combining these two aspects of energy conservation - namely heat and water recovery.)

Similar opportunities exist for heat and water recovery in other dairy processes, as well as in breweries and other bottling plant. Refrigerator units may also be regarded as a heat source, and possibilities exist for integrating this with the heat pump unit, either as a heat source for a

The Energy Intensive Industries - 2

separate heat pump circuit or as a heat supply, where it is acting as the condenser in the refrigerator (heat pump) circuit.

Use of a heat pump in dishwashers: The washing cycle of conventional large dishwashing machines is similar to that of bottle washers, and it is natural that the heat pump is applied here also. In this case the rinse water is heated to 90°C, somewhat higher than that in bottle washing machines. The waste heat is discharged in part as hot vapour, and it is this which is used as the heat source in the heat pump system. The resulting cooling of this vapour improves comfort conditions in the vicinity of the washer, as well as saving energy.

The heat pump, operating with a coefficient of performance of 4 (see Chapter 7) raises the rinse water to 60°C, and affects the power input to the washer as follows:

<u>Conventional</u>

	Input		Output	
Motors		5.5 kW	Heat dissipation	5.5 kW
Calorifier		67 kW	Waste water	26.5 kW
Tank Heating		54 kW	Dishes	41.5 kW
Total		126 kW	Vapour	53.0 kW

<u>Heat Pump Design</u>

	Input		Output	
Motors		5.5 kW	Wasted heat	5.5 kW
Heat for final rinse		25 kW	Drain	26.5 kW
Heat pump compressor		24 kW	Dishes	22.5 kW
Total		54.5 kW	Recovered heat	96 kW

The capital cost of this unit is recovered in thirty one months.

4.4 The Textile Industry

The textile industry is particularly vulnerable to increasing costs of labour, raw materials and energy because of its relative position with respect to overseas manufacturers in countries where labour is cheap or where subsidies may be given to boost exports. Energy costs in the industry are not insignificant - in 1973 the energy bill for Courtaulds Limited was £50 million and this company is probably more energy conscious than many of the smaller organisations in textile manufacture. The aim of Courtaulds was to save 10 per cent, i.e. £5 million, on the energy bill within one year (ref. 4.22).

Textile manufacture involves a number of processes which begin as being rather diverse, the production of materials for synthetic fibres being a completely different procedure to the preparation of wool and cotton, but later tend towards similarity in finishing processes; indeed it is in the area of finishing (dye-houses, dryers and stenters) that energy is a major portion of the total cost involved.

It is proposed to concentrate in this section on the finishing processes, as they involve in part equipment common to a large number of industries, namely dryers. Dye-houses use large amounts of hot water, and again heat recovery here has much in common with other industries and items of plant, not least bottle washing machines discussed earlier in this chapter. Mention will also be made of the work of two Research Associations that are closely involved in energy conservation in the textile industry.

4.4.1 Dye-Houses The recovery of heat from the hot effluent leaving dye-baths and similar plant in the textile industry is a technique which is becoming much more popular. However there are still instances where waste heat and, equally important, waste water is dissipated in substantial quantities, and while a number of developments in the dyeing process have decreased energy consumption, replacement of all plant requires considerable capital expenditure. In many situations the limited capital available might well be invested if used to purchase heat and water conservation equipment. Alterations in procedures, particularly with respect to planning of work loads to ensure maximum steady equipment utilisation, is also important.

Boiler efficiency is improved if fluctuations in steam demand can be minimised, either by re-scheduling or by incorporating some form of thermal storage device (see Chapter 8). Start-up of plant at the beginning of a shift, followed by a period of full utilisation, makes particularly high demands on steam consumption. If there is an insufficiently steady flow of work to maintain full utilisation throughout the shift, staggered start-up and operation would be preferable, possibly in conjunction with an accumulator (ref. 4.23). Of course any improvements of this type affecting steam utilisation should be carried out with consideration being given at the same time to correct steam trap functioning, boiler burner optimisation etc. as discussed in Chapters 5 and 6.

The dye-house should not be treated in isolation from the remainder of the plant. Waste heat from dryers, stenters and washers may also be considered for heating dye-house water, if not recirculated within the dryer units themselves. Reference 4.23 cites heat recovery from boiling effluent from an open-width washing machine, which, when used to preheat supply water from ambient to $65^{\circ}C$, saved 35 per cent of the machine steam consumption. In cases where raw steam is injected directly into fluids to promote rapid heating, operatives should be made aware of the need to control injection rate so that all the steam condenses before reaching the liquid surface. As emphasised later in this book, the cost of boiler feedwater is such that direct steam injection heating should only be used as a last resort.

Gillies (ref. 4.24) highlights the problems of dye-house waste disposal brought about by the restrictions imposed by some local water authorities on the maximum temperature at which waste can be discharged. In some plants water had to be stored in settling tanks prior to final discharge, to permit natural cooling to take place. In a case where 91 000 litres of effluent are discharged per day from a dye-house, heat recovery could save up to 9000 kg steam per day if the effluent was cooled to $38^{\circ}C$ before discharge. Over an annual period, this would save approximately £6000. Shell and tube heat exchangers of the simplest type could be used, although care would be necessary in selecting the materials to ensure resistance against any corrosive matter in the effluent. Fibrous matter is a common constituent of dye-house effluent, and filtration would be advisable upstream of the heat exchanger to prevent fouling, which would in time reduce heat transfer effectiveness.

Considerable interest is evident in the textile industry in techniques for speeding up the dye process, (ref. 4.25). Much can be accomplished by the introduction of precise process control, in conjunction with accurate knowledge of the properties (weight and moisture content) of each batch of yarn being dyed. Computer control systems have been used to implement these requirements.

One method which offers energy and time savings is package dyeing. A package dyeing machine comprises a cylindrical pressure vessel with a cover, often pneumatically operated. Dye liquor is generally circulated using a high capacity centrifugal pump, (ref. 4.26). Heating of the dye liquor is implemented via a heat exchanger, thus saving boiler feed-water, which could affect the dying process if injected directly. By carrying out the process in a pressure vessel, a greater range of temperatures can be achieved, increasing the versatility of the machine. An advance on the basic package dyeing system has been incorporated in units produced by Obermaier & Cie, as shown in Fig. 4.14. This machine differs from standard models in that the dyestuff is injected into the machine only after the maximum dyeing temperature has been reached.

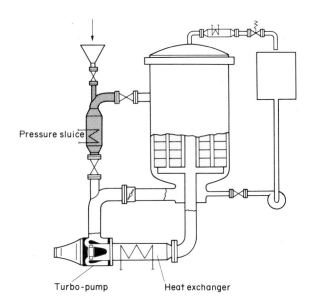

Fig. 4.14 A turbostat vertical pressure package dyeing machine incorporating heat recovery

This arrangement eliminates the controlled heating-up phase, shortening the dyeing cycle and reducing dyeing times. Less heat is needed, peaks in heat demand are reduced and liquor quantities used are less than in conventional pressurised package dyers. Dyestuffs are heated to the required temperature in the pressure sluice, and the turbo-pump then rapidly distributes the contents uniformly throughout the material in the vessel. Obviously waste heat can be used to full advantage in the two heat exchangers on the machine.

Fabric printing is followed, particularly in the case of synthetics and polyester/cotton materials, by a fixation process which must be carried out at a high temperature, of the order of 180°C. Normally the complete process is implemented using superheated steam as the heating medium. Work carried out by Artos in Germany has shown that an externally heated standard loop steam-ager (fixer) with a working width of 2.6 m and a fabric capacity of 200 m would require 1700 kg of steam per hour. Based on a steam cost in Germany of approximately £4.75 per 1000 kg, the total operating cost per annum would be £21 500, based on two shifts per day. An internally heated universal loop machine of identical capacity would require 560 kg of steam per hour, costing £7000 per annum. Although this type of machine is more expensive, steam savings indicate a payback period of one to one and a half years. With the internally heated system, Artos suggest that some of the steam heating may be replaced by hot air heating, with no loss in quality, providing that correct dyes are used. Steam consumption is reduced by over 65 per cent, and the annual running costs drop to only £2500 (ref. 4.27).

The textile industry, particularly dyeing, has much in common with laundry processes, and a considerable amount of effort has been put into the conservation of heat and water in large laundries (refs. 4.28, 4.29, 4.30). Available sources of heat in a laundry which are also present in textile plant include air conditioning units (0.9 to 1.5 kW per ton of refrigeration), heat from waste water (discharged from laundries at 57°C), heat from flash steam or hot condensate (unless returned to the boiler) and exhaust from dryers. (In the latter case, preheating of drying air is the preferred use, as

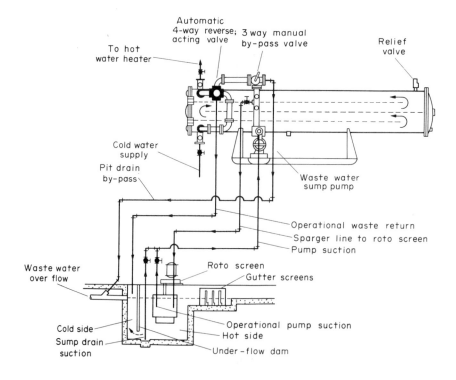

Fig. 4.15 Waste water sump pit recovery system
(Reprinted by permission of the American Society of Heating, Refrigerating & Air-Conditioning Engineering Inc., from ASHRAE Transactions)

discussed below).

One recovery technique is illustrated in Fig. 4.15, using a waste water sump pit, (ref. 4.28).

Waste water discharged from the washers flows by gravity through a screened gutter to the sump pit. The pit is sized to accept the full volume of all machines discharging at the highest required level. The water level in the pit remains constant and is governed by the elevation of the overflow line to the sewer. The bottom of this line should be below the inlet gutter to the pit, as it will guarantee complete drainage of the gutters. The pit is divided into two sections: hot inlet side from the gutter and cold overflow side to the sewer. The dividing partition or baffle, open at the bottom, provides a level, equalising underflow between the two sections. The intermittent discharge from the wash wheels to the hot side of the pit displaces an equal volume of cooled waste water under the dam and out through the overflow line to the sewer. The pressurized closed waste water heat recovery unit is a special multipass shell-and-tube heat exchanger with quick-opening hinged doors at each end. The waste water is pumped through the straight tubes and the fresh water passes through the cylindrical baffled shell surrounding the tubes.

A second system, which can be operated as a primary heater, rather than relying on waste recovery, is shown in Fig. 4.16. Also operated in laundries (ref. 4.30), it is designed to cater for intermittent requirements for hot water in cyclical processes such as washing. The idea of this arrangement is to minimize energy input rate consistent with meeting demands which could normally only be satisfied with high energy surges. Heated water is kept in

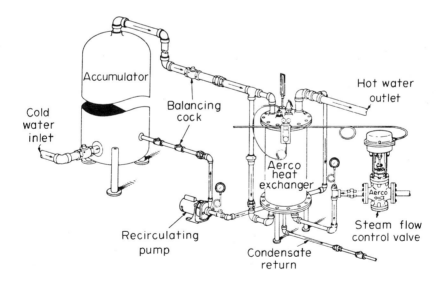

Fig. 4.16 Hot water heating system for optimum energy use in laundries (Reprinted by permission of the American Society of Heating, Refrigerating & Air-Conditioning Engineering Inc., from ASHRAE Transactions)

a storage tank upstream of the heaters, and the heated water is drawn into the hot water supply piping directly from the heater. Cold water is injected into the storage tank as make-up, and this is supplemented by a flow of hot water back from the heater outlet. Balancing cocks are so adjusted as to keep the flow through the heater constant regardless of demand, any surplus being returned to the storage tank, where it will rise to the top (outlet). Thus the warmest water in the storage tank will be the first supplied to the heater.

4.4.2 Drying in textile plant

It is becoming increasingly evident that the use of heat to evaporate water as a means for the drying of fabrics is an expensive process, and more effective heat recovery, in conjunction with prior mechanical treatment using mangles or vacuum slots to reduce the water content to a minimum, are necessary components of efficient mills. Greater control of drying processes, based on humidity and temperature measurements, also contribute to efficiency, as discussed in Chapters 5 and 6.

Drying can be carried out using hot air directly, or by passing the material over heated cylinders, similar to the process used in the paper industry described earlier. Commonly known as contact drying, the evaporation of moisture from the textiles is brought about by heat conducted through the walls of the cylinders, which are steam-heated internally. Uniform and good heat distribution depends on rapid removal of condensate, the maximum amount of condensation occurring in the first two or three cylinders. Heat losses from the ends of the cylinders and the feed lines can be minimized by thermal insulation. As with paper drying, the supplementation of cylinder heating with forced air across the top of the material assists with the removal of humid air, and accelerates the drying rate.

Apart from return of the condensate to the boiler, the opportunities for direct heat recovery in cylinder dryers are not so great as in stenters, where air heated by steam is applied directly to the cloth. There are also greater opportunities for accurate control of drying conditions to influence the total energy consumption in stenter-type dryers. Adjustment of recirculation and exhaust dampers can, states Sundaram (ref. 4.23) have a marked effect on energy consumption, and in some stenters the exhaust fan can be left inoperative with some benefit to the energy consumption. In the above reference stenter steam consumption of 1.9 kg per kg of water evaporated was reduced to 1.5 kg per kg, when these factors were taken into account in setting up the machine.

The most important fact to consider when selecting a heat recovery unit for textile dryers, and indeed for most other drying applications, is that the humidity of the exhaust air must not be given up to the incoming air. Thus the heat exchanger must be of the recuperative type, where a solid wall separates the exhaust and inlet air flows. Economiser-type units are cheap and operate with efficiencies of up to 50 per cent (ref. 4.22), in one example this type of heat exchanger cooled 4.7 m^3 per second of exhaust air from 175°C to 90°C, while raising the temperature of 3.25 m^3 per second of fresh incoming air from ambient to 90°C, a recovery rate of almost 300 kW.

Heat pipe heat exchangers, described in depth in Chapter 7, have been used extensively in the United States in the textile drying field. Stenter dryers, incorporating these units with electrostatic precipitators to minimize fibre contamination in the exhaust, have benefited from their use. Spiezman Industries Inc., claim that savings per dryer range from £2000 per annum on a

2 m³ per second unit to £16 000 on a 9.9 m³ per second system. Payback periods are comparatively short if one relates this to the cost of heat pipe heat exchangers given in Chapter 7.

The hot air exhausted from stenters which are processing synthetic fibres can contain organics which are a source of atmospheric pollution. Cartwright (ref. 4.22) reports on an interesting application of catalytic combustion which, when applied to the fumes in stenter exhausts, reduces the pollutants to carbon dioxide and water. In order to sustain the reaction, the exhaust must be raised in temperature to 360 to 380°C, and this is the function of the catalytic burner. In the example quoted a supplementary oil burner is used to raise the temperature of the exhaust before it reaches the catalyst.

Two possibilities exist where the relative humidity of the exhaust air is at an acceptably low level. It could be reintroduced into the stenter once it has been cleaned by the catalyst to transfer heat to the air entering the stenter; in this case it may be acceptable to use a regenerator if the air is dry. Alternatively recuperators of the type discussed above may be used, and recovery efficiencies of 70 per cent have been achieved.

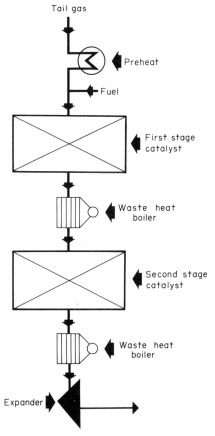

Fig. 4.17 An example of heat recovery downstream of a catalytic reaction (Courtesy Engelhard Industries Ltd.)

106   Industrial Energy

The purification of exhaust gas using catalysts in conjunction with waste heat recovery is not unique to the textile industry. Fig. 4.17 shows a two stage Engelhard catalyst installation used to purify nitric acid tail gas, with the catalysts operating at temperatures of up to 800°C. Heat is recovered using waste heat boilers.

4.4.3 The activities of the Textile Research Associations

There are a number of Research Associations in the United Kingdom which are in varying degrees active in promoting energy conservation in the textile industry (see Appendix 3). The Wool Industry Research Association (WIRA) can carry out surveys on behalf of companies, and are very active in machinery and instrumentation development. Included in their development programme have been the following projects:

- (i) Development of an automatic dissolving unit which can produce large quantities of sodium carbonate solution using cold water. Previous methods required steam.

- (ii) Improved scouring set which has reduced the hot water consumption in washing greasy wool by 50 per cent.

- (iii) System for controlling humidity in loose wool dryers to minimize heat loss.

- (iv) System for the control of cloth regain to minimise heat usage in dryers.

- (v) Effluent treatment techniques to enable re-use in processes.

- (vi) Flash steam heat recovery system - pay back periods of one year are typical.

- (vii) Development of radio frequency (r.f.) package dryers, leading to a 30 per cent reduction in drying costs.

The Textile Institute in Manchester are also active in the promotion of energy conservation. Shortly after the commencement of the 'oil crisis' a number of presentations were made on ways of reducing energy consumption in the textile industry. Recommendations ranged from 'in-house' electricity generation (with heat recovery) to the use of low-energy content textiles. In general finishing processes came in for the greatest attention, and an attempt was made to quantify the savings possible. These are listed in Table 4.6, (ref. 4.31).

4.5 Conclusions

All of the industries discussed in Chapters 3 and 4 are concious of the need for energy conservation, and, particularly among the larger organisations there are numerous ways in which it has been implemented. Apart from 'good housekeeping', which is discussed in the next chapter and is common to all industries, and some of the more specialised processes unique to the chemical or non-ferrous metals industries, it is hoped that the discussion will have 'sown the seed' for ideas which may be applied in the medium term by energy managers responsible for processes, and indeed industries outside those mentioned.

TABLE 4.6 Check List of Energy Conserving Measures
in the Textile Industry

(a) Water Using Processes

Action	Typical Saving (therms/100kg cloth)
1. Avoid underloading	Proportional to loading
2. Re-use of hot water	1.8
3. Use counterflow	1.3
4. Limit idling losses	0.9
5. Reduce rinsing water	1.3
6. Reduce water temperature	0.9
7. Recover heat where possible	35% of total

(b) Drying and Heating Processes

Action	Typical Saving (therms/100kg cloth)
1. Maximize mechanical drying	0.2
2. Avoid over-drying	0.45
3. Minimize setting and dye-fixation times	0.2
4. Reduce exhaust air to practical minimum. Monitor humidity of drying processes	} 25% of total
5. Examine thermal efficiency of forced convection systems	Small
6. Reduce idling losses, e.g. by switching off exhaust air	0.45
7. Recover heat from exhaust air	25% of total

To the author, it seems that the recovery of waste itself, typified by the recovery in bottle washing machines, in addition to the heat recovery, is most significant. It is in this area that interest is possibly most rapidly growing, and this extends to incorporate the use of waste to generate power and/or heat, discussed in Chapter 6.

REFERENCES

4.1 Anon. Energy conservation in the UK - achievements, aims and options. National Economic Development Office, London HMSO (1974)

4.2 Over, A.J. and Sjoerdsma, A.C. Energy conservation - ways and means. Future shape of Technology Publications, No. 19, The Netherlands (1974)

4.3 Anon. Technology of efficient energy utilisation. Report of a NATO Science Committee Conference, France. (NATO Scientific Affairs Division), Oct. 1973

4.4 Newton, G.E.H. Fuel and productivity in the pulp and paper industry Paper 11. Fuel and Productivity Conference, London 1963.

4.5 Webzell, A.B. Energy utilisation in the paper industry. Combustion 44, 8, 36 - 42 (1973)

4.6 Frei, D. Gas turbines for the process improvement of industrial thermal power plants. Sulzer Technical Review, 56, 4, 195 - 200 (1975)

4.7 Frei, D. and Holik, H. Total energy supply systems for paper mills. Sulzer Technical Review, 55, 3, 189 - 194 (1973)

4.8 Minett, P. Radio frequency and microwave heating saves labour and energy. Electrical Review, 5 Dec. 1975

4.9 Anon. Energy saving: The fuel industries and some large firms. Energy Paper No. 5, Department of Energy, HMSO, London (1975)

4.10 Meyers, P.G. The potential for energy conservation in the pulp and paper industry. Paper Trade Journal, pp 68 - 71, 17 -24 Feb, 1975.

4.11 Reding, J.T. and Shepherd, B.P. Energy consumption: Paper, stone, clay, glass, concrete and food industries. Dow Chemical Company Environmental Protection Technology Series. Report EPA-650/2-75-032-c, April, 1975

4.12 Annual Report, 1975. The British Glass Industry Research Association, Sheffield (1975)

4.13 Budov, V.M. and Seskutov, Yu.V. Use of waste gas heat in waste heat boilers. Glass and Ceramics (USA) 31, 7 - 8, 567 - 568 (1974)

4.14 Anon. Dry compressed air in glass manufacturing. Glass Industry (US) 40, 689, (1959)

4.15 Fuller, R.A. Lehr design for energy conservation. Ceramic Bulletin (US), 54, 3, 277 - 279 (1975)

4.16 Gibbs, M.G. et al. Thermal storage for industrial process heating. Electricity Council Research Centre, Capenhurst Report ECRC/M504, Sept. 1972.

The Energy Intensive Industries - 2

4.17 Hawkins, A.E. and Cowper, J.E. Research at Unilever Research Laboratory, Colworth/Welwyn. *Chemistry & Industry*, 45, 1426 - 32 (1970)

4.18 Anon. Radio telemetry equipment. Food Research Association Leaflet No. 8, June 1973

4.19 Steele, D.J. Radio telemetry; the development of this technique to measure product temperatures in closed or continuously moving equipment. Food Research Association, Research Report. No. 186, undated.

4.20 Starkie, G.L. Some aspects of energy conservation in dairy process plant. *J. Society of Dairy Technology*, 28, 3 (1975)

4.21 Bunt, B.P. The energy used by packaging and its minimization. *J. Society of Dairy Technology*, 28, 3 (1975)

4.22 Cartwright, K. Modern air conditioning for the textile industry. *Textile Month*. May 1975.

4.23 Sundaram, S. Effective use of heat energy in textile finishing. *Colourage*, April 18, 1974

4.24 Gillies, J. Heat storage and waste heat recovery. *The Plant Engineer* 14, 11, (1970)

4.25 Ward, J.S. How fast can you dye? International Textile Machinery, Supplement to Textile Month, 1973.

4.26 Anon. Package dyeing - the future way. *Textile Month*, Jan. 1975.

4.27 Anon. Reducing steaming costs in fabric print fixation. *Textile Month*, August 1975

4.28 Coleman, J.J. Waste water heat reclamation. *Trans. ASHRAE*, Pt 1, 370 - 384 (1974)

4.29 Killebrew, J.B. State of the art of laundry water heating. *Trans. ASHRAE*, Pt. 1, 363 - 369 (1974)

4.30 Angelery, H.W. Meeting maximum energy demands with minimum average energy input rate in water heating systems. *Trans. ASHRAE*, Pt. 1, 385 - 389 (1974)

4.31 Anon. Implications of higher energy costs for the textile industry. *The Textile Institute and Industry*, 12, 2 (1974)

Common Items of Plant - 'Good Housekeeping'

So far in this book we have dealt with the broader aspects of energy conservation, emphasising the need for economies, and highlighting areas of industry where energy intensive processes exist. The potential for energy conservation has also been indicated.

This and the subsequent Chapter deal in more detail with the individual items of plant used in these industries. Many of the processes used are common to a wide range of industries, some of which are not considered separately in this book. As a result Chapters 5 and 6 in effect widen the scope of the treatment to almost every industry, whether it employs, for example a boiler, a small furnace or batch drying machines. This Chapter discusses 'good-housekeeping' - factors such as preventative maintenance, the use of temperature control systems, insulation and similar low cost activities which can save energy. A section at the end of the Chapter gives information on fuel additives.

Where the treatment of a particular aspect of energy conservation on an item of plant requires detailing in some depth, or where the energy conservation technique is common to a wide range of different processes, full discussion is in some cases reserved for later Chapters. This applies in particular to heat recovery techniques, which are fully covered in Chapters 6 and 7. Obviously there are several features which cannot be considered in depth in a book of this type, and more comprehensive treatment, particularly in the form of theoretical analyses, is given in the references cited in each Section, and in Appendix 5.

There are a number of organisations which are fully equipped to offer advice on energy conservation in any plant, and these are listed in Appendix 3. Similarly, equipment manufacturers who can supply systems which contribute towards more efficient fuel utilisation are tabulated in Appendix 4.

5.1 Air Conditioning

Air conditioning, as applied to buildings in commercial and manufacturing plant, relates to the control of the working environment to maintain temperature and humidity within limits appropriate to the type of activity being carried out. In most cases the environment has to be maintained at comfort conditions for the personnel working within the building, but air conditioning is also necessary in plant where the particular process being carried out necessitates temperature and humidity control. Storage of goods is another example where conditioning can affect the 'shelf life' of the material being stored.

An air conditioning system has to compensate for a large number of variable energy inputs to and outputs from a building, and these are illustrated in Fig. 5.1. Unless the system is operating efficiently and is well-maintained, the energy balance can be easily upset and the resulting losses can have a significant affect upon the system operating costs. While it is becoming increasingly obvious that conditioning of office and plant spaces in winter can be particularly expensive, it may not be equally realised that the air conditioning duty in summer can also be raised to unacceptable limits by the flow of heat into the building.

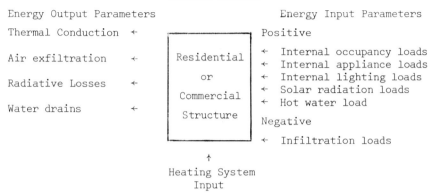

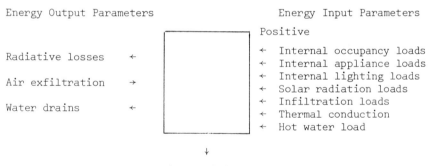

Fig. 5.1 Building energy inputs and outputs to be considered in air conditioning analyses

Much can be done to minimise the energy requirements of a building by careful design. The structure itself can have a considerable influence on the seasonal thermal variations. Increasing the thermal storage capacity of a building helps to damp down seasonal variations, and this can be done either by using a heavy structure or by the incorporation of some form of thermal energy store centrally located, using heat of fusion salts or some comparable medium. These energy storage techniques are discussed in detail in Chapter 8.

A technique which can also be effectively fitted to existing buildings is the use of solar radiation screens. Solar radiation during the summer months can be responsible for a large part of the air conditioning load, and solar glare also leads to uncomfortable working conditions. Coatings for windows, in the form of an adhesive film, are now available (see Appendix 4) which can transmit visible radiation with only a very slight loss, but reflect infra-red radiation. One film, manufactured by the 3M Company, can cut solar heating during summer by over 75 per cent. Film coatings such as aluminium, indium oxide and zinc oxide have all been used to demonstrate this phenomenon.

Heat inflow is one aspect of air conditioning which can be controlled using these films. Various combinations of double glazing, low conductivity gases in between the two glass layers, and the application of low conductivity and low radiation transmitting coating can have a marked effect on the thermal conductivity of windows, as illustrated in Table 5.1.

TABLE 5.1 Thermal Conductivity of Glass/Coating Combinations

Type of Window	Conductivity (kJ/m^2 h $°C$)
Single glass	21
Double glass with air	11
Double glass with one coating	6.6
Double glass with one coating and krypton gas	3.8
Double glass with two coatings and krypton gas	3.3
Single glass with one coating	11

Obviously when an air conditioning plant is installed in a building, it is tested and balanced to meet the expected load requirements. However the loading may be changed as building utilisation varies, and a regular programme of testing, balancing and adjustment, if necessary, should be implemented. A wide range of accurate instrumentation for flow distribution, velocity and pressure drop measurement is available for this purpose. One part of the system which is frequently neglected is the filtration. In any ducted air system the filter can, if not cleaned or replaced, create a large pressure drop, leading to excessive fan power consumption. A dirty filter also leads to unpleasant office conditions. While replacement filters of the type originally used by the manufacturer of the air conditioning plant may be used, the development of filters is such that new types having a better performance may be available, and the user should be aware of improved accessories in all fields.

A number of air cleaners are now available which can be installed in air conditioning ducts to remove many types of contamination which would otherwise be distributed throughout the system. Most operate on the principle of electrostatics, and some manufacturers are listed in Appendix 4.

Similar maintenance should be carried out on the heating and cooling coils. These normally employ extended surfaces to aid heat transfer, and cleanliness of these surfaces is reflected directly in operating costs and efficiency.

Insulation of air conditioning ducts can also contribute to more efficient operation. Where ducts may span areas between buildings, or where they pass through sections of a building which do not require air conditioning, excessive heat gain or loss can occur, and it is in these areas where some form of insulation should be applied. Thermal insulation performance is dealt with in more detail in Chapter 8. However, the insulation of industrial buildings is, in the United Kingdom, actively encouraged by a provision in the 1975 Finance Act, which provides a 100 per cent first year allowance for expenditure incurred on adding insulation against loss of heat to an existing building. This covers installation and material costs, for roof lining, double glazing etc. Also covered are devices for preventing heat loss from loading bays and doors, which are discussed later in this Chapter.

It is sometimes forgotten that air conditioning implies regulation of humidity as well as temperature. In some cases it can be less expensive as far as energy consumption is concerned to vary the humidity than to control the temperature. This is applicable only when small corrections are required to satisfy comfort conditions.

Air conditioning units are now being installed more frequently with heat recovery units built in at the design stage. The various layouts possible are detailed in Chapter 6, and the types of heat recovery units which can be used in buildings are fully described in Chapter 7.

5.2 Air Conditioning Control Systems

The control of the air conditioning system is most important both from the point of view of ensuring satisfactory comfort conditions and the efficient use of energy throughout the building.

One of the most common failings of air conditioning systems is their inability to react quickly to changes in external conditions. A number of control systems, some which are solid state, are now available which can minimise energy wastage by overheating or overcooling by compensating much more quickly to ambient changes (ref. 5.1).

Turning equipment off and on to meet the requirements dictated by the occupancy of the buildings can save energy. Restarting of the system should be delayed until the last possible moment. Reset controllers are available and are widely used where the air conditioning load requirements are reduced at night or during the weekend. Controllers can be obtained with a built-in weather reset for night and/or weekend cutback. These units can pay for themselves within a few months, and some feature automatic early start-up for mornings when the weather is particularly cold. It should be remembered that even if the existing air conditioning system has what appears to be a satisfactory control system, this is probably out-of-date and would have been installed when energy was comparatively cheap. It should at least be adjusted to meet today's requirements, or replaced with more versatile controllers of the types described above.

When the air conditioning load is reduced during the heating season in off-peak hours, it is important to keep ventilation air dampers closed as long as possible, consistent with minimum ventilation requirements. A similar argument applies during restarting of the system. This encourages a more rapid return to normal comfort conditions within the building. Towards the end of the working day the dampers and exhaust ducts can again be closed to 'store' the heat already within the building. In many instances, an acceptable ventilation level can be achieved even if the dampers in the outside air intake duct are closed for a part of every hour during peak utilisation periods.

Facilities with low pressure air distribution systems can use reset controllers to cycle the units on 100 per cent return air while maintaining lower air temperatures during the 'off-peak' periods. High velocity air distribution systems can have exhausts closed off during periods of unoccupancy, in conjunction with a reduction in the fresh air intake and heating rate.

A large number of the air supply units to air conditioning systems employ mixed air control, using warm air from the building mixed with cold air from outside, to obtain a fixed design air supply temperature. If this supply temperature is only $12^{o}C$, say, an excessive amount of cold air may be brought into the mix, and this has to be reheated. In some instances it may be worth raising the design air supply temperature to $16^{o}C$ or thereabouts to allow a greater proportion of the return air to be recirculated.

An exercise carried out by the Fuel Economy Unit of the Property Services Agency, part of the UK Department of the Environment, (ref. 5.2) has shown that the energy consumption in buildings can be considerably reduced by the appropriate application of control systems and other techniques, and the major areas of energy conservation are illustrated in Table 5.2. The PSA estimate that up to a 30 per cent reduction in consumption and cost within a five year programme is possible.

The Table presents data on lighting and boiler efficiency, and these are discussed in the context of 'good housekeeping' later in this Chapter.

Computers in Air Conditioning The computer has two primary functions in the context of air conditioning, one of which may be of interest to users of large air conditioning plant who would like to examine their options for energy conservation. The second function is the control of an air conditioning system in a large building using a computer. Although perhaps out of context in this Chapter, one example of such a system is given to illustrate the scale and duty of some of the large air conditioning systems.

Naturally enough, the use of computers to analyse air conditioning systems has been most extensive in the United States. One program available, known as AXCESS, forms part of the energy management service of the Edison Electric Institute in New York. AXCESS is a computer analysis program which enables the user to determine the most efficient use of electric energy in a building by comparing various methods and systems of heating, ventilating and air conditioning, and other influences on internal electric loads. The program also provides information on the capital and operating costs of each arrangement. The program is capable of simulating most types of air conditioning systems in use today, and can cater for heat recovery installations and ceiling induction units utilising lighting cavity heat.

The minimum requirements and output data for an energy analysis program are shown in Fig. 5.2 (ref. 5.3). The procedure for applying such a computer program to an existing building is as follows:

(i) Study building plans and check actual building conditions against these, and collect data for program.

(ii) Run a load program calculation

Plant - Good Housekeeping

TABLE 5.2 Predicted Effect of Energy Conservation Measures in UK Government Buildings

Type of Conservation/ Financial Benefit	Potential savings in % of sectorial consumption (years)	Costs/ benefits	Type of policy and action required
Running boiler plants at optimum efficiency	5% of fuel	Small cost compared to savings	Information, training, encouragement, monitoring performance, accountability
Less non-productive idling of boiler plant	5% of fuel	Small cost compared to savings	Information, training, encouragement, monitoring performance, accountability
Introduction of optimum start control to Defence Estate	10% of fuel	Return on investment 1 - 2 years	Identifying suitable buildings for conversion contract, monitoring performance
Adjustment to existing heating controls - Defence & Civil Estate	10% of fuel	Small cost compared to savings	Monitoring performance to highlight large wastage. Corrections to controls. Training staff in control technology. Accountability
Installation of heating controls to inadequate system - Defence and Civil Estate	10% of fuel	Return on investment <1 year in most cases	Monitoring performance to highlight large wastage. Installations of controls. Training of staff in control technology. Accountability
Improvements to electrical power factor	2% of electricity	Return on investment - $1\frac{1}{2}$ years	Identifying poor power factor. Fitting power factor correction devices Monitoring performance - accountability
Reduce excess lighting by staff action	20% of electricity	Small cost compared to savings	Encouragement and publicity. Monitoring performance
Reduce excess lighting by technical improvements in switching and controls	20% of electricity	Return on investment 1 - 4 years	Development of new design and control concepts. Implementation and monitoring performance
Improve thermal insulation and temperature control	20 - 25% of sectorial consumption	Return on investment 5 - 10 years	Revised insulation and control standards. Implementation of schemes giving greatest cost benefits
Automatic controls on catering equipment	20 - 25% savings on modified equipment	Return <4 years	Development of fuel saving devices and implementation of the scheme after British Gas approval.

(iii) Use data from (ii) with the energy program and all mechanical system data to obtain an output of energy consumption.

(iv) Check this ouput against records of consumption given in fuel bills etc.

(v) If there are differences between calculated and actual consumption, find reasons. (e.g. damper leakage, improper calibration of controls).

(vi) Once the differences have been satisfactorily explained, input modifications to the air conditioning system and assess capital and running cost changes for each modification.

(vii) Prepare a course of action based on the economic analysis, to implement improvements.

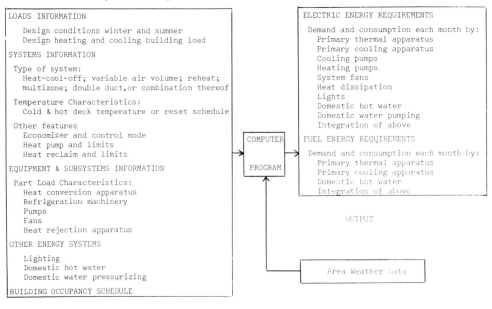

Fig. 5.2 The minimum requirements and output data for an energy analysis program

An impressive example of the use of a computer to control the heating, ventilating and air conditioning system in a large building is typified by the Chicago IBM office. With 52 floors, and a base of 38 m by 64 m, the building has a volume of 6.8×10^5 m^3 and a glass area of 32 500 m^2. The design conditions are as follows:

Heating: Loss   15 000 kW
           Ventilation   188 m^3/s
           Design condition 22°C outside
                            21°C inside

Cooling: Gain   20 500 kW
           Ventilation   188 m^3/s
           Design condition 35°C dry bulb, 24°C wet bulb outside
                            25.5°C, 50% relative humidity inside

The total air conditioning power is 22 500 kW, and the additional connected loads in the building amount to a further 22 000 kW. The computer control system is continuously updated as experience is gained, data being obtained in some instances from personnel and then stored on the computer data bank. Such has been the success of the system that in the winter of 1973/74, the IBM building used 42 per cent less energy per square foot of office space than the average of thirteen large buildings surveyed, (ref. 5.4).

5.3 Boilers

The boiler, by its nature, is used in conjunction with a number of other items of equipment, such as steam traps, which all can benefit from preventative maintenance, insulation and other low cost measures to save energy. These are detailed in a number of other sections of this Chapter.

If the boiler system is producing steam or hot water, the insulation of at least all main lines is recommended. Leaks should be repaired as soon as possible, and steam trap operation regularly checked. The use of storage tanks to receive hot water during off-peak operation can assist in energy conservation.

Maintenance of the boiler itself should receive emphasis. Cleaning of the shell and tube sides is important, and the burner combustion efficiency can affect fouling as well as lowering heating rates. Humidification of the combustion air supplied to the boiler can assist in preventing boiler fouling. Typically 12 - 20 kg water are recommended for each 1000 kg air (ref. 5.5).

Boilers can waste considerable quantities of heat. The stack temperature should be regularly monitored; if it is too high, an excessive amount of heat is going to waste in the atmosphere. Heat recovery from boiler flue gases can save 10 per cent of the boiler fuel input, even when stack temperatures are comparatively low.

Where possible, boiler condensate should be returned to the boiler as feedwater. If this is impractical, the heat should be recovered from the condensate before it is discharged. Boiler feedwater is normally of high quality, as is the condensate, and the higher the feedwater temperature, the lower the quantity of fuel required to raise the steam.

Boiler blowdown is also a source of heat, and should be recovered in a blowdown heat exchanger (see also Chapter 6). This will prove economical in instances when poor quality feedwater necessitates frequent blowdown. Flash steam from high pressure processes should also be considered as a candidate for recovery.

If the boiler provides both process heat and space heating services, it is worth remembering that it is likely to be producing a comparatively small steam load during summer months, hence operating at low efficiency. Selection of smaller package boilers for each duty may be economically advantageous.

A number of devices, some of which are discussed in detail in Chapter 8, are available for isolating boiler flues during burner off-periods, to take full advantage of the thermal storage capacity of the boiler. During normal boiler operation the burner firing time is reduced, resulting in substantial fuel economies. During night or weekend shut-down periods, heat loss via the stack is minimised, again saving energy. Isolators are available which open and close the flue automatically as the burner cuts in and out.

5.4 Chimneys

The functions of a chimney are primarily to conduct waste gases to a sufficient height where they can be discharged to the atmosphere, and to maintain the gas velocity, or draught, through the system. A chimney can operate using natural or mechanical draught, the latter being either forced or induced.

In natural draught systems, which are most susceptible to growing contamination, the interior and the condition of the top are particularly important. If the top of the chimney is encrusted with tar, grit and other contaminants, it can lose over 50 per cent of its effective area, seriously affecting the ability to maintain a sufficient draught. Thus it is necessary to carry out regular inspections of the condition of any chimneys in the plant, cleaning where necessary.

If an increase in draught is required, raising of the height of the chimney can improve performance. This is best carried out on chimneys which have a relatively large cross-sectional area, with low gas velocities, otherwise the pressure drop could become prohibitive.

A second technique for improving the performance of natural draught chimneys is to install barometric draught controls. Alternatively, draught inducers mounted at the top of the chimney can increase capability.

Mechanical draught comes into its own as the size of the installation increases, when a greater output is required from the plant, or when heat recovery in the stack, leading to an increased pressure drop, is contemplated.

The ideal system comprises a combination of induced and forced draught, known as a balanced draught flue. The forced draught fan supplies primary air for the combustion process and secondary air for the burning of volatile matter. The induced draught fan has the sole function of evacuating products of combustion from the plant. Fan-driven systems tend to operate with chimney base temperatures about 40 per cent below those of a good natural draught system, with a proportional reduction in flue heat loss. Up to 25 per cent of the heat in the system can be dissipated up the flue in a poor natural draught installation. If the chimney top is caked and the inside of the stack contaminated, heat losses can be 50 per cent greater than when the natural draught system is properly maintained.

Plant - Good Housekeeping

5.5 Combustion Systems

Fuel combustion efficiency is of the utmost importance in maintaining satisfactory operation in boilers, furnaces and any other heating plant which uses coal, oil or gaseous fuels. The techniques for improving combustion efficiency discussed in this Section fall into three major categories:

(i)   Improvements to existing systems
(ii) New burners available
(iii) Automatic control systems.

These are discussed in turn below.

5.5.1 Improvements to existing systems:

The 'good housekeeping' aspects of combustion systems are largely associated with measurements which enable efficiency to be checked and improved where necessary. This may also point towards replacement of the burner with a more modern type, or the installation of a control system which can adjust air/fuel ratios as required by the load and ambient conditions.

Combustion test kits may be obtained for flue gas analysis. Results, if applied, can lead to improved combustion and resulting energy savings. In boilers, furnaces and similar equipment, benefits will also result from less fouling of air preheaters, economisers and other heat transfer surfaces. Flow meters can also be employed to check that optimum conditions exist in combustion systems which use pressurised combustion air.

It is sometimes not appreciated that flame geometry and burner design can have a significant effect on overall performance. The flame must be directed to where it will be used most effectively and a number of burners are available to meet the requirements of applications where this is important

Fully automatic hydraulic or electronic flow control systems are available for large gas furnaces. These simultaneously monitor combustion chemistry, and enable the operator to cater for slight changes in ambient temperature or humidity.

If poor mixing is suspected as a cause of low efficiency, it may be advisable to switch to a power burner, in which instead of the gas inducing the air into the burner, a blower is used for the supply air.

A technical service to customers who wish to improve combustion performance is available from British Petroleum (BP). BP have established a number of mobile combustion laboratories (MCL), so that investigations associated with fuel oil firing could be readily undertaken on operational plant. The first MCL was commissioned in the UK in 1960, and seven such laboratories are now in operation throughout the world.

They were originally conceived for technical service duties and advice to industry on combustion efficiency and flue gas corrosion problems. Now they also encompass pollution control. The MCL's fulfil multi-purpose roles on such diverse plant as large and small industrial boilers, power plant, industrial gas turbines, lime kilns, blast furnaces, dryers, office heating equipment and other types of continuous combustion equipment.

BP have also carried out studies on chimneys and furnaces, and claim to be well placed to advice industry on efficient furnace operation (see Section 5.11). Recent work has concentrated on the cement and lime industries, and in particular the iron and steel industries. BP were among the pioneers of oil injection into furnaces to increase iron production (now an unattractive technique because of the high oil prices), and have studied several techniques for improving furnace thermal efficiency.

Even the viscosity of the oil used in burners (and in internal combustion engines) can influence efficiency. As shown in Fig. 5.3, changes in viscosity associated with temperature variations of only 5°C can reduce burner efficiency by up to 2 per cent. A noticeable reduction in efficiency is also seen in diesel engines.

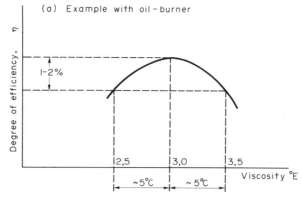

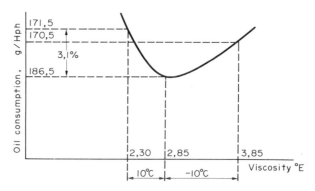

Fig. 5.3 The effect of fuel oil viscosity on the performance of (a) an oil burner and (b) a diesel engine (Courtesy Euro-Control Ltd.)

Fuel additives, which can affect regions remote from the burner, as well as improving combustion efficiency, are discussed separately in Section 5.20.

5.5.2 New burners available:

There are many types of burner which are designed for particular tasks, and the flame geometry is often the feature which enables one to identify these burners.

One such type used in furnaces produces a flat flame which follows the contours of the inside wall of the furnace. As the air enters the burner body it passes over a side orifice plate which fixes the air capacity and also initiates a spinning action which carries through the burner to produce a flat flame. Applications are those where a very even heat distribution is needed, and the system enables burners to be placed closer to the load without the fear of direct flame impingement. The flame pattern and hot gas flow is illustrated in Fig. 5.4.

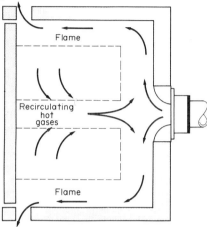

Fig. 5.4 Flame and hot gas pattern in a Stordy Hauck burner designed to follow the furnace wall
(Courtesy Stordy Combustion Engineering Ltd.*)

A burner developed to meet the requirements of high heating rates in furnaces, resulting in significant fuel savings, employs the concept of very rapid recirculation of the heating medium within the furnace chamber, heat transfer to the load being by forced convection at all stages of the cycle. It is claimed that this breaks down the layer of still air around the load which can inhibit heating in less vigorous surroundings. The effect of this 'high velocity' burner is shown in Fig. 5.5 and compared with a conventional burner arrangement.

Burners which can produce large quantities of hot gas often employ pre-heating of the combustion air. If the heat in the products of combustion can be recirculated in the combustion zone, rapid vapourisation of the fuel (be it atomised oil or a mixture of oil and gas) occurs, assisting combustion.

Submerged combustion burners are not new, (ref. 5.6) but their application in heating of process liquids is worthy of emphasis because of their high efficiency and the opportunities for energy conservation which their design offers. A submerged combustion unit manufactured by Hygrotherm is illustrated

---

*Manufacturing licencees to the Hauck Manufacturing Company, Lebanon, PA, USA.

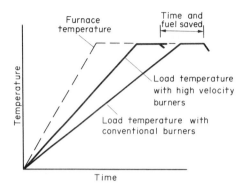

Fig. 5.5 Reductions in furnace load warm up rate
using a Hotwork high velocity burner

(By kind permission of Hotwork Development Ltd.
Combustion Engineers, Dewsbury, Yorkshire)

in Fig. 5.6. The burner fires through a tube, similar to a radiant burner tube, which is immersed in the liquid to be heated. Gaseous products of combustion are discharged into the liquid at the base of the tube. Hot gases then break up into small bubbles and rise to the surface between the combustion tube and a draught tube. This promotes turbulence in the space between the two tubes, thus keeping the burner tube cool. Heat transfer from the bubbles to the liquid is rapid because of the very large amount of surface area created by the vast number of bubbles formed. The advantages of such a system are many. Combustion takes place where it is required to use the heat,

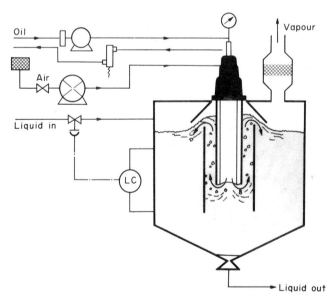

Fig. 5.6 A submerged combustion burner
(Courtesy of Hygrotherm Engineering Ltd., Manchester)

and no heat exchanger is needed, reducing capital and maintenance costs. In acid pickling plant, submerged combustion burners can be used either to reduce fuel requirements - the turbulence generated enables pickling to be carried out at 10 - 15°C lower than using conventional bath heating - or to increase throughput.

A number of new burner systems are discussed in references 5.7 and 5.8. Of particular interest is the recuperative burner (ref. 5.8) which is fully described in Chapter 6.

5.5.3 Automatic control systems:

Combustion control systems are standard equipment on most new furnaces and similar combustion-heated systems. The primary aim of the control system is to ensure that the correct air/fuel ratio is used in the burner. This maintains high efficiency as the fuel/air mixture is prevented from becoming too lean, excess air taking too much heat up the stack, or too rich, resulting in the waste of unburnt fuel. A control system also contributes towards plant safety as a very rich mixture can lead to an explosion, and a burner which is allowed to err outside its stability limits and is extinguished can affect product quality and process functioning.

Good control may also be needed in the process because a reducing atmosphere may be required at times, or a lean atmosphere for oxidation may be more suitable. Changes in the atmosphere as the process progresses may also necessitate an automatic control system for best results. A direct result of elimination of rich and lean mixtures during normal operation is a reduction in atmospheric pollution (ref. 5.9).

There are four main types of automatic control systems for combustion units (ref. 5.10). The simplest type, known as area control, relies on modulation of the areas of the flow passages for the fuel and air to maintain the correct ratio. Unfortunately the area is not the sole criterion affecting flow rate, such variables as pressure having an additional influence. Pressure control of the air/fuel ratio eliminates this possible error, and is versatile and accurate. In large systems, a hydraulic flow control system can be used instead of simple pressure control, helping to give faster response, particularly where large loads may be needed to move heavy dampers or valves.

The most sophisticated methods of combustion control are based on the use of electronics. Their use will be illustrated using as an example the AFR Series Combustion Control Equipment, manufactured by Hamworthy Engineering Limited. The basis of this control system lies in the fact that the fuel and air flows are computed to a balanced state, which readily enables automatic corrections to be made for pressure, temperature, and programmed excess air requirements. The unit can be used for oil and/or gas fired burners, and incorporates flow meters for air, gas and oil flow rates.

Features include:

(i) If the load is increased, the fuel flow will only rise when an increase in actual air flow, initiated by the control system, occurs. Similarly a reduction in load, and a turn-down in the fuel rate, will only lead to a reduction in air flow when the fuel rate reduction has been seen to occur.

(ii) Offset adjustment is a facility which allows a required excess air quantity (in terms of a percentage) to be maintained regardless of variations in load. Slope adjustment permits the excess air ratio to be automatically increased on turn-down. Offset and slope adjustment can be implemented without affecting one another.

(iii) Output signals from the system can be used to actuate pneumatic or electric systems for control.

(iv)   The control system will detect excursions in mixture outside predetermined conditions, and flow meters are continuously checked for faulty operation.

The significance of correct air/fuel ratios is illustrated in Fig. 5.7, and this is sufficient to emphasise the savings accrueing to the use of control systems.

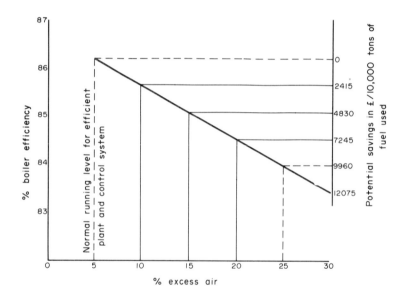

Fig. 5.7 Potential savings accrueing to the use of combustion control systems. (Courtesy Hamworthy Engineering Ltd.)

5.6 Compressed Air Systems

Compressed air is often used in a plant with little appreciation of the fact that it is a comparatively expensive form of energy. The compressors used to supply this air consume electricity, or in some cases oil or gas fuel, and the product is too expensive to be used as a seemingly 'inexhaustible' power supply.

There are a number of checks and improvements which can be carried out to reduce the operating costs of compressed air systems. All leaks should be repaired as soon as possible, and the system pressure should be the lowest acceptable for the applications where it is used. It should be borne in mind that the applications may have changed since the system was last overhauled, and a change in pressure may be beneficial, if only for this reason. A critical analysis of the applications should be carried out to see if the system retains worthwhile economic advantages over other methods. Never use compressed air for forced convection cooling if it can be avoided; cheaper methods are available.

Heat from the compressor aftercooler may be used to supplement space heating. Compressors often run when no demand for compressed air exists, and it is prudent to question whether this is necessary, based on the current load and utilisation schedule.

Filters and dryers are integral components of most compressed air systems, and regular cleaning or replacement of filter elements should be a part of the maintenance programme. Increased pressure drop through a dirty filter necessitates increasing the compressor power input to maintain equal power at the compressed air tools etc. The importance of drying is illustrated by the experience of a firm specialising in PTFE coatings. Pinholes in the surfaces, caused by moisture in the compressed air used to spray PTFE, were leading to a 5 per cent rejection rate. Installation of drying and filtering plant reduced the reject rate to almost zero, and for a capital outlay of £622, increased quality production worth £100 per week was achieved.

By driving the compressor with a gas reciprocating engine, the engine waste heat can be used for space and/or process heating.

5.7 Cooling Towers

The primary function of a cooling tower is to cool water which has been heated up in a process so that the water may be re-used for further cooling instead of being discharged. A good cooling tower will only lose at most 5 per cent of the water throughput by evaporation and carryover.

Obviously by saving water, which may cost something like 10p/1000 litres, the cooling tower can pay for itself in a relatively short time. However, a cooling tower, by its nature, wastes heat. Although this heat is generally comparatively low grade, it has a value, and in some instances the heat may be recovered either by direct heat exchange or by using a heat pump (see Chapter 7). It is therefore recommended that in instances where water is being fed to a cooling tower at temperatures in excess of $30^{\circ}C$, alternative uses for the heat be investigated. The heat pump will, if the water is used as the heat source, cool it still further, thus reducing the cooling tower load.

With regard to the normal functioning of a cooling tower itself, efficient operation is dependent on the condition of the packaging, and it may be advantageous to replace existing packing with newer plastic types (ref. 5.12). This can improve performance and duty within the same basic shell. Checks should be made of water loss and mist eliminators can be used to prevent excessive carry-over. Any fouling occurring in the tower can affect performance, and preventative maintenance should ensure that satisfactory operation is maintained.

ICI Petrochemicals Division have analysed in detail their plant which involves cooling towers in the circuits. The power required for circulating the cooling water is high and on some plants may account for as much as one-third of the total consumption of electricity. To minimise the cooling water flow in such systems, it is important to monitor the cooling water exit temperature. Thus ICI have installed temperature recorders at strategic points, and thermostatically controlled valves on cooling water exits from large condensers and coolers are used for automatic flow control.

On one large plant ICI have three separate cooling water circulation systems, each employing several large constant speed pumps. By linking together two of the systems and installing a variable speed pump, they were able to reduce pumping costs significantly. Normal fluctuations in the demand for cooling water are being catered for by the variable speed pump, and larger changes in demand are met by switching the constant speed pumps on or off. On this particular plant, all conservation measures to be taken are expected to save an estimated 2 MW.

5.8 Dryers

Dryers are an integral part of a large number of industrial processes, and the importance has already been indicated in Chapters 3 and 4. While gases frequently require drying, and several types of gas dryer are available (an air conditioning system is under certain conditions a gas dryer), this Section is principally concerned with the drying of solids.

Solid drying is normally associated with the removal of water, but in general drying is the separation of any liquid from a solid. Separation is carried out either mechanically, for example by centrifugal separation processes, or thermally, where hot air may be blown through the medium to be dried. The potential for energy conservation is much greater in thermal drying processes, and most of the discussion on dryers in this book is restricted to this type. (In the food industry more specialised drying techniques such as freeze drying are being applied to an increasing number of products, but these techniques are not in general used elsewhere).

Thermal drying, be it direct or indirect, batch or continuous, can only be regarded as having been successfully carried out when the moisture content of the material leaving the dryer has been reduced to that required before the material is passed on to another process, or leaves the plant altogether. In general, this leads to overdrying; for example in textile manufacture, under-drying causes mildew, which in turn creates problems later in the production. Similarly the moisture content of powdered ores is important as it affects crushing strength and pellet size.

The importance of the moisture content as far as energy conservation is concerned cannot be over-emphasised. Sira Institute (see Appendix 3) have established a Moisture Measurement and Control Centre which can assist United Kingdom industry to apply on-line moisture measurement instrumentation. Several projects have been concerned with the control of dryers where Sira have successfully applied commercial moisture meters and produced substantial fuel savings. During the course of this work Sira developed the temperature difference technique (described below) for determining the moisture content (or more correctly the equilibrium relative humidity) of material at the exit from a dryer. As well as reducing fuel consumption, improved control accelerates throughput, product quality and uniformity. Sira have found that the elimination of over-drying can, in most cases, reduce fuel consumption by a minimum of 10 per cent. In the textile and pharmaceutical industries, savings of 20 per cent have been obtained, and in one application energy costs in drying were halved.

The moisture measurement technique is based upon the fact that when material is being dried in a hot atmosphere, its temperature is depressed below the dry-bulb temperature of that atmosphere. If the material is saturated, its temperature will be approximately the wet-bulb temperature of the atmosphere; if less than saturated, the temperature is not depressed that far. The difference in temperature between the material and the wet bulb is related directly to the moisture content of the material. The method is therefore suitable for measurement of any moisture content below saturation point. Fig. 5.8 shows the percentage moisture contents plotted against temperature difference for a variety of materials which undergo drying.

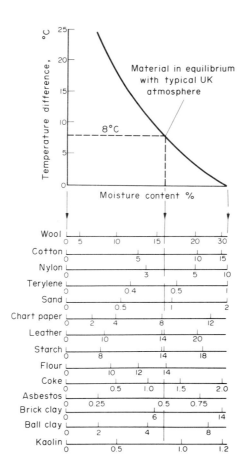

Fig. 5.8 Temperature difference versus percentage moisture content for various materials. (Courtesy Sira Institute)

Probes are required to measure two temperatures; the temperature of the material being dried and the wet bulb temperature of the atmosphere within the dryer. For continuous processes these measurements are made near the dryer exit, and the control system is illustrated in Fig. 5.9. (An alternative to speed control is control of the thermal input).

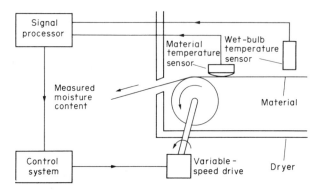

Fig. 5.9   The Sira moisture control system applied to a textile dryer. (Courtesy Sira Institute)

A number of other measurement techniques have been developed by Sira, and the results of their application, including payback periods, is given in Table 5.3.

An interesting assessment of energy conservation techniques in aggregate drying which gives pointers to other drying processes has been carried out by the United States National Asphalt Pavement Association (NAPA). Their conclusions concerning dryer types and heat recovery are discussed fully in Chapter 6, but NAPA have also highlighted several areas where 'good housekeeping' minimises dryer heat load (ref. 5.13). A more detailed analysis carried out on behalf of NAPA also includes useful recommendations which may be implemented on existing machinery (ref. 5.14). (The United States asphalt industry used the equivalent of 30×10^8 litres of fuel oil for aggregate drying and heating in 1974.)

Of particular significance is the effect of the atmosphere on the aggregate during storage prior to drying. It is recommended that the material be covered when stockpiled if in a damp atmosphere, or arranged so that some natural drying can take place. NAPA calculated that only a 1 per cent reduction in aggregate moisture content as it entered the dryer could save up to 10 per cent of the fuel needed in the dryer itself.

It is proposed that as the fine aggregate tends to hold the most moisture, effort should be expended on that type. A lower moisture content on entry to the dryer also reduces the fan power needed to shift the vapour, which may represent up to 50 per cent of the total flow (including combustion gases) in this particular process. Selection of aggregate from the outside of the stockpile is also a recommendation made by NAPA. By digging deeply into a stockpile, the material with the highest moisture content is removed before it has had a chance to dry naturally.

The volume of air used in a dryer is directly related to fuel consumption, as this air is heated, and in turn removes heat from the drying compartment. If the dryer operates at less than its maximum production rate, or if the aggregate has a lower than average moisture content, the burner can be

TABLE 5.3 Sira Case Studies – Drying and Humidity Control

Industry: Why measurement is important	How measurement and control was achieved	Accuracy (% moisture content, m c) and pay-back period	Real or estimated benefit
Textiles Underdrying causes mildew and trouble at subsequent processes. Operators therefore overdry wasting fuel and reducing throughput	*Temperature difference technique* An automatic feedback system controlled throughput of cloth on production tenter	Measurement accuracy depends on fibre: ±2% wool, ±1% cotton, ±0.5% nylon. Pay-back period: less than 12 months	Textile manufacturers estimated control of dryers could mean savings exceeding £4000 pa per dryer
Ceramic tiles Correct moisture content produces green tiles of maximum strength reducing breakages and cracking and speeds up rate of tile production	*Infrared reflectance technique* Instrument installed, over belt carrying clay flake controls speed of steam drum dryers	Measurement accuracy: clay dust ±0.16% m c clay flake ±0.57% m c Pay-back period: 2 months	Company's estimate of savings from installation of dryer control system £40 000 pa
Calcium-silicate bricks Correct moisture content reduces cracked, distorted and blistered bricks	*Mouldability controller* Control system based on Ridsdale Dietert mouldability controller installed in a brick plant	Control accuracy ±0.5% m c Pay-back period: about 2 months	Control reduces number of defectives and saves about £2000 pa. Improvement in compressive strength saves about £10 000 pa
Minerals (eg iron ore and copper ore concentrates) Moisture content of powdered ore important because it controls crushing strength and size of pellets	*Infrared reflectance technique, working at longer water-absorption wavelength and greater intensity* Measurement system installed over conveyor belt	Measurement accuracy: preliminary trials indicate ±0.3% m c	Depends on application
Pharmaceuticals (fluid-bed dryers) In tablet making the moisture content of the powder determines crushing strength dimensions and colour	*Temperature-difference technique* Consistent results on fluid bed dryer with starch/lactose material in plant. Trials with other powders also successful. Now being applied to control drying of sugar coatings on tablets	Measurement accuracy: preliminary results indicate ±0.16 m c. Pay-back period: results not yet available	Estimated savings from accurate control of moisture content: £1000 pa per fluid-bed dryer. For special products figure substantially higher
Flour milling Correct moisture content gives optimum extraction rate from wheat grains	*Microwave absorption technique* Measurement system fed from wheat screw feeder. Plant trials over several months	Control accuracy: ±0.25 m c Pay-back period: 4 months	Comparison of average extraction figures with and without automatic control, showed increase in flour yield 1 – 2.4% (depending on wheat) with automatic moisture control

readily throttled, but control of the airflow through the drying compartment is more difficult. Damper control may not be sufficiently accurate, and leakage into the chamber via other openings can lead to an extra burden on the fan, (ref. 5.15).

As with other combustion systems, checks should be made on the stack temperature - too high an exhaust temperature means heat is being wasted. Heat recovery can be carried out on many types of dryer (see Chapter 6).

One of the problems in aggregate drying is combustion noise. Stordy-Hauck have developed a combustion system which, as well as offering noise reductions, modulates both fuel and airflow rates to suit the load, hence minimising excess air. A recuperative combustion chamber is also provided, increasing thermal efficiency. The cost savings resulting from use of such a system are shown in Fig. 5.10. NAPA, in their assessment of the combustion region in the dryer, indicated that a layer of insulating firebrick lined with refractory brick would increase efficiency in conventional direct fired rotary dryers by raising the inner wall temperature, helping radiation and conduction heating of the aggregate and reducing heat loss through the dryer wall in this region.

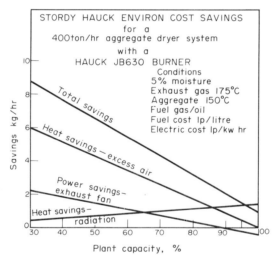

Fig. 5.10 The performance of the Environ Combustion System, developed to minimise running costs when operating at less than full capacity.
(Courtesy Stordy Combustion Engineering Ltd.)

In Chapter 6 a number of recent developments in drying are discussed, involving new plant, and the potential of heat recovery equipment is assessed.

5.9 Electricity Generation (In-Plant)

The in-plant generation of electricity, or more specifically the use of total energy plant, has already been discussed in some detail in Chapter 2. There it was emphasised that electricity generation should be carried out in conjunction with full utilisation of the waste heat resulting from the

Plant - Good Housekeeping

operation of the prime mover, to gain maximum benefits.

There are a few points which should be considered when attempting to obtain maximum efficiency (in addition to plant maintenance etc.). Transmission losses are often believed negligible, but these do include losses in transformers, which should be de-energised wherever possible. It is also important to maintain transformer heat exchanger surfaces in good condition.

The inefficiencies accrueing to the use of plant on continuous low loads cannot be over-emphasised, and this applies as much to generating plant as to a small boiler. Demand in the factory should be accurately scheduled, and if the power factor is found to be low, a professional evaluation is recommended.

The power factor in an electrical circuit is the ratio of the product of voltage and current when they coincide, to the product of the maximum instantaneous values of voltage and current. It may be alternatively expressed as the ratio of working power to total kVA, (ref. 5.16). The total kVA in a circuit can be greater than the working power because current is required to produce the flux necessary to make the electric equipment operate. Voltage and current therefore achieve greater instantaneous values than the coincidental values normally associated with the running power of the equipment. If a large motor, say 100 kW, is used to drive a fan which may only require 70 kW, the resulting power factor of the motor operating at only 70 kW may be 0.8. If a 75 kW motor had been selected the power factor at 70 kW would be higher, say 0.85. In the first case the kVA is 70/0.8, i.e. 87.6. In the second example the corresponding kVA is lower, only 82.3.

Thus correct matching of electrical equipment is important to keep the size of power distribution equipment down, and also to reduce the cost of electricity to the consumer, (ref. 5.17).

Power plant auxiliaries can use a substantial proportion of the output of the plant itself. It can be beneficial to operate these auxiliaries using high pressure steam instead of electricity, if a use for the waste low pressure steam can be found in the factory.

5.10 Electric Equipment

As mentioned in the preceeding section, correct sizing of electrical equipment, in particular motors, can assist in reducing running costs and, as a larger motor than necessary is obviously more expensive, capital outlay.

Variations in voltage can be a contributory factor in lowering the life of electrical equipment, as well as reducing the power factor and increasing starting current. Voltage stabilisation equipment should be considered if it is discovered that voltage fluctuations are significant.

All electrical equipment (and this includes office equipment such as electric typewriters) should be switched off when not in use.

Although relevant to all services, including gas, water, oil and communications, a tariff rate is also applied to electricity used by a company.

Tariff structures are particularly complicated in many cases, and because of the numerous factors involved in fixing the tariff - these include meter rental, off peak supplies, restricted hour rates, maximum rate of supply and maximum annual demand - are many and varied. It is therefore recommended that a company checks that the tariff rate charged for its services, including electricity, is the lowest possible. A number of specialist companies are able to advise on tariffs, and are experienced in putting forward cases for reductions and in negotiating new tariff rates. One of these companies, National Utility Services, (ref. 5.18), operates on a minimum 5 year contract with a firm, recouping their costs by way of a retainer and a proportion of the savings resulting from their recommendations. NUS claim that companies can usually produce evidence of savings within 9 months of initiation of tariff investigation exercises.

5.11 Furnaces

A furnace consumes vast amounts of energy, and as one of the greatest users of energy in industry, it has been singled out by many organisations as a prime candidate for the application of energy conservation techniques.

Heat recovery systems for furnaces have probably received the most attention, and are fully discussed in Chapter 6. It suffices to mention here that the waste heat can be used for many purposes, including preheating of combustion air, preheating cold stock, steam or hot water production, or thermal fluid heating. The high grade heat available in a furnace installation is second to none in industry, and should be used to its full extent. It is estimated that in the United Kingdom alone approximately £100 million is lost in the form of furnace waste heat per year.

Another exercise which in general involves less capital expenditure than waste heat recovery apparatus is the insulation of the unit. The heat loss depends upon the condition of the furnace linings, insulation and doors, and proper maintenance of these features is essential. Probably the most important development as far as insulation is concerned is the manufacture of low thermal mass (LTM) refractories, consisting of ceramic fibres, generally based on alumina or silica. They are available in a variety of forms ranging from wool and rope to large blankets. Such is the effectiveness of this refractory that a layer 100 mm thick is equivalent to a 300 mm layer of fireclay brick backed with high and low temperature insulation brick (ref. 5.19). Some ceramic fibre blankets are available which will withstand temperatures of up to $1600^{\circ}C$.

LTM refractories have been widely applied in electric furnaces, particularly in batch processes. Savings of over 50 per cent in the energy used by these furnaces, compared with conventional brick lined types, have been recorded, and similar improvements have been reported with gas-fired furnaces.

An example of a furnace being converted to 'Kaowool' ceramic fibre blanket insulation was provided by the United Kingdom Steel Castings Research and Trade Association, (SCRATA), (see Appendix 3).

A gas-fired bogie hearth heat treatment furnace constructed of ordinary fire-brick was relined with 'Kaowool' ceramic fibre blanket to improve the thermal efficiency of the furnace. Both the installation time and the cost were considerably reduced by anchoring the lining directly on to the existing casing brickwork, thereby eliminating the need for the usual steel shell 'back up'. This was achieved by a special technique using adaptor bosses for supporting the anchor in the brickwork structure.

The installation of the lining on the 4.6 m x 3 m bogie hearth furnace was completed in five days at a total cost of £2300 (£1000 material cost plus £1300 installation charges). Gas consumption measurements were made over a wide range of heat treatment cycles both before and after the conversion. These showed that an average saving of 30 per cent in gas usage was achieved due to the new insulation.

The conversion of three heat treatment furnaces has resulted in a saving of about £1000 per month in fuel costs. The heating and cooling phases of the various heat treatment cycles were greatly reduced due to the low heat storage capacity of the lining, and for certain special annealing cycles, it was found necessary to maintain a low heat input to the furnace during the cooling phase to prevent too rapid a cooling rate.

SCRATA point out that the furnaces in question were poorly insulated prior to conversion, and these savings are unlikely to be obtained were the conversion applied to a well-insulated installation. However, the pay-back period is still likely to be well within two years, and the conversion time is very short.

Another example of effective insulation, this being applied in a gas-fired kiln used for first stage firing in the pottery industry, uses Saffil fibres manufactured by ICI. The kiln operates at temperatures of up to 1250°C and has an external steel shell 4.54 m long, 2.18 m wide and 3.02 m high. Table 5.4 shows the performance achieved with fibrous insulation, and the results are compared with an equivalent conventional brick-lined kiln.

TABLE 5.4 Comparison of Brick and Saffil Fibre Insulation in a Pottery Kiln

Kiln	Run	Stages (h) Heating	Soaking	Cooling	Total (h)	Fuel consumption Joules x 10^{-8} (Therms)	Fuel Savings %
			HOLLOWARE				
Brick	A	$15\frac{1}{2}$	1	11	$27\frac{1}{2}$	251(238)	35
Fibre	B*	$7\frac{1}{2}$	1	8	$16\frac{1}{2}$	162(154)	
Brick	C	$8\frac{1}{2}$	1	11	$20\frac{1}{2}$	202(192)	20
Fibre	B*	$7\frac{1}{2}$	1	8	$16\frac{1}{2}$	162(154)	
			FLATWARE				
Brick	D	$18\frac{1}{2}$	3	11	$32\frac{1}{2}$	332(315)	26
Fibre	E*	$10\frac{1}{2}$	3	8	$21\frac{1}{2}$	247(234)	
Brick	F	11	3	11	25	298(283)	17
Fibre	E*	$10\frac{1}{2}$	3	8	$21\frac{1}{2}$	247(234)	

*Fastest heat-up rate consistent with integrity of ware in kiln

The proper scheduling of furnace work loads reduces the energy expended in reheating, but if the furnace is inoperative for extended periods, the holding temperature should be kept as low as possible. Similarly the preheating of furnaces should be carefully observed to see that it is only carried out for as long as is necessary. Pressure control in large furnaces ensures minimum egress of flames and also minimum intake of cold air through any openings. A slight internal positive pressure is generally maintained, and good pressure uniformity in the furnace aids even heating of the load.

Monitoring of the essential parameters in a furnace (and other combustion equipment) and good maintenance are essential for efficient operation. With the help of an efficiency monitoring diagram in which excess air and stack temperature are related to performance, operators can quickly reset the air flow to achieve efficient operation. Monitoring diagrams used by ICI are shown in Figs. 5.11 and 5.12 (ref. 5.20), and in most cases this company has found that an improvement in excess of 5 per cent is possible. A more obvious indication of overfiring of a furnace is the appearance of flames in the flue. Many furnace flues are oversized: this leads to excessive heat loss and the inability to pressurise the chamber, unless dampers are used. Flues should be reduced in size if the furnace operates below maximum design load for most of the year.

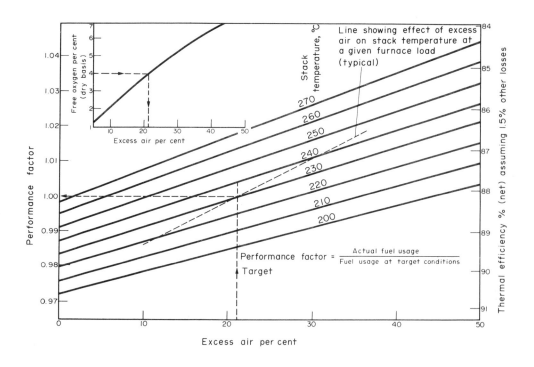

Fig. 5.11 Furnace efficiency monitoring graph

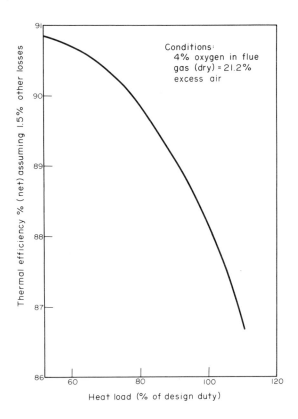

Fig. 5.12 Effect of heat load on furnace thermal efficiency

With regard to furnace maintenance, scale on the furnace heating system can decrease efficiency, and the control of the oxidation can assist in eliminating this. Furnace life, in particular that of the refractories, can be shortened if the temperature of the furnace is frequently altered, and a maintenance programme should take account of this, which may be alleviated by rearranged scheduling.

If a melting furnace is used, it may be advantageous to load via the flue, as this directly uses exhaust heat to preheat the load (as an alternative to recovering exhaust heat for this task via a heat exchanger).

The use of high gas velocities in furnaces (as opposed to indirect or low velocity firing) can improve heat transfer. However, the performance of natural draught low velocity burners on furnaces leaves a lot to be desired. British Petroleum have found that the main causes of poor performance are badly maintained burner tips, or poor tip alignment. Air registers can also become jammed by fuel oil deposits. BP have initiated a replacement and maintenance programme for burner tips at most refineries, and short term studies at Grangemouth should lead to the elimination of the jamming of air registers.

As well as the furnace itself, much of the equipment needed to load and unload the chamber may be a source of energy loss. Furnace rails, which are often water-cooled, can act as a substantial heat sink. It is advisable to insulate these to prevent continuous excessive heating of the water which passes through them. Of greater importance are the conveyors in continuous process furnaces and heat treatment plant. These should be returned through the heated enclosure, or preheated with hot exhaust gas, whenever possible to save the need to use additional furnace energy to reheat them when they re-enter the furnace zone.

Another study carried out by SCRATA on furnace efficiency was concerned with the insulation of ladles used for carrying liquid metal. The majority of the heat lost from liquid metal in a ladle occurs by radiation and convection from the surface.

Insulating compounds reduce these heat losses by cutting down on the radiation and by preventing cold air flowing over the metal surface. Consequently the heat can be tapped at a lower temperature, i.e. the heat time can be reduced, or the metal can be held for longer without skulling, the metal can be transported greater distances or castings poured for a longer time. To be most effective the insulator must be added to the ladle before the slag has the chance to solidify, so that the slag remains liquid and the whole mass flows down the ladle as pouring proceeds without allowing the hot interface to come into contact with cold air.

Two proprietary compounds, K-lite A and Exothermic Ladle Cover, and the conventional foundry practices of no cover, slag cover, and vermiculite were investigated. The programme of work was carried out on a 50 kg induction furnace, using molten steel superheated to 1650°C. Cooling curves were obtained, shown in Fig. 5.13, and these reveal that all the materials used reduced the rate of heat loss by producing a cooler air/cover interface and a consequential reduction in radiation. Of the covers investigated, the vermiculite and slag were similar in performance because of fusing of the vermiculite to form a slag-like layer. The proprietary materials were the most effective. The cost of the covers could be offset against the reduced superheat necessary to complete the pouring or the longer pouring times that could be employed.

5.12 General Process Heating

While most aspects of 'good housekeeping' are discussed in the context of particular items of plant, there are a number of aspects common to most process heating applications, and these are listed below:

(i)   If thermal control of a process is extended to its external environment, the process should be thermally separate from the rest of the factory.

(ii) Use reflective shielding to ensure that heat is retained within the process area

(iii) If a continuous heating process can be justified by the demand, replace batch processes with this system. Productivity is increased and less energy is needed to reheat the unit as each load is inserted.

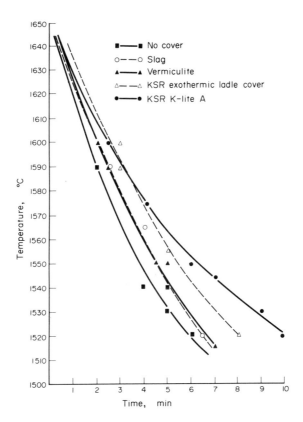

Fig. 5.13 The cooling curve of a 0.30% carbon steel in an induction furnace using various ladle covers (Courtesy Steel Castings Research and Trade Association: Journal of Research No.26, Sept. 1974. The use of ladle insulating compounds, by M.W. Hubbard)

(iv)   Consider the possibility of automatic control for processes if this would increase efficiency

(v)   Use lagging wherever possible. The insulation material should be waterproof.

(vi)   Dual firing can save money if full advantage is taken of fluctuating fuel prices and tariffs.

(vii) In some processes heat in endo- or exothermic reactions can eliminate the need for 'external' heat and can act as a heat source.

(viii) In all cases, make sure that the correct grade of fuel is being used.

138   Industrial Energy

(ix)   If process requirements change, or the heating load is no
       longer required, check that any pipes carrying steam, hot
       air or hot water to the process area are isolated. Often
       the heat loss from pipes servicing areas where there is no
       longer a need for heat is considerable, and these pipes tend
       to be 'forgotten'.

(x)   Continuous processes such as heat treatment or drying can
       involve considerable heat loss from the inlet and exit to
       the treatment chambers. Air curtains (see Section 5.15) can
       be used to prevent this loss, improve operator comfort, and
       increase effective oven length. Fig. 5.14 shows an installation
       of this type.

Fig. 5.14   An air curtain installed on a heat treatment oven
                     (Courtesy Minikay Ltd.)

5.13 Incinerators

It is tempting to neglect the operation of an incinerator because it is used to burn waste products and therefore may not be regarded as an integral part of the apparently more constructive elements of the production process.

However, incinerators are becoming very sophisticated pieces of equipment, capable of dealing with solid, liquid and gaseous waste, and their economic operation can only be ensured if guidelines similar to those laid down for

Plant - Good Housekeeping   139

plant such as boilers and furnaces are followed. An increasingly common application of incineration is in the pollution control field, and fume incinerators are particularly appropriate to this area. Fume incinerators using catalytic processes for effluent control are now available.

All waste treated by incineration produces heat, and waste heat recovery methods are being applied in incinerators, this being discussed in detail in Chapter 6. With respect to 'good housekeeping', similar practices as those common to other combustion plant apply. Preheating of combustion air, adequate thermal insulation, burner design rated to enable poor grade fuel to be used, and control of the temperature to ensure that the minimum incineration temperature is not greatly exceeded, are all features which can contribute to reduced operating costs. If an incinerator is purchased without a heat recovery facility, it is also worth ensuring that a heat exchanger could be added at a later date if there is a potential use for the waste heat, or to counter further rises in fuel cost. Also, before incinerating waste, consider whether the waste can be used as a fuel (see Chapter 6).

5.14 Lighting in Factories and Offices

Lighting is another neglected area as far as efficiency and energy conservation are concerned. However a number of publications are now available offering advice on lighting selection and maintenance with a view to cost-effectiveness, bearing in mind the need for energy conservation. The Electricity Council, the Lighting Industry Federation and the Illuminating Engineering Society (with its equivalent body in the United States having the same name) are all involved in assisting industry with lighting problems (ref. 5.21), (ref. 5.22). The lighting industry is also now well aware of the benefits of emphasising energy conservation, and this is reflected in their product publicity data. Addresses of organisations offering advice in this field are given in Appendix 3.

The characteristics of the more common types of lamp are described in Table 5.5 (ref. 5.23). The current emphasis on energy conservation has led to a greater interest in the light output of the various types of lamp, generally expressed in lumens per Watt, or efficacy, as shown in Fig. 5.15.

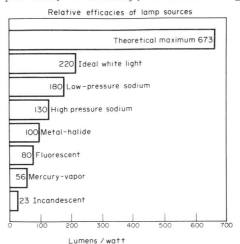

Fig. 5.15 Relative efficacies of lamp sources

TABLE 5.5 Characteristics of Basic Lamp Types
(Courtesy IEEE Spectrum, (ref 5.23))

Characteristics	Incandescent (including Tungsten Halogen)	Fluorescent
Wattages (lamp only)	15 to 1500	40 to 219
Life (hours)	750 to 12 000	9000 to 30 000
Efficacy (lumens per Watt lamp only)	15 to 25	55 to 88
Color rendition	Very good to excellent	Good to excellent
Light direction control	Very good to excellent	Fair
Source size	Compact	Extended
Relight time	Immediate	Immediate
Comparative fixture cost	Low because of simple fixtures	Moderate
Comparative operating cost	High because of relatively short life and low efficacy	Lower than incandescent; replacement costs higher than HID because of greater number of lamps needed; energy costs generally lower than mercury - vapor

Characteristics	High-Intensity Discharge		
	Mercury-Vapor	Metal Halide	High-Pressure Sodium
Wattages (lamp only)	40 to 1000	400, 1000, 1500	75, 150, 250, 400, 1000
Life (hours)	16 000 to 24 000	1500 to 15 000	10 000 to 20 000
Efficacy (lumens per Watt lamp only)	20 to 63	80 to 100	100 to 130
Color rendition	Poor to very good	Good to very good	Fair
Light direction control	Very good	Very good	Very good
Source size	Compact	Compact	Compact
Relight time	3 to 5 minutes	10 to 20 minutes	Less than 1 minute
Comparative fixture cost	Higher than incandescent, generally higher than fluorescent	Generally higher than mercury - vapor	Highest
Comparative operating cost	Lower than incandescent; replacement costs relatively low because of relatively few fixtures and long lamp life	Generally lower than mercury-vapor; fewer fixtures required, but lamp life is shorter and lumen maintenance not quite as good	Generally lowest; fewest fixtures required

Incandescent and self-ballasted mercury lamps have relatively low efficacies. Fluorescent lights have a much better performance from this point of view, with further improvements being obtained using sodium and metal-halide lamps. High-intensity discharge lights are popular because of their long life. The white fluorescent tubes now available have efficacies of over 70 lumens per

Watt, and discharge lamps also combine high efficacy with greatly improved colour-rendering, and are particularly useful where high mounting is necessary.

Maintenance is very important if the correct standard of lighting is to be maintained. A fall-off in illumination due to improper maintenance can be countered, and often is, by the wasteful installation of extra lighting to compensate for losses due to dirt and time-expired light fittings. The effect of the maintenance period on performance is shown in Table 5.6

TABLE 5.6 Typical Maintenance Factors for Industrial Lighting Installations

Locations	Months between cleanings				Luminaires
	3	6	12	24	
"Clean-room"	1.00	.99	.98	.94	Special
	1.00	.99	.98	.94	Ordinary
Particular clean	.98	.97	.96	.90	Special
air-conditioned area	.95	.84	.82	.76	Ordinary
Average factory,	.97	.90	.88	.82	Special
e.g. assembly dept.	.90	.82	.74	.57	Ordinary
Dirty factory,	.90	.82	.79	.71	Special
e.g. engineering.	.87	.73	.65	.48	Ordinary
Particularly dirty,	.89	.78	.73	.64	Special
e.g. foundry	.84	.67	.59	.42	Ordinary

Notes: These figures are typical only of, and are based on BZ3 luminaires in, fairly large rooms with walls of 30% average reflectance; other data are averaged. The Maintenance Factors (MF) shown for "ordinary" luminaires refer to the common types of open industrial luminaires. The factors for "special" luminaires refer to luminaires having effective through-draught construction, giving a "self-cleaning" property, with or without being fully enclosed. Some ventilated open-construction luminaires produce better MF than some enclosed ones.

The light output of any lamp decreases with time, and the rating in lumens per Watt quoted by manufacturers is generally a figure corresponding to that which the lamp will produce after about 30 per cent of its operating life (i.e. after 2000 hours for a 7500 hour lamp). Fluorescent tube lamps and high pressure discharge units can continue to function well beyond their rated life, unlike incandescent filament lamps, at the expense of a considerable reduction in light output. Thus a replacement programme should be implemented.

As well as the lamp, cleaning of its holder (collectively called a luminaire) is also recommended. While some lights can be 'self-cleaning' to a degree by virtue of convection air currents, all reflectors tend to get dirty after some time. Wall surfaces and windows should also be maintained in a clean state, as they can absorb useful light.

Some allowance is made for loss of light due to dirt by the designer, by use of a Maintenance Factor. The value of this factor can be agreed between the supplier and purchaser, and the desired illumination will only be achieved if this is done, and a regular maintenance and replacement programme is implemented.

As well as losing performance, lamps become out-of-date. It is therefore advisable to regularly review (say every 5 to 10 years) the types of lighting available to enable assessment of the use of new types to be made. Codes defining lighting levels are available, and these are produced by the Illuminating Engineering Society.

As well as selection of the luminaire and its correct maintenance, there are several other ways in which energy can be saved, and these are listed below:

(i) Lights should be turned off when not required. Most indoor lighting in areas close to windows can be turned off in daytime. This should be taken into account when designing lighting circuits for large open-plan offices.

(ii) Time switches on lights can be beneficial.

(iii) The use of light colours on walls and ceilings lowers the electric lighting levels required.

(iv) Dimmers can be used to adjust lighting levels as required.

(v) Lighting should be selected on the basis of the activity which it is serving. If the activity in that particular part of the building changes, it may save money to change the lighting to suit as well.

(vi) Heat generated by lighting adds to the air conditioning load. However this heat can also be used to advantage in winter for space heating (see Chapter 6).

Shell UK Ltd. have estimated that the best economies in their headquarters building, Shell Centre in London, could be obtained by replacing tungsten lighting by fluorescent lights. This would result in a 3 per cent energy saving (£6000 per annum), with a pay back period of two to three years. More efficient lighting such as metal halide sources, with improved diffusers, show considerable savings in boiler and plant rooms. Lighting manufacturers are also offering more efficient fluorescent tubes, and trials in the Shell Centre are in progress, using a more efficient luminaire with three tubes in each fitting, allowing alternate fittings to be switched off. Shell estimate that if universally applied in the Shell Centre, a 25 per cent saving in energy consumption and maintenance costs would result, with a present pay back term of the order of five years.

Shell are also considering extensive use of a ripple control system for lighting in this building. They argue that while timeclocks and time delay switches are essential for local lighting and power consumption, it is more economical to control general lighting in the Shell Centre using one time clock and an electrical distribution network. A 200 Hz impulse is injected onto the neutral at medium voltage and this operates contactors on the lighting switches in corridors, toilets, conference rooms etc. Shell believe that it is now economical to extend this form of control to cover garage areas and all circulation corridors. An investment of £3000 would pay back in two years, (ref. 5.20).

At another plant, twenty three 1 kW tungsten lamps were replaced by an identical number of high bay 400 W high pressure sodium lamps. The luminance was trebled to 390 lux and the load reduced from 23 kW to 10.35 kW, resulting in an annual running cost reduction from £540 to £172. Maintenance costs fell from £525 to £200 and the total capital cost of the installation, £1996, was recovered in 2.6 years.

5.15 Plant Buildings

Energy conservation in factory buildings has much in common with the measures recommended for air conditioning systems and other plant, discussed in other sections of this Chapter. However there are some features unique to assembly shops and areas serving similar functions which warrant separate treatment.

As most air from outside entering a plant requires adjustment to inside conditions, and exhaust air represents a heat loss, it is always worth investigating the possibility of heat recovery units to preheat incoming air, and to keep air changes to a minimum consistent with comfort and safety conditions. In particular, where forced ventilation is used, excessive air use means increased fan running costs.

Most industrial buildings have large doors, loading bays, etc. Even small openings such as unused fan ducts and gaps in the roof can account for significant heat loss. Warm air curtains can be used to protect doorways which are open for long periods. If loading or unloading is being carried out, the vehicle may be covered to reduce heat loss. Radiant space heating is probably the most efficient where heat losses cannot be stopped, as personnel are least affected by the loss. A convection system tends to have to reheat the whole building before full comfort conditions return.

The amount of heating required to compensate for heat losses through open doors is considerable. Fig. 5.16 shows the quantity of fuel wasted per hour in heating incoming air, as a function of wind speed and the area of the oepening. For a door 3 m by 3 m in area, based on an outside air temperature of $-1^{\circ}C$ and an inside temperature maintained at $18^{\circ}C$, it would cost approximately £1900 over a three month period to compensate for heat losses through the open door. (It is assumed that the building is heated with oil fired units at an 80 per cent efficiency, for 8 hours per day, five days per week). The installation of an air curtain would cost considerably less than this. Operating costs of the air curtain are low - in some cases ambient air being sufficient to prevent losses if correctly directed across the doorway. Existing plant heating, or an independent heat source for the air curtain may also be used, the former being the most common solution. Air curtains can also be applied to cold stores, to reduce refrigeration load, and to ovens, (see Section 5.12).

Non-electric control valves are available for use with radiators and convectors. These may be fitted on existing equipment and do not require remote thermostat wiring, thus making installation simple. Use of these valves is said to result in fuel savings of up to 40 per cent in some instances, although they are more applicable to small spaces such as offices.

If gas-fired convection heating is used, all surfaces should be regularly

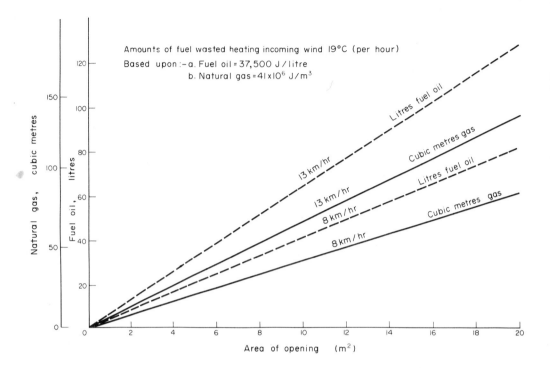

Fig. 5.16 The amount of fuel per hour wasted heating air flowing into a building through an open door (Courtesy Minikay Ltd.)

cleaned, and burner adjustment checked for maximum efficiency. The cleaning of radiant heating surfaces is also a pre-requisite for efficient performance.

Space heating systems should be switched off in areas which are unoccupied for long periods. Radiant heating response is fast, so the personnel need not be uncomfortable for long if the heating is switched on as they enter a particular area. In confined spaces direct convection heaters are most efficient.

If staff start and finish work at a different hour, this may necessitate heating being on longer than if all staff were to start and finish at the same time.

Avoid having open doors and windows on opposite walls of a building, as this aggravates heat loss by creating a through-draught. The use of self-closing doors is recommended, and these should be well maintained to prevent them jamming open. In winter windows should preferably not be used to control temperature.

In some buildings a particular process may require good ventilation to remove noxious fumes. If this is the case, ensure that the ventilator is local to the process, retaining where possible normal air change rates in the rest of the building. All ventilation equipment should be switched off when the particular process is not in operation.

5.16 Plant Domestic Hot Water Systems

There are a number of alterations which can be readily implemented to save heat loss in domestic hot water systems in factory and office buildings, and these include the following:

(i) Reduce the maximum temperature of washing water by a few degrees. This is unlikely to cause inconvenience, and can easily be done if the system is separate from the main space heating or plant process water heaters.

(ii) Use taps which automatically shut off when released. These save both hot and cold water.

(iii) Insulate hot water storage tanks and long pipe runs to hot taps.

(iv) Check that the hot water system does not have a capacity much greater than that needed. The requirement for hot water may have changed since the system was installed.

(v) Consider the use of process or space heating waste heat for preheating of the domestic hot water

(vi) Relocate hot water heaters and storage tanks closer to the point of use.

5.17 Useful Aids to Energy Conservation Implementation

There are several types of instrument available for helping the energy conservation manager check the performance of plant and equipment, and some of these are listed here. Manufacturers of these instruments are listed in Appendix 4.

(i) Infra-red cameras and scanning systems. These can be used to locate points of excessive heat leakage in furnaces etc., and enable the user to check on the effectiveness of insulation. While the most sophisticated systems are expensive, the user is advised to check on hire facilities for this equipment, unless its use as a permanent feature of an R & D programme can be justified.

(ii) Gas analysis equipment. Analysers can be obtained for sensing a single gas, or a large variety of different types. These may be used for checking combustion efficiency, which in turn can lead to pollution, as well as fuel control.

(iii) Flowmeters. Flows of air, water and other gases and liquids is of interest in process control and energy conservation. A flowmeter for use in a water main can indicate when excessive wastage is occurring. Oil flow rates measured in an oil central heating system over a 24 hour period may reveal excessive fuel use when the building is unoccupied or ambient temperatures are high.

(iv) Temperature control equipment. Thermostatically controlled heating, be it in space heating or process applications, can save excessive energy use. Control equipment related to ambient conditions, building occupancy etc. is also available, and is discussed in Section 5.2.

(v) Humidity measurement. Humidity is important in air conditioning, and also in processes such as drying (see Section 5.8). Humidity measurement and control can assist in saving energy.

All the above types of instrument are generally applicable in industry. A number are available which are unique to particular industries, and these are discussed in the sections relevant to those particular industries. Remember that these instruments only remain accurate if they are adequately maintained and regularly calibrated.

5.18 Steam Traps

A steam trap is an automatic valve installed in a steam line to remove condensate and air. Steam traps can be actuated in a number of ways, thermostatic traps responding to temperature differences between steam and the cooler condensate, and thermodynamic traps reacting to changes in phase. Mechanical or float traps are actuated when the density of the medium changes, indicating the presence of condensate or air.

Good reliability of steam traps is essential. Condensate in heaters reduces their effectiveness by decreasing internal surface for heat transfer. Air and other non-condensable gases have a similar effect. Carbon dioxide and oxygen in the steam can cause corrosion in any pipework carrying it, and of course can also shorten the life of the heater.

Steam trap failure can have serious implications. One that fails closed prevents steam flow and the heating process is terminated. This may be countered using a steam bypass system, but if the bypass valve is left open for extended periods, the functions that the trap was performing are not available, and heating efficiency gradually deteriorates. Unless repairs are quickly implemented, the fall-off in performance leads to excessive steam use or low process efficiency. If a trap fails open, the failure is more difficult to detect. However the cost of an open trap can be very high. A single steam trap on a 30 to 150 kN/m^2 system with a 9 mm to 12.5 mm orifice can waste between £500 and £1000 per month in fuel. Even if the trap is functioning, but is dirty and not operating at peak efficiency, this can lead to a 20 per cent increase in fuel consumption (ref. 5.24).

Standard Oil of California have implemented a steam leakage prevention and steam trap maintenance programme, (ref. 5.25). As a result the steam consumption at their Richmond refinery was reduced by over 30 kg/s. At their Perth Amboy plant, which contains 2500 steam traps, a maintenance programme on the traps, involving checks and maintenance every month, accounted for 50 per cent of all the steam saved in the plant. At Richmond traps were previously checked only once every 3 months and the failure rate was still 10 to 15 per cent, emphasising the need for regular checks at relatively short intervals.

ICI Petrochemicals Division (ref. 5.20) are successfully using ultrasonic

Plant - Good Housekeeping

detectors for examining steam traps. The Division has also produced a video tape to explain the operation, correct installation, and fault finding procedures of steam traps. This is being used in seminars organised for supervisors and maintenance engineers.

5.19 Vats and Hot Storage Tanks

If a material is being stored at a temperature above (or below) ambient, or if it is being heated in a container, a number of tasks can be carried out to reduce heat loss (or gain), or improve heating rates. These are as follows:

(i) Vats and storage tanks should be insulated and covered with a lid.

(ii) If the vat produces a significant amount of hot exhaust, as in a copper in a brewery, some waste heat recovery techniques may be beneficial.

(iii) The heating of vats is more effectively carried out using immersion heating than by applying heat underneath or outside the container, although where insulation proves inadequate, some good guard heaters may be employed. Submerged combustion may be appropriate for vat heating (see Section 5.5)

(iv) Immersion heaters should be regularly checked to ensure that surfaces are clean. A heavily coated tube could lead to burn out.

(v) Thermostats should be added to heater systems on storage vessels to ensure that the temperature does not rise above the minimum necessary.

(vi) If hot effluent is available, this could be passed through a coil immersed in a storage tank to provide at least some of the heat input required.

5.20 Fuel Additives

An increasing number of companies are offering fuel additives for use in fuel oil burners in boilers, furnaces and other process equipment. The manufacturers claim that these additives can overcome a wide variety of problems, leading to improved combustion efficiency, lower maintenance costs and reduced pollution. One product, when introduced to the fuel oil storage tank offers the following benefits:

(i) Prevents sludge formation in the storage tank and minimises burner coking.

(ii) Improves combustion efficiency.

(iii) Reduces low and high temperature fouling and corrosion.

(iv) Controls acid smut emission and excessive smoking.

Other additives are available which can be added to oil or natural gas fuels to prevent scaling of metal during heat treatment (ref. 5.26).

A short but comprehensive review of the properties of burner fuel additives was given in 'Processing' in 1975 (ref. 5.27). The problem areas were identified and cost data given on some of the additives. Viscosity was pinpointed as one problem associated with the use of some residual fuel oils. A high viscosity can cause comparatively rapid build-up of sludge in both the fuel storage tank and the lines feeding the burners. Fuel additives can be obtained which overcome this problem by treating the sludge-forming components of the oil to make them combustible. This improves burner performance and reduces maintenance costs.

High and low temperature corrosion, brought about directly by vanadium pentoxide and sulphuric acid respectively, can be reduced by using additives which prevent the formation of these compounds, or neutralise their action.

Burner performance itself may be improved by additives. By including forms of nitrogenous amine compounds, which can modify the interfacial tension of the oil molecules, the atomisation of the fuel becomes more effective, improving combustion efficiency and reducing fuel costs (ref. 5.28).

A typical cost range for fuel additives appears to be 0.1 to 0.2p per gallon of fuel oil treated. Obviously this is only a very small proportion of the total fuel oil cost, and a fuel additive must be regarded as a good investment, provided of course that the combustion system is set up correctly in the first place.

Table 5.7 summarises the properties of a number of fuel additives, some of which were listed in ref. 5.27, other data having been obtained by the author.

5.21 Sources of Other Data and Suggestions

The reader will readily be able to obtain a wide range of publications which summarise ways in which energy can be saved by short, medium and long term measures. A selection of this literature is detailed in the Bibliography. Other Appendices list equipment suppliers and organisations specialising in energy audits etc.

Energy saving techniques detailed in this and subsequent chapters can yield substantial savings, NIFES estimating that United Kingdom Industry as a whole could save £500 million per year if it used energy as effectively as the most efficient firms. As a first step it is necessary to be able to quantify the consumption of each item of plant. Following this, short term measures of the type described in this chapter can be implemented, and a more detailed economic analysis will precede any investment in heat recovery plant of the type detailed in Chapters 6 and 7.

Plant - Good Housekeeping

TABLE 5.7 Proprietary Fuel Additives and their Properties

Supplier	Additive	Application	Composition
Arrow Chemicals Ltd.	Solvenol	Reduction of sludge formation	Solution of various additives in hydrocarbon base
	Carbonex	Cold-end treatments	Powdered alkali nitrates for direct injection into boilers
Combustion Chemicals Ltd.	Alumag CH22	Control of acid smut emissions and V h.t. corrosion	MgO and alumina powders suspended in a light oil carrier
	CCL Concentrate	General purpose corrosion inhibitor	Manganese/amine containing additive
	Triad MM	Reduction of sludge. Anti-corrosion, improved combustion Anti-acid smut and smoke	Manganese/magnesium and solvents
Industrial (Anti-Corrosion) Services Ltd.	Fernox-Rolfite 404 (refs.5.26,5.28)	General applications from sludge prevention to acid smut elimination	Added to storage tank in 6000:1 ratio
Nalfloc Ltd.	A810 and A825	Fireside applications to reduce fouling and soot deposits	Powder type additives
	A880 and N158D	For sludge prevention and reduction of burner fouling	Liquid for direct addition to fuel
Tribol Ltd.	Maxiflo	Sludge emulsifier	Added to storage tank where sludge is converted into finely dispersed particles
Triple-E(UK) Ltd.	R4000 and B1000 (refs.5.29,5.30)	General combustion improvers	Chemical control of combustion by non-metal-bearing additive. Fuel savings of 5 to 10% & cleaner systems
United Lubricants Ltd.	Carburol Power Plus	Fuel booster/ combustion improver	Liquid additive containing organo-metallic compounds which absorb asphaltic and resinous impurities in oil to make them combustible
	Carburol S.D.	Sludge disperser	Similar action to Power Plus and also improves oil atomization
Also UK agents for Du Pont Pet. Chemicals	DMA-F	Multipurpose additive	A 49% w/w solution in kerosene of organic compounds, to preserve stability, control haze, etc.

REFERENCES

5.1   Schoenberger, P.K. Energy saving techniques for existing buildings. Heating, Piping & Air Conditioning, pp 98 - 105, Jan. 1975

5.2   Loten, A.W. Energy conservation in the Property Services Agency. Building Serv. Eng. Vol. 43, Aug. 1975

5.3   Coad, W.J. The computer as a tool for energy analysis. Heating, Piping & Air Conditioning, pp 46 - 50, Jan. 1975.

5.4   Anon. Computer-controlled HVAC system helps cut energy usage in glass-walled office building. Edison Electric Institute Energy Management Case Study 12, New York, 1974.

5.5   Lyle, O. The Efficient Use of Steam, HMSO, London, (1956)

5.6   Kurz, G. and Carnegie, I.L. Submerged combustion burners. Process Engng., Jan. 1969

5.7   Anon. New burners for old. Processing, pp 38 - 39, May 1975.

5.8   Bryan, D.J. et al. Applications of recuperative burners in gas-fired furnaces. Inst. Gas Engineers, Communication 952, Proc. 40th Autumn Research Meeting, London, Nov. 1974.

5.9   Reed, R.J. Fuel/air ratio control systems. Industrial Process Heating pp 14 - 17, April, 1972

5.10 Lindsley, D.M. Automatic combustion controls for smaller or older installations. The Heating & Air Conditioning Journal, pp 24 - 27, May 1974.

5.11 Weinberg, F.J. The first half-million years of combustion research and today's burning problems. Progress in Energy and Combustion Science, 1, 17 - 32, (1975).

5.12 Anon. Enlarging the capacity of cooling towers. The Heating & Air Conditioning Journal, Sept. 1974

5.13 Foster, C.R. and Kloiber, F. Fuel conservation, National Asphalt Pavement Association Report, 1974

5.14 Dickson, P.E. Heating and drying of aggregate. National Asphalt Pavement Association Report, April 19, 1971.

5.15 Foster, C.R. Heating and drying of aggregates - Btu requirements and exhaust volumes. NAPA Information Series 47, 1973.

5.16 Silverthorne, P.N. Power factor correction for energy conservation. ASHRAE Journal, pp 28 - 32, May 1975

5.17 Coombs, V.T. The possibility of energy saving by the correct sizing of electric motors. Electrical Review, pp 744 - 746, Dec 5, 1975

5.18   Sanders, P.K. Are you paying too much for your energy supplies ?
       Bradford Chamber of Commerce Journal, 1, 4 (1974)

5.19   Beatson, C. How to save power in furnaces with lightweight refractory
       The Engineer, pp 48 - 49, April 24, 1975

5.20   Anon. Energy saving: The fuel industries and some large firms.
       Energy Paper No. 5, Department of Energy, HMSO, London, (1975)

5.21   Anon. Essentials of good lighting. A simple guide for industry. The
       Electricity Council, London (1974)

5.22   Anon. Manage to save it - Efficient lighting. The Lighting Industry
       Federation, London, 1975.

5.23   Beardsley, C.W. Let there be light, but just enough. IEEE Spectrum,
       12, 12, pp 28 - 34 (1975)

5.24   Ulrich, O.E. Using steam traps effectively. Plant Engineering (US)
       Feb. 21, 1974

5.25   Humphrey, R.C. Standard Oil of California stresses steam savings,
       improved firing, in all refineries. Power, Nov. 1973.

5.26   Henson, C.G. Furnace technology - Increasing the efficiency and
       quality of heat treatment. Metallurgia and Metal Forming, Aug. 1972

5.27   Spear, M. Burner fuel additives. Processing, pp 35, 37, May 1975

5.28   Henson, C.G. Furnace technology - Fuels and fuel economy. Metallurgia
       and Metal Forming, July 1973

5.29   Anon. Fuel additive can save 5% in costs. Maintenance Engineering.
       pp 24 - 27, Oct. 1974.

5.30   Anon. An experimental evaluation of Rolfite fuel oil additive in
       domestic heating systems. Report of the Rolfite Company, May 1975.

Common Items of Plant - Energy Recovery

As discussed in the introductory remarks to Chapter 5, much of the preceeding data has covered areas of energy conservation from the point of view of the particular industry concerned, with descriptions of processes which at first sight may be regarded as unique to that industry. However, in reality many of the processes have components common to a wide variety of industries, and in Chapter 6 the common components are discussed from the point of view of energy (in particular heat) recovery.

Some have already been discussed in some detail. Steam and gas turbines, and reciprocating engines have been covered in Chapter 3, with particular emphasis on heat recovery in the concept of the total energy plant. It has been indicated in the appropriate sections of Chapter 5 which of the various items of plant can usefully employ waste heat recovery equipment, but in this chapter we will also be concerned to some extent, with recovery of fluids and other 'effluent' which in many instances is really wasted energy.

In most cases the improvements in plant efficiency which can be made using techniques recommended in this chapter will, unlike many of those in Chapter 5, involve some capital expenditure, often necessitating a quite considerable investment. However, the returns on the investment are normally high.

Most of the equipment for heat recovery is described in detail in Chapter 7, and the manufacturers of this equipment are listed in Appendix 4.

6.1 Air Conditioning

An air conditioning system is probably the most complex 'item of plant' which we will investigate from an energy recovery point of view, because it can involve a large number of components in its own right. One air conditioning system can supply several hundred rooms, and in a commercial building the requirements in different locations may be many and varied. There may be a simultaneous need for heating and cooling in different areas, either due to poor system design, which can create overheating or a heating deficiency, or due to the operating requirements of particular plant in the building, such as computers etc.

As air conditioning systems are in most cases based on the use of electrical energy, it is the national electricity organisations which have shown the greatest interest in energy recovery and general efficient energy usage in air conditioned buildings. Two publications, one distributed in the United Kingdom by the Electricity Council (ref. 6.1) and the other published in the United States by the Electric Energy Association (ref. 6.2), present a most useful account of the ways in which energy recovery can be applied in air conditioned buildings with emphasis on the economics as well as the technical considerations.

6.1.1 Heat sources and sinks in a building
In order to be able to assess

the economic aspects of an air conditioning system, and the likely improvements brought about by adding energy recovery units, it is necessary to carry out an accurate heat balance. The heat losses from the building are:

(i) Fabric heat loss - the heat loss according to the building structure, which normally includes half an air change every hour to allow for the inflow of air through doors etc.

(ii) Ventilation air heat loss - the heat required to be added to the air entering the building to provide comfort conditions.

The heat gains within a building are numerous, and may be difficult to quantify accurately. The most important are:

(i) Lighting - all the energy consumed by lights may be regarded as a heat gain, although an accurate statistical analysis of lighting use will be needed to assess this.

(ii) Metabolic heat - occupants of a building have a resting metabolic heat loss of approximately 100 W, which is given up to the atmosphere.

(iii) Computers - a large computer can generate considerable quantities of heat, although much of this may arise intermittently.

(iv) Exhaust air - heat contained in the stale air which is exhausted from the building may be recovered.

(v) System gains - heat is generated by motors driving fans and pumps as part of the general air conditioning services.

(vi) Process plant - other items of equipment having no association with the air conditioning system may generate heat. These may be regarded as heat sources from the air conditioning point of view.

The heat losses from the building are a function of the external temperature, and may be plotted as a straight line, with heat loss increasing as external temperature decreases. In general the internal heat gains may be regarded as independent of external temperature, and are plotted in Fig. 6.1 as a straight line parallel to the x axis. The theoretical balance point, where the heat loss equals the heat gained, is given by the intersection of these two lines. In practice the theoretical heat balance point is never likely to be achieved without a redistribution of heat in some parts of the building.

The heat balance for a building in the United Kingdom, the Merseyside and North Wales Electricity Board Headquarters at Chester, is as follows. The figures are based on an outside ambient of $-4°C$ and an internal temperature of $21°C$, (ref. 6.1).

	Gain kW	Loss kW
Fabric heat loss	-	378
Ventilation loss	-	490
Lights	550	-
People	100	-
Fans, pumps, compressor	218	-
	868 kW	868 kW

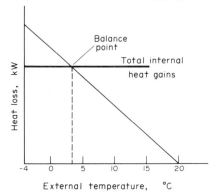

Fig. 6.1 Graph indicating balance of internal heat gains and heat losses from a building, as a function of outside ambient temperature. (Courtesy The Electricity Council)

6.1.2 Methods of heat recovery Assuming that a number of basic requirements can be met, namely that there is sufficient heat to be economically recovered (both in terms of capital and running costs) and that an acceptable use can be found for the recovered heat, the methods for recovery can be investigated. The recovery of heat in air conditioned buildings may be divided into three categories; exhaust air heat recovery, recovery from lighting systems, and refrigeration-type heat recovery.

Exhaust air heat recovery: The recovery of heat contained in the exhaust from an air conditioning system (or, in the summer, the transfer of heat from incoming fresh air to the exhaust cool conditioned air), is the easiest form of heat recovery to implement, and can be fitted to existing air conditioning systems.

There are four basic systems available for heat recovery in air conditioning exhausts. These are:

(i)   Rotating regenerators

(ii)   'Run-around' coils

(iii) Simple air to air heat exchangers

(iv)   Heat pipe heat exchangers.

All of these systems are described in detail in Chapter 7.

Plant- Energy Recovery

The performance of a rotating regenerator (commonly called a 'heat wheel') in an air-conditioning application is illustrated in Fig. 6.2. It can increase the temperature of cold make-up air by up to 15°C in winter, and in summer it is able to reduce air inlet temperatures by 4 - 8°C, (ref. 6.3). As with the other air conditioning exhaust heat recovery systems, with the exception of the run-around coil, the rotating regenerator can only function if the exhaust and intake ducts are located adjacent to one another at some point in the system. This drawback may affect 'retrofitting' of these systems.

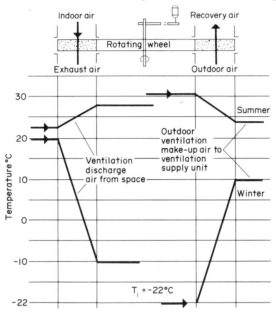

Fig. 6.2 Performance of a rotating regenerator used to heat or cool building make-up air (Courtesy IEEE Spectrum, ref. 6.3)

A run-around coil is in general inferior in performance to a rotating regenerator, as can be seen by comparing Figs. 6.2 and 6.3. In the system illustrated, water is circulated through the two heat exchanger coils, and as a pump is used, the two coils can be located some distance apart. Both the run-around coil and the rotating regenerator employ moving parts (a pump and an electric motor drive respectively), and this sets them apart from the air to air and heat pipe heat exchangers. However the rotating regenerator is unique in that some cross-contamination can occur in the ducts. Dirt can be deposited on the wheel, which then carries it over to the incoming duct. Most wheels now have purging sections which minimise carry-over, and the performance of these is quantified in Chapter 7.

The simple air to air heat exchanger, illustrated diagrammatically in Fig. 6.4, represents a static means for transporting heat between exhaust and inlet air streams, which pass through it in counterflow pattern. It resembles an open-ended steel box with a rectangular cross-section that is compartmented into a multiplicity of narrow passages in a cellular format. Alternate passages carry exhaust and make-up air and energy is transferred from one air stream to the other simply by conduction through the passage wall. No cross-contamination can occur, and as a considerable amount of surface area can be packed into a compact space using this technique, the efficiency obtained is reasonable.

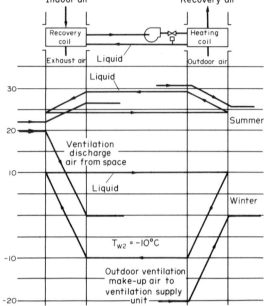

Fig. 6.3 A 'run-around' coil used to transfer heat between air streams. Performance is inferior to the regenerator shown in Fig. 6.2. (Courtesy IEEE Spectrum, ref. 6.3)

The heat pipe heat exchanger, detailed in Chapter 7, may be regarded as a logical extension of the above, where instead of bringing the two air streams into compartments, they retain their completely separate identity, being connected instead by an array of finned heat pipes, which transfer heat from one duct to the other. Although the wall of the heat pipe may be considered as an additional thermal resistance, the efficiency of heat transfer within the heat pipe itself, which uses an evaporation/condensation cycle, is so great that these heat exchangers can recover up to 70 per cent of the exhaust heat. One of their major advantages over the run-around coil and the thermal wheel is their reliability. The failure of a few of the heat pipes will only

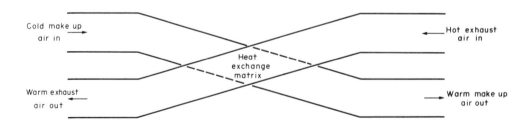

Fig. 6.4 Diagrammatic representation of a static air-to-air heat exchanger

slightly reduce the efficiency of the unit, rather than leading to total shutdown of the recovery system.

With all of the above heat recovery systems, it is important to bear in mind the fact that their use when applied in air conditioning is most beneficial and economically advantageous in countries where extremes of temperature are greater than in the United Kingdom and some European areas.

As with systems such as heat pumps, high utilisation in the summer, as well as in the winter heating season, is desirable. In plant where ambient conditions are by necessity different to 'normal' comfort conditions, possibly due to the requirements of a particular process being carried out, these arguments may not apply, and significant temperature differences between the inside climate and the external ambient could well exist. In these cases heat exchangers of one type or another would undoubtedly prove advantageous.

Heat recovery from lighting systems: A ducted air system in a building may take full advantage of the heat given off by lights, which are a major source of internal heat gain. In order to do this, special types of light fittings, or luminaires are needed, of the geometry shown in Fig. 6.5. In this luminaire, return or exhaust air is drawn over the lamps and sheet metal of the lamp fittings before passing into plenum chambers in the ceiling. In this way the air can pick up as much as 80 per cent of the heat given off by the lights. Although some of the heat transferred in this manner into the ceiling space is lost, either back into the room below or through the floor into the offices above, between 50 and 65 per cent of the heat is retained within the ceiling plenum. Fig. 6.6 shows a portion of a single duct heat recovery system used in the interior zone of a building which may require

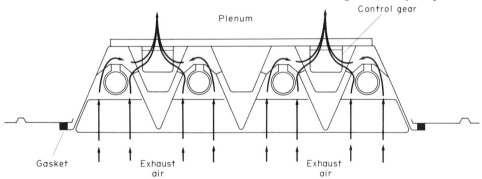

Fig. 6.5 An air handling luminaire for fluorescent tube lights
(Courtesy The Electricity Council)

simultaneous winter heating and cooling. The induction box with thermostatically controlled dampers is the most important element in this system. When full cooling is called for, 100 per cent cold primary air is delivered to the space. When warmer air is needed, warm air from the ceiling cavity is gradually induced into the unit and is mixed with a proportion of cold air. Alternatively warm plenum air can be used to heat areas closed to the building perimeter, with supplementary heating supplied as required.

In summer the use of air-cooled luminaires is particularly beneficial, as the lighting heat can be removed and dissipated to the external atmosphere,

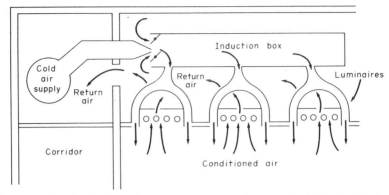

Fig. 6.6 A single duct heat recovery system applied to luminaires
(Courtesy US Electric Energy Association)

thus leading to a reduction in the cooling load of the air conditioning system. An advantage resulting from their use, not always appreciated, is the fact that lighting efficiency is improved. By operating at a lower temperature, the lamp can give up to 13 per cent more light output for the same power, compared to an uncooled lamp. (Conversely an over-cooled lamp will give a lower output. For a conventional fluorescent tube, the maximum light output is achieved when the tube wall is at $40^{\circ}C$.)

The performance, in terms of heat and light output, of a heat transfer luminaire manufactured by Osram-GEC is detailed in Fig. 6.7. These results

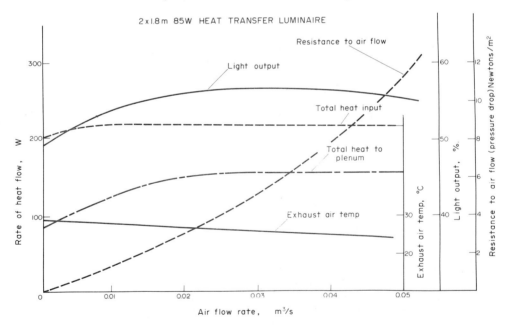

Fig. 6.7 Light and heat performance of a heat transfer luminaire, (2 x 1.8 m, 85 W units). (Courtesy Osram-GEC Ltd.)

were obtained using a purpose-built calorimeter at the GEC Hirst Research Laboratories (ref. 6.4). Room temperature was 22°C, and the pressure drop is that measured across the luminaire. The effect of over-cooling can be seen in this figure, the light output decreasing as the airflow rate rises above 0.035 m^3 per second.

The discussion has so far concentrated on air-cooled luminaires. It is also possible to recover the heat from light fittings using water cooling (refs. 6.2, 6.3). One method of doing this, in a refrigeration-type heat recovery system, is shown in Fig. 6.8. The luminaires used have special aluminium reflector housings that are formed in such a way as to provide integral water passages. Using this method, up to 70 per cent of the heat generated may be transferred to the water.

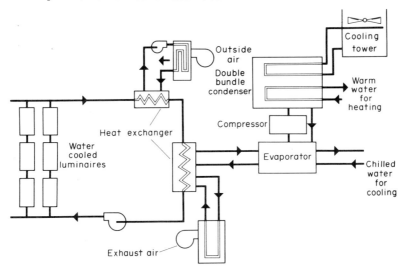

Fig. 6.8 Water cooled luminaires linked to a comprehensive refrigeration-type heat recovery system
(Courtesy US Electric Energy Association)

In one design, the fluid from the luminaires is pumped through an evaporative heat exchanger, minimising internal heat gain. If lighting accounts for 50 per cent of the heat gain in a building, then absorbing 70 per cent of this heat directly from the fixtures can lead to a 35 per cent reduction in the total air conditioning load, a not inconsiderable quantity. In the winter, this system would be turned off to allow the heat to be used for comfort conditioning.

Refrigeration-type heat recovery (heat pumps): The principle of the heat pump and its potential in industrial applications is covered in detail in Chapter 7. However, as emphasised there, the heat pump will come into its own in industry over a period of several years, whereas for space conditioning in commercial buildings, the technology has already been successfully applied on a considerable scale, (see Appendix 5).

Typical heat pump systems which can be applied to space conditioning are illustrated in Figs. 6.9 and 6.10. The arrangement shown in Fig. 6.9 utilises

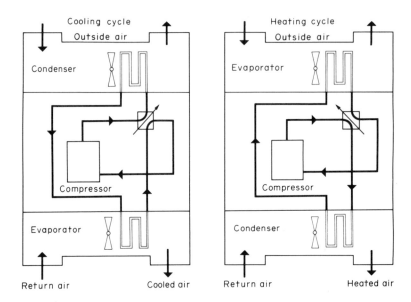

Fig. 6.9 A simple air-to-air heat pump system, using outside air as a heat source in winter and a heat sink in summer (Courtesy US Electric Energy Association)

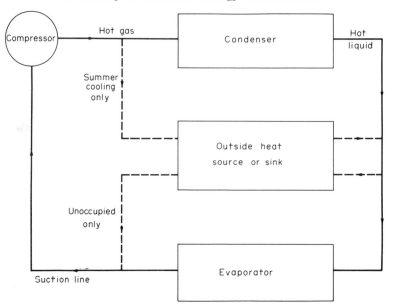

Fig. 6.10 A heat pump applied to simultaneous heating and cooling duties within one building shell

Plant - Energy Recovery

low grade heat recovered from the air for heating in winter, and uses this same air as a heat sink in summer. Figure 6.10 shows an air-to-air heat pump acting as a heat recovery unit in a double duct air conditioning system capable of simultaneously heating and cooling. In this mode the compressor delivers the hot vapour to the condenser heat exchanger, where heat is given up for space heating. The condensed refrigerant then passes to the evaporator, where it takes up heat, cooling the area around it. When there is no internal heat in the building to be recovered, for example during periods of unoccupancy, heat is taken from the outside air instead, or, as in Fig. 6.9, the outside may be used as the heat sink.

As an alternative to air, water may be circulated through the evaporator and condenser for heating and cooling respectively, in conjunction with a cooling tower, as illustrated in Figs. 6.11 and 6.12. Use is made of a double-

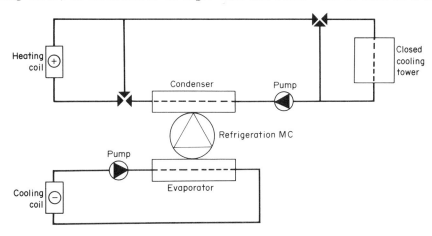

Fig. 6.11 Heat recovery system with a closed circuit cooling tower
(Courtesy The Electricity Council)

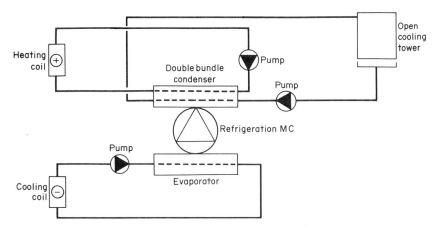

Fig. 6.12 An alternative to the system shown in Fig. 6.11,
- use of a double bundle condenser
(Courtesy The Electricity Council)

bundle condenser in the second of these circuits, with two separate water circuits enclosed in the same shell. The heat given up by condensation of the heat pump working fluid may be put into either or both of the water circuits, depending on requirements at that time. Also, by splitting the condenser the possibility of contamination of the building water circuit by water from the open cooling tower is eliminated.

If the heat is not needed for direct re-application, it may be stored. Heat generated during the day from lighting, operating equipment and human occupation may be collected and stored for use during the night. It can then be used to compensate for the building heat loss through the walls and windows. This may be accomplished by using the stored water as a heat source for a heat pump. A system of this type, having a water storage capacity of 90 000 litres, stored at $52^{o}C$, would be capable of providing 600 kW for approximately seven hours, with water being delivered to the heat pump evaporator at $15^{o}C$, (ref. 6.5). The use of supplementary heating in the storage tank may assist in overcoming an extended demand period, or intermittent peaks in demand. The tank may also be charged using off-peak electricity, and if the storage temperature is sufficiently high, the tank may be able to directly supply building heat, thus reducing machine running costs. (Heat storage systems are also discussed in Chapter 8).

6.2 Boilers

Indication of the possible ways in which heat recovery can assist the performance of boilers has already been given in Chapter 5. This section will deal with three topics associated with boiler operation which involve completely different aspects of energy conservation, all of which are, however heat recovery techniques. They are:

(i)   Boiler air preheaters

(ii)   Boiler condensate return

(iii)   Recovery of boiler blowdown.

Of these, only the air preheater is described in detail in Chapter 7, because of its applications in many other items of plant. Condensate return and blowdown are unique to boilers, and in general either do not involve heat exchangers, or, in the case of blowdown recovery, necessitate the use of only the simplest type of heat exchanger.

6.2.1 Boiler air preheaters An analysis has been carried out (ref. 6.6) of the effect on boiler performance and economics of the installation of a rotating regenerator for preheating combustion air. Rotating regenerators are discussed in detail in Chapter 7, and their application in a boiler, in this case a horizontal shaft air preheater, is illustrated in Fig. 6.13. The data on the oil fired boiler, which already contains an economiser which raises the feedwater temperature from $150^{o}C$ to $260^{o}C$, and a steam air heater which heats combustion air from ambient to $70^{o}C$, is given in Table 6.1.

The selection of rotating regenerators may in the first instance be carried out by the user, on the basis of manufacturer's data of the type illustrated in Section 7.4.1. However it may be necessary to carry out an optimisation exercise to take into account factors such as capital cost, when weighed against the quantity of heat recovered and the pressure drop through

Plant - Energy Recovery

TABLE 6.1 Data on Oil-Fired Boiler

Steam production rate	38 kg/sec
Fuel consumption	2.5 kg/sec
Combustion air quantity	41 kg/sec
Flue gas quantity	43.5 kg/sec
Oxygen in flue gas	3%
Ambient air temperature	15.6°C
Gas temperature leaving economiser	450°C
Net/gross calorific value (cv) of fuel	41 400/44 500 kJ/kg
Fuel sulphur content	2.5%
Fuel cost	£24/tonne
Power cost	0.75p/kW hr.

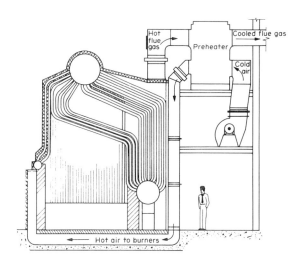

Fig. 6.13 A James Howden rotating regenerator employed
as a horizontal shaft boiler air preheater
(Courtesy Howden Group Ltd.)

the heat exchanger, which, of course means a variation in running cost. Capital costs also must allow for necessary modifications to ductwork and other plant to provide for incorporation of the heat exchanger.

In this case it was decided to select a gas leaving temperature of 155°C, before leakage dilution, (the leakage arising because of the imperfect seal between gas and air ducts). This gives an average metal temperature of 116°C and a gas temperature to the stack of 149°C. The gas and air quantities are now re-calculated taking into account fuel savings, which with the above gas outlet temperature amount to about 3.8 per cent. The preheater design conditions therefore become:

Gas entering quantity	41.85 kg/sec
Air leaving quantity	39.5 kg/sec
Gas entering temperature	232°C
Gas leaving temperature	155°C
Gas leaving temperature (diluted)	149°C
Air leaving temperature (by heat balance)	157°C

As mentioned above, a trade-off must be carried out to obtain the optimum heat recovery unit to cater for the design conditions. The optimisation exercise by necessity involves detailed cost analyses, and will normally be carried out by the manufacturer of the heat recovery equipment. The form of the data obtained for rotating regenerators manufactured by James Howden is shown in Table 6.2.

TABLE 6.2 Optimisation of Air Preheater Size

Preheater Size	19HS700	$18\frac{1}{2}$HS700	18HS750	$17\frac{1}{2}$HS750	17HS800
Gas Side Resistance mm Hg	3.03	3.57	4.66	5.44	7.04
Air Side Resistance mm Hg	3.09	2.46	3.22	3.75	4.85
Total Resistance mm Hg	5.10	6.04	7.90	9.20	11.85
ID Fan Head mm Hg	11.5	12.1	13.15	13.95	15.55
FD Fan Head mm Hg	17.6	18.0	18.8	19.3	20.3
ID Fan Head Capitalised at £60/kW	£38205	£40056	£43451	£46228	£51474
FD Fan Head Capitalised at £60/kW	£38616	£39435	£41073	£42301	£44554
Total Capitalised Running Cost	£76821	£79491	£84524	£88529	£96028
Selling Prices Preheater including erection and spare elements and seals 'A'	£49010	£45340	£42490	£40150	£38220
Replacement of fan wheels, couplings and motors, including erection 'B'	£17280	£17280	£17280	£17280	£17280
Ducting, expansion joints, dampers and controls, including erection and insulation 'C'	£35520	£33870	£32370	£30890	£29400
Total Capital Cost 'A'+'B'+'C'	£101810	£96490	£92140	£88320	£84900
Total Capital Cost + Capitalised Running Cost	£178631	£175981	£176664	£176849	£180928

The results of the analysis show that in this case the $18\frac{1}{2}$HS700 model Howden rotating regenerator has the lowest overall cost, although it is not the cheapest in terms of simple capital cost. The capitalised running cost takes into account the cost of power used in overcoming the pressure drop through the heat exchanger. In order to complete the analysis unit performance must be described in terms which can be presented before the management of a company who are contemplating investing in the equipment, and several accounting methods are available for doing this. The calculation of the payback period, i.e. the length of time it takes for the net fuel savings to

Plant - Energy Recovery

equal the equipment capital cost, is a popular presentation method when energy conservation equipment is being assessed. Other methods are discussed in Appendix 2.

The calculation is best carried out for a range of fuel and electricity costs, so that the effect of projected increases can be estimated. In the data below oil is assumed to cost £28/tonne and electricity 0.75p/kW hr. It has been indicated previously that the fuel saving amounts to about 3.8 per cent, and therefore based on a utilisation of 8000 hours per annum, the total fuel saving will be £75 630. The additional fan power needed to overcome the resistance of the heat exchanger must be subtracted from this value to obtain the net fuel saving, and the power needed to rotate the regenerator, normally supplied by an electric motor, must not be neglected. The total power requirement is 35.7 kW, giving an annual cost of £2140. Thus the net energy saving in one year is £73 490.

The payback period is obtained by dividing the capital cost (£96 490) by the energy cost (£73 490), resulting in a payback period approaching 16 months. (This would probably be considered attractive to industry, where payback periods of two years or less seem to be generally regarded as acceptable, depending also on the amount of capital involved).

6.2.2 Boiler condensate return It has already been emphasised in Chapter 5 that the return of boiler condensate to the boiler is, assuming that the condensate remains in good condition, a simple method of saving heat and therefore money. This is best illustrated with an example of the type of saving which can be made, based originally on United States data (ref. 6.7), and it can generally be stated that 10 to 30 per cent of the amount of fuel used for steam raising can be saved if condensate is re-used as boiler feedwater.

For convenience, take the value of steam as £2.20 per 1000 kg. In a plant saturated steam was delivered to a building at a pressure of 1380 kN/m^2 and a flow rate of 3.4 kg per second for an average of 8000 hours per annum. The steam pressure was reduced through control valves and condensed in heating coils at an average pressure of 172 kN/m^2. The condensate was returned to the boiler as feedwater. The value of the heat recovered may be calculated using steam tables. 1380 kN/m^2 saturated steam has a heat content of 2786 kJ/kg, and the heat content of make-up water which would be used if the condensate had not been returned to the boiler is 88.4 kJ/kg (calculated at an ambient of 20°C). Thus the net heat value of the condensate is 2698 kJ/kg. The proportion of the heat remaining in the condensate may also be obtained from the steam tables, and this is illustrated in Fig. 6.14. From this figure it can be seen that 17 per cent of the heat remains in the 172 kN/m^2 condensate from the 1380 kN/m^2 saturated steam.

Thus the total amount of heat recovered in the condensate is:

$$0.17 \times 2698 \text{ kJ/kg} \times 3.4 \times 3600 \text{ kg/hr} \times 8000 \text{ hr per year}$$
$$= 4.5 \times 10^{10} \text{kJ per year}$$

The value of the heat recovered is then calculated to be £42 600.

In this calculation it is assumed that there is no heat loss in returning the condensate to the boiler. The heat loss is in practice dependent on line length and insulation. However, it must also be appreciated that by

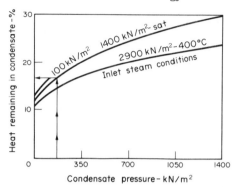

Fig. 6.14 Heat remaining in steam condensate, as a function of condensate pressure and inlet steam conditions

recovering condensate, one is saving the cost of additional water and chemicals for treating this water to bring it up to feedwater quality. The cost of a condensate return system should of course, be taken into account in any analysis.

6.2.3 Recovery of boiler blow-down

It is necessary to regularly clean out a boiler, even though water treatment is normally carried out, because most feedwater will contain suspended or dissolved solids which will be deposited in the boiler as steam is generated. The cleaning procedure is either intermittent or continuous, and involves removing water which has become highly concentrated with impurities from the boiler, and replacing it with clean feedwater. The criterion for determining the amount of water removal, or 'blow-down' required is based on the measure of the total dissolved solids (TDS) level. The TDS level in a modern packaged boiler is generally recommended to be in the range 2000 to 3500 ppm, (ref. 6.8).

It has been common practice on most boilers to 'blow-down' the unit intermittently. This enables the cleaning process to be carried out at the most convenient time, when TDS level is high and generation rates are low. It also ensures that precipitate near the blow-down outlet is carried away with the sudden water flow.

However, intermittent blow-down is becoming less desirable for a number of reasons. In the United Kingdom the grid water distribution system has led to a reduction in quality of feedwater, meaning a higher TDS and a greater blow-down requirement. Also in modern packaged boilers intermittent blow-down, with the associated thermal shocks and complete fall-off in pressure, can create problems. Local authority regulations concerning maximum temperatures at which waste water can be dumped into the sewerage system also work against intermittent blow-down.

The alternative is continuous blow-down of the boiler, which has the attraction of considerable potential for waste heat recovery. This system can also be fully automated, obviating the possibility of neglect which could lead to corrosion associated with intermittent manually directed operations.

It has been recommended (ref. 6.8) that continuous blow-down be applied if the average boiler requirement for blow-down exceeds 10^{-2} kg per second. Equipment is available for implementing this (see Appendix 4), and it is

Plant - Energy Recovery

claimed that the use of a continuous blow-down system does not interfere with the boiler steam raising rate or pressure.

Continuous blow-down, by its name, implies that a continuous supply of hot water is being discharged from the boiler. The heat contained in this water may be recovered using a flash vessel or a heat exchanger, the recovered heat being used for feedwater heating, or preheating of fuel oil or combustion air.

The quantity of heat which can be recovered from the blow-down depends largely upon the use to which it is put, and this is illustrated in Fig. 6.15.

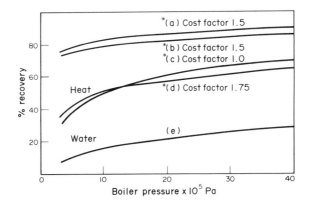

Fig. 6.15 Heat and water content recovery from boiler blow-down
(Courtesy Gestra (UK) Ltd.)

The most efficient recovery processes, and the most expensive, are represented by the upper two curves in this figure. These involve flash steam recovery by directly injecting the water into the feedwater. This is combined with sensible heat recovery into cold make-up water using a conventional heat exchanger, and the layout is illustrated in Fig. 6.16. Alternatively the flash vessel may be omitted, and all the sensible heat recovered used for

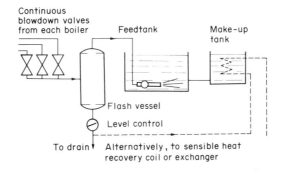

Fig. 6.16 Flash steam recovery, with alternative sensible
heat recovery for feedwater heating
(Courtesy Gestra (UK) Ltd.)

make-up water heating. The lowest line on the graph illustrates the amount of water which can be recovered when a flash vessel is used. This is obviously becoming increasingly attractive in view of rising water costs.

A more sophisticated system is shown in the flow diagram, Fig. 6.17. The flashed water here may be further used in a number of processes, including the neutralisation of cracked gasoline and treatment of oil sludge, chemical properties of the blow-down making it ideal for these functions (ref. 6.9).

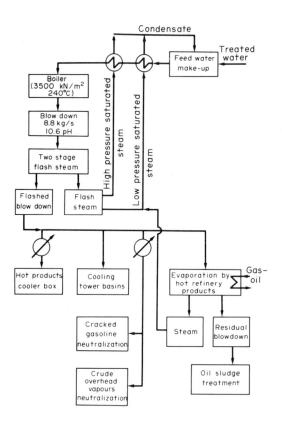

Fig. 6.17 A complex boiler blow-down recovery system in an oil refinery

With regard to capital cost of the equipment needed to implement a continuous blow-down system plus heat recovery, an outlay of £1000 to £5000, depending on the degree of automation and other 'optional extras', appears typical. The user may expect to recover this expenditure within one year, and thereafter to save several thousands of pounds every year. In addition, of course, the user benefits from a reliable blow-down system which will increase boiler life and protect associated plant.

6.3 Burners - Heat Recovery to Improve Efficiency

As discussed later in Chapter 7, recuperative heat exchangers, used for preheating combustion air in furnaces and other plant, have been applied primarily where vast quantities of heat are used. The recuperator is commonly regarded as being a massive structure located downstream of the main heating chamber, and hence economically viable only in large plant. The recuperative burner, with its own self-contained recuperator, now offers the user of furnaces of any size an opportunity to save considerable quantities of fuel.

Recuperative burners fired by gas have been the subject of a significant development programme undertaken by the British Gas Corporation (ref. 6.10). It was argued that while air preheating using recuperators or regenerators in large continuous process furnaces was fully acceptable, smaller units, in particular batch furnaces, were not ideal vehicles for these types of heat exchanger. British Gas contended that these latter furnaces were the most uneconomical, an average batch forge furnace having a thermal efficiency of about 10 per cent, with approximately 70 per cent of the heat supplied being lost up the flue. In order to overcome this, British Gas concentrated on the development of a recuperative burner which incorporated the functions of burner, flue and recuperator in a single compact unit. It was also stipulated that installation of the new type of burner was to be as simple as that of a standard non-recuperative burner.

The recuperative burner developed as a result of this study at the Gas Corporation Midlands Research Station (MRS) is illustrated in Fig. 6.18. It consists of a high velocity nozzle-mixing burner surrounded by a counterflow

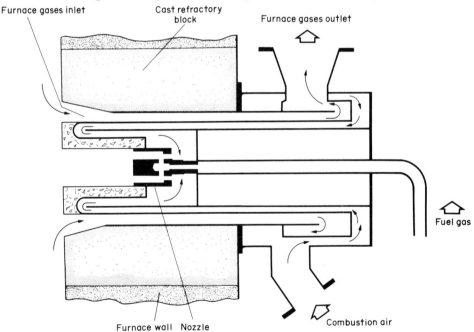

Fig. 6.18 A recuperative burner developed at the Gas Corporation Midlands Research Station
(Courtesy British Gas Corporation and The Institution of Gas Engineers)

gas-to-gas heat exchanger which supplies the hot combustion air. The outlet for the spent furnace gases is located on the top of the burner enclosure, and all recuperative components are made of heat resisting steel. The basic design of the burner is discussed in detail in ref. 6.11 which also lists two patents which have been granted.

The series of burners designed covers a range of thermal inputs from 100 kW to 900 kW, and exit velocities from the burner exceed 50 m per second. Combustion is essentially complete before the gases pass through to the furnace chamber. Control systems have been developed for use with this type of burner, these being similar to those for conventional burners, with the addition of a governor to compensate for the increase in back pressure resulting from higher air temperatures.

There are a number of applications where this burner has demonstrated its good efficiency and fuel economy. These include forge furnaces, reheating furnaces, intermittent kilns, (see Chapter 4), soaking pits and heat treatment furnaces. The improvement in performance obtained may be illustrated with data from an aluminium crucible furnace. It is normal practice to hold the metal in these furnaces at its casting temperature, which is approximately 720°C, prior to flowing to dies, and rapid melting is often required. A representative series of tests were carried out, consisting of three basic trials:

(i) Melting from cold - here the time required and the amount of heat supplied to heat 180 kg of the metal to 720°C was recorded

(ii) Holding rate - measurement was made of the heat required to maintain 180 kg of aluminium alloy at 720°C in the crucible

(iii) Melting from a hot start - this test is similar to (i), with the exception that the crucible is preheated.

The results obtained with a 100 kW recuperative burner were compared with those achieved using a standard nozzle mixing burner and a packaged burner, and are given in Table 6.3. Furnace A was the same type as that used for the recuperative burner tests, but furnace B, although broadly similar, had a lower capacity.

It had been intended to use a 150 kW burner, but a 100 kW unit was immediately available. This accounts for the longer melting rates, and the MRS claimed that a larger burner capacity would not adversely affect the results of the holding test.

In spite of this, however, it can be seen that the results obtained, especially when compared with the packaged burner, are particularly impressive. A 150 kW unit would overcome the greater times necessary to achieve the control temperature at the expense of some fuel, but the savings are still likely to be of the order of 40 to 50 per cent.

TABLE 6.3 Comparison of Recuperative and Non-Recuperative Firing of an Aluminium Crucible Furnace, (ref. 6.10)
(Courtesy The Institution of Gas Engineers)

	MRS recuperative burner	Furnace A nozzle-mixing burner	Furnace B packaged burner
Alloy	LM13	LM13	LM4
Charge (kg)	180	160	140
Control temperature (°C)	720	720	700
1. Cold Melting Test			
Time to control temperature (min)	173	135	145
Heat input (kW)	110	175	248
Total heat input (MJ/kg)	6.37	8.81	16.66
Fuel saving over nozzle burner	27.8%	–	–
Fuel saving over packaged burner	61.5%	–	–
2. Holding Test			
Heat input (kW)	16.35	45	69
Heat input to hold 1 kg for 1 hr (MJ)	0.32	1.03	1.83
Fuel saving over nozzle burner	68%	–	–
Fuel saving over packaged burner	82%	–	–
3. Hot Melt Test			
Time to control temperature (min)	107	–	85
Total heat input (MJ)	690	–	1214
Heat input (MJ/kg)	3.83	–	9.1
Fuel saving over packaged burner	57.5%	–	–

6.4 Dryers - Heat Recovery and Fluidised Bed Systems

As indicated in Chapters 4 and 5, the drying of products is a process which requires accurate control of conditions if energy consumption is to be minimised. With many industrial dryers, this control will probably prove sufficient to save a considerable amount of heat energy. The user may well feel that once he has achieved this goal, the need for further economies is unnecessary. However, should the energy content remain a significant part of the production cost of the material being dried, the use of heat recovery systems can reduce the fuel bill even further.

Obviously the air exhausted from a dryer has a high water content, and therefore it cannot be re-used directly in the dryer. The humidity of the air must be reduced, or, more commonly, a heat exchanger must be used to recover the heat from the exhaust before it is finally rejected to the atmosphere.

A fact which is not always appreciated is that the dried product itself also contains a significant amount of heat. A system has been developed, (ref. 6.12) for using the heat contained within the product to preheat drying air, and this is described later in the section.

6.4.1 Indirect heat exchange in dryers

Most types of air-to-air heat exchanger described in Chapter 7 may be applied to dryers, with the stipulation that the exhaust and incoming air streams must not be brought into contact with one another. These include finned heat exchangers of the recuperative type, heat pipe heat exchangers, run-around coils and, more recently, heat pump systems.

In most dryers, the inlet and exhaust ducts are at opposite ends of the plant, and therefore the run-around coil offers a convenient way for transferring heat from one end to the other without the expense of additional ductwork. An example of this heat exchanger applied to a continuous dryer is illustrated in Fig. 6.19. The coils in the inlet and exhaust duct take the form of extended surface heat exchangers, and in this particular example, (ref. 6.13) the duct dimensions were 1.2 m x 0.75 m. The mean air velocity in the air inlet was 5 m per second and approximately 5 kg per second of air was passing to exhaust. Savings of up to 60 per cent in the heat needed to be provided by inlet burners were possible using this method. As in other applications of run-around coils, the circulating fluid may be water in most instances, but high temperature organic heat transfer fluids may be used if the operating temperature is high.

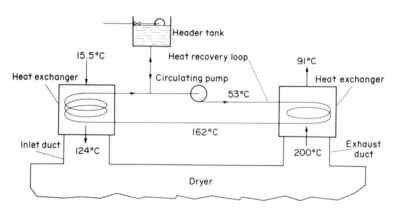

Fig. 6.19 A dryer incorporating a run-around coil-type heat recovery system

The run-around system, by permitting remote location of the inlet and exhaust duct, is also worth considering for preheating dryer air using waste heat from other processes. It is possible to recover sufficient heat in this manner to provide all the requirements of the dryer, depending of course, upon the dryer size and load schedule.

More conventional finned heat exchangers can also give an impressive performance as far as dryers are concerned. The problems of aggregate drying, introduced in Chapter 5, can be alleviated using this type of heat exchanger. Its use in conjunction with a rotary dryer for aggregates is illustrated in Fig. 6.20.

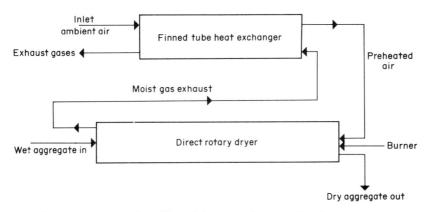

Fig. 6.20 Use of a finned heat exchanger for heat recovery
in a rotary aggregate dryer
(Courtesy US National Asphalt Pavement Association)

In this application, an exhaust gas flow of 6.2 m^3 per second will consist of 3.4 m^3 per second of excess air (230 per cent excess) 1.6 m^3 per second of combustion gases and 1.2 m^3 per second of exhaust water vapour. The relative humidity of the exhaust is of the order of 10 per cent, there being 0.197 kg water per kilogram dry air in this gas flow. A heat exchanger applied to the rotary dryer, which utilises air at 110°C having an inlet flow of 5.04 m^3 per second could cool the exhaust air from 120°C to 66°C. The heat available to preheat the inlet air amounted to 630 kW. The recuperative heat exchanger, discussed in reference 6.14, carried the exhaust gases on the inside of the tubes, the preheated air flowing over the finned outer surface. (It was considered that fins on the inside could assist heat transfer, but contamination of the fins would have made cleaning more difficult). The above quantity of heat recovered amounts to 18 per cent of the total input to the dryer. In addition, 10.5 per cent of the water vapour is condensed. It was estimated that the cost of the heat exchanger required was equivalent to £13 000 and the fuel oil saving was of the order of 68 litres per hour. Thus the payback period was comparatively short.

Turbo-tray dryers: One type of dryer which employs direct heat recovery from the exhaust gases is the turbo-tray dryer. An example of this unit, which can dry crystals, pulp, powders, pellets, and a wide range of granular materials, is illustrated in Fig. 6.21. This turbo-tray dryer, manufactured by Buell, incorporates a series of super-imposed annular shelves mounted in a rotating steel framework. These shelves consist of segmental shaped plates spaced so as to provide a slot between adjacent trays. Turbine fans mounted on a central vertical shaft ensure a uniform flow of drying air over the material on the trays. Wet material is fed to the top shelf by a special feeder and the material is then carried in a thin layer on the rotating trays. After each tray has completed one revolution the material is removed from the segment plate by scrapers or rakes so that it falls through the slot on to the segment plate below where it spreads uniformly over the full shelf area.

This process is repeated on each layer of shelving in the dryer, until the material is delivered dry into the outlet chute.

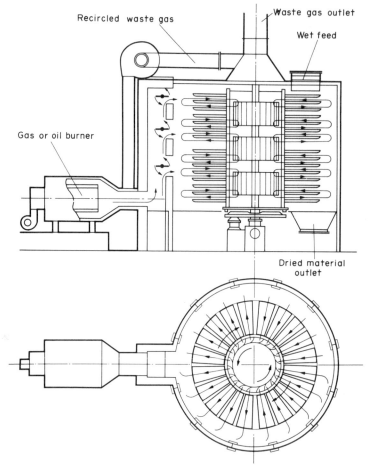

Fig. 6.21 A Turbo-Tray dryer which employs recirculation of waste hot gases. (Turbo-Tray dryer by Buell Ltd. England)

The speed of rotation of the shelving system determines the drying time of the material and a variable speed drive is provided to enable the drying time to be adjusted without stopping the machine.

The transfer of material on each shelf is of value in exposing fresh surfaces of the material to the drying medium while avoiding any violent agitation and dust formation.

The heat source for turbo-tray dryers can be indirect via a steam/air heat exchanger or direct from combustion gases using a separate chamber burning fuel oil, coal, gas or other suitable fuels.

Rotary air jet dryers: In most dryers the air is generally introduced at one point and extracted at another location, coming into contact with the material being dried only once on its path through the equipment. The rotary air jet dryer made by Cannon Air Engineering and used principally for drying bricks and similar materials, recirculates the bulk of the air, as shown in Fig. 6.22. Because of the nature of the product being dried, a batch drying process is necessary, and this is carried out in the chamber shown.

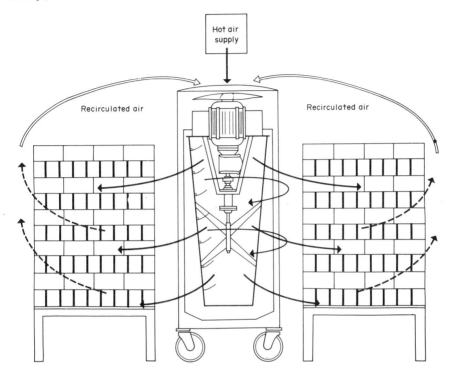

Fig. 6.22 A rotary air jet dryer which uses direct recirculation of warm air

Clay for rapid drying must be subjected to air movement which will remove the thin film of heavily moisture laden air which forms on and clings to the surface of the ware. This layer prevents evaporation from the clay body and requires a considerable effort for its removal. If however, the clay is subjected to a continuous air current capable of removing the moisture laden film, drying is far too rapid at the surface, which becomes over-dried, while the interior remains wet and cracking is the result. This difficulty is overcome by the 'Rotary Air Jet', which projects an air blast over the ware to be dried, lasting a few seconds, and the clay is then left alone for some time to allow moisture to travel from the interior to the surface before another air current sweeps the place to be dried.

The Rotary Air Jet consists essentially of an axial flow fan mounted at the top, handling a large volume of air. This volume is composed of a controlled amount of hot dry air from a heat exchanger or system recovering residual heat from the kiln*. At the same time the fan draws in a large amount of recircu-

---

*Here we have an example of heat from another process being used to dry a product.

lated air from the body of the dryer, this volume being many times that of the hot air supply. The whole volume in a thoroughly mixed state flows downwards into a slowly rotating cone which incorporates a louvred slot, through which the air is discharged.

These dryers normally incorporate heat exchangers or direct fired air heaters, both of which may be oil fired, or if steam is available suitable heater batteries can be provided, which may include residual heat recovery systems.

6.4.2 Regenerative dryers

As mentioned earlier in this section, heat is contained within the product being discharged from the dryer, as well as in the exhaust airflow. In the United States there has been considerable emphasis on improved drying techniques in the food industry (ref. 6.12). Two occurrences were in part responsible for this; the coincidence of a very wet harvest season and a national energy shortage in 1972, and the recommendation that all animal feeds and cereal grains be pasteurised to counter the growth of salmonella. Pasteurisation, which necessitates heating up the material, was initially found to be economically unattractive because of the low cost of the product and the high energy content of the process.

In order to make the drying or pasteurisation process for granular dry materials more feasible economically, the concept of regeneration was incorporated in a dry materials heat exchanger, and performance evaluations carried out. The regenerator and heat exchanger are shown in Fig. 6.23. The hot air

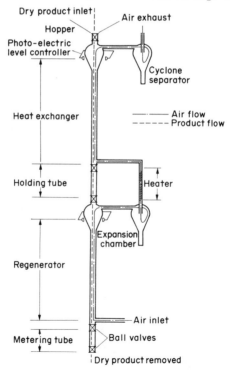

Fig. 6.23 Schematic of dry materials heat exchanger with regeneration

Plant - Energy Recovery

passes in counterflow past the product, which flows vertically downwards under gravity. Four main sections make up the device:
(1) a heat exchanger where the product is raised to temperature by the hot air,
(2) a holding tube where the material is maintained at temperature,
(3) a regenerator where heat is recovered from the material and given up to preheat the air, and
(4) a metering tube for controlling material flow rate.
A heater is provided to supplement regenerative heat input.

The graph in Fig. 6.24 shows the relationship between regeneration efficiency and moisture content of the product as it enters the heat exchanger. As expected regeneration efficiency increased as moisture content decreased, but even a regenerative efficiency of 30 per cent for very high moisture contents indicates an equivalent saving in the energy cost of processing.

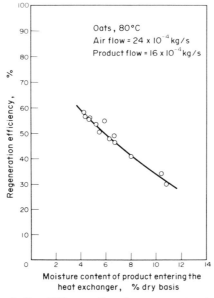

Fig. 6.24 Effect of moisture content on the regeneration of oats at 80°C

Based on the results obtained from a series of runs made with oats, cracked corn and black peppercorns, the development team were able to make the following conclusions:

(i) Regeneration efficiency is a function of the relative flow rates of the air and the product. Best operation occurred when the ratio of air-to-product heat capacity fluxes was approximately unity.

(ii) Higher regeneration efficiencies could be obtained by lowering the process temperature, at the expense of drying efficiency.

(iii) The apparatus dries sorptive matter effectively. The

regenerator provides additional drying for all products with
with moisture contents above 5 per cent before processing.

It was emphasised that for the types of materials tested, the functions of drying and regeneration could not be separated, and the final design would in each instance necessitate a trade-off between regenerative efficiency and drying efficiency. While the experiments to date have concentrated on foodstuff, there is no reason why this system should not be applied elsewhere, particularly in cases where the product has a high specific heat.

6.4.3 Heat pumps in drying
The application of the heat pump in a dryer is illustrated in Fig. 6.25, where it functions as a dehumidifier and a preheater. Moist air from the dryer exhaust flows over the heat pump evaporator coil, where it is cooled, giving up its heat to the fluid flowing in the closed heat pump circuit. As a result of being cooled, the water in the air condenses and is removed. The heat taken up at the heat pump evaporator is circulated to the heat pump condenser, where it is given up at a higher temperature to preheat the dry air, which then re-enters the drying chamber.

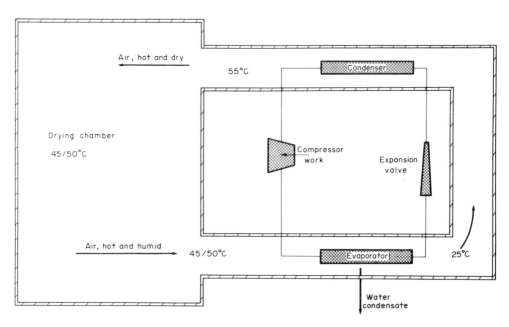

Fig. 6.25 Use of the heat pump in a dryer, resulting from developments at the Electricity Council Research Centre, Capenhurst.

The heat pump is one of the systems being advocated by the United Kingdom Electricity Council Research Centre at Capenhurst (see Chapter 7) as a more efficient way of using electricity, which has to date been a relatively extravagant consumer of primary energy resources. A heat pump is able to provide three times as much heat energy at the condenser as is put in as work equivalent at the compressor (i.e. its Coefficient of Performance is 3). Thus a heat pump having a compressor driven by an electric motor will have an efficiency of approximately 100 per cent in terms of primary energy resource utilisation, (on the basis that a power station is 33 per cent efficient).

Development at Capenhurst is currently directed at dryers running with chamber temperatures of up to 85°C, but it is intended to extend this range in the future to encompass temperatures in excess of 100°C, (ref. 6.15).

6.4.4 Fluidised bed dryers Many small or medium sized companies requiring dryers for processing one of their products are not geared to continuous production, either because the nature of the process may not permit this, or it may not be justifiable on economic grounds. In these cases the user demands a dryer which can process batches of the product efficiently, and one system which meets this requirement is the fluidised bed dryer, (ref. 6.16).

The drying process can be considerably accelerated if the material to be dried has a large surface area compared with its mass. Thus if a material can be separated into independent particles, the drying process can be assisted. This is the principle behind spray drying, (used normally in large plant), and fluidised bed drying. In fluidised bed drying, the aim is to suspend the wet particles in a vertical airstream, thus exposing as much surface area of the particles to the forced convection effect of the air as possible.

For batch processes, where one dryer having a capacity of around one tonne per hour may treat a variety of products, the fluidised bed dryer of the type illustrated in Fig. 6.26 meets these requirements. (The layout in the figure and some of the data below is based on the dryer manufactured by Arthur White Process Plant Limited).

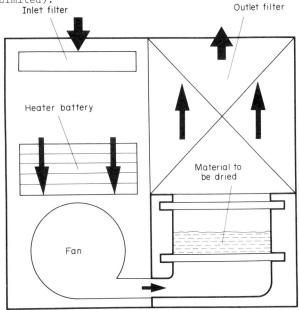

Fig. 6.26 A fluidised bed dryer, showing adjacent location
of inlet and exhaust ducts
(Courtesy Arthur White Process Plant Ltd.)

Of particular use in the pharmaceutical, chemical, food and plastics industries, this type of dryer may use steam coils for heating the air, or is available with direct gas heating. Operating temperatures of up to 120°C are standard on some models, and secondary filtration and carbon adsorption systems are available for added environmental protection.

The layout of this type of dryer is particularly attractive from the point of view of heat recovery. As can be seen in Fig. 6.26, the inlet and exhaust ducts are located adjacent to one another. Thus provided that any condensation resulting from cooling of the exhaust air can be prevented from falling back into the product being dried, this unit appears ideal for application of heat pipe heat exchangers (see Chapter 7). Installed as shown in Fig. 6.27, the energy consumption of this type of dryer, which in larger units can be 7.5×10^{-2} kg per second of steam, could be more than halved. A typical payback period of one year should be achievable.

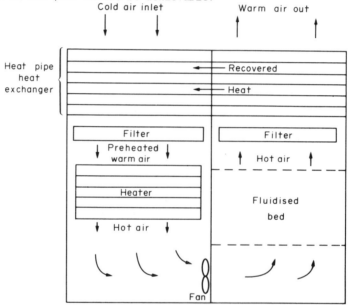

Fig. 6.27 Location of a heat pipe heat exchanger above a fluidised bed dryer

As in most instances the boiler supplying the steam to the heating coils in a fluidised bed dryer will be serving other parts of the plant, the considerable quantities of waste heat in the boiler flue could also be used in the dryer. A run-around coil would be an appropriate form of heat exchanger if this heat source was selected.

From the above discussion, it may be concluded that many types of drying equipment are suitable media for heat recovery systems, and dryer manufacturers should offer these as options on their products, if this is not currently standard practice.

6.5 Furnaces

The requirements of furnaces vary considerably from one industry to another in terms of size, temperature range, firing rate, fuel and operating cycle. It is therefore perhaps misleading to discuss 'ad nauseam' features of

Plant - Energy Recovery   181

furnaces such as heat recovery and new melting techniques which may be
relevant to only one section of industry. For example, we have discussed the
use of ceramic recuperators in the steel industry in Chapter 3, and these
units are currently unique to this industry where furnace fuel use is particularly significant. Heat treatment furnaces have special requirements,
depending on the product being treated, one instance in Chapter 4 being concerned with lehrs in the glass container industry. Furnaces like this are
often of the continuous type, the product being fed on a conveyor from one
end to the other. Batch production furnaces may be required to process
material at as fast a rate as possible, and similar, although smaller, holding
furnaces are used to keep material in a molten state for a long period.

In many instances, therefore, the reader will find data on furnace energy
conservation in the sections dealing with the industries where the particular
type of furnace is employed. Topics discussed below will in some cases, be of
more relevance to one industry than another, but the use of electric heating
in furnaces has broader implications, as detailed below.

6.5.1 Heat flow paths in furnaces   The heat flow in batch and continuous
flow furnaces may be described with reference to Figs. 6.28 and 6.29,
(ref. 6.17). In the batch furnace the stock is heated up 'in situ', with the

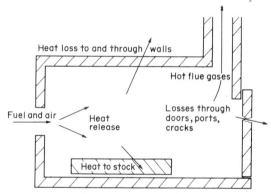

Fig. 6.28   Schematic furnace showing major heat sinks

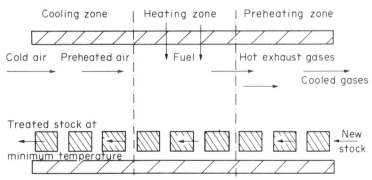

Fig. 6.29   Layout of a continuous furnace,
showing heat saving possibilities

major heat loss from the furnace being via the stack in the form of hot flue
gases. (Other heat losses shown on the diagram, and techniques for
minimising them, are discussed in Chapter 5). The flue gases therefore offer
the greatest potential for heat recovery, either for preheating combustion air
using recuperators or for raising steam using a waste heat boiler (see
Chapter 7). Alternatively this gas can be used for preheating scrap metal
prior to it being fed into the furnace. This system is currently unique to
BOS steel furnaces and is discussed in Chapter 3, where a more exotic energy
recovery system for furnaces, based on a turbine to utilise the blast furnace
top pressure, is also detailed.

   The continuous furnace, used primarily for heat treatment, offers more
scope for heat recovery and utilisation within the furnace itself. Some of
these are indicated on the figure. Cold air entering at the discharge end of
the furnace may be used to cool the emerging stock, providing preheated air
for the burners. The hot exhaust gas from the burners themselves may be
passed upstream to preheat the stock and it is obviously not too difficult to
ensure that optimum use of energy is obtained in this type of furnace.

6.5.2 Batch furnaces - cupola blast heaters

Ironically, in many iron
foundries the move has been away from recuperators to direct firing of the
air blast using coke or natural gas. As discussed in Chapter 7, the nature
of the exhaust gases from a process can affect the life of a recuperator, in
some cases making ceramic materials necessary. Maintenance costs can be high
when the environment is unattractive, and it has been found that the use of
recuperators in cupola installations, (the cupola being the main melting unit
used in the iron-founding industry, accounting for over 85 per cent of molten
metal output), has proved less economical than a direct fired blast heater.
In other words the cost of fuel for heating the blast is more than offset by
the savings in maintenance; (based on current fuel and labour costs).

   Figure 6.30 shows a section through a cupola blast heater manufactured by
Wellman Incandescent Furnace Co. Ltd. (ref. 6.18). The output from the combustion chamber is accelerated through a nozzle into the mixing duct. Flow
through this nozzle induces a proportion of the waste gases approaching in a
counterflow direction along the outer annulus, keeping the gases in the main
heat exchanger down to an acceptable temperature. The blast air passes

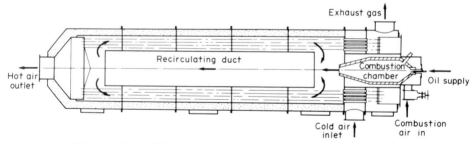

Fig. 6.30 A Wellman incandescent cupola blast heater
(Courtesy Wellman Engineering Corporation Ltd.)

through the tubes contained within the annulus, and the hot products of combustion passing over these tubes heat the air, which is then passed to the
furnace. The blast heaters in this installation use U-shaped tubes to counter
thermal expansion problems. Cupola blast heaters of the above type have

efficiencies of about 80 per cent, with air outlet temperatures approaching 500°C.

The major economy measure in cupola installations has resulted from the use of the divided blast system, developed by the British Cast Iron Research Association (BCIRA). In this system (refs. 6.19, 6.20), which is used with coke-fired units, the air needed to burn the coke is blown into the cupola through two rows of tuyeres correctly spaced apart, instead of via a single tuyere. A divided blast (each row taking 50 per cent of the hot air produced by the blast heater) has reduced coke consumption in some cupolas by 30 per cent, and has in addition increased molten iron output by 25 per cent.

The capital cost of converting existing cupolas to divided-blast operation is low compared to the savings obtained. BCIRA quote one British foundry where the payback period was only fourteen weeks. A large Canadian foundry saved $170 000 in one year for a conversion cost of only $18 000. Added benefits resulting from a lower coke utilisation per tonne of iron include a lower sulphur content, thus saving on desulphurisation plant and giving a higher quality iron.

Cupola replacement with electric melting: The Electricity Council have recently given publicity to the electric arc and coreless induction furnaces as replacements for cupolas in large iron melting plant (ref. 6.21). In an arc furnace electrodes are inserted into the chamber via the furnace roof; when the arcs are struck, heat is generated directly by the current flowing through the charge, and radiation from the arc and the electrodes also contributes to the heat input. The arcs may also be submerged in the charge, and in some installations preheating of the charge is carried out before it is introduced into the furnace (ref. 6.22). These furnaces are reliable and can accept the cheapest type of charge material for melting.

An alternative method of electric melting is via induction. In this case the furnace is surrounded by a copper coil, and a high current is induced in the furnace charge if alternating power is applied to the water-cooled coil.

6.5.3 Reheating and heat treatment

Reheating and heat treatment furnaces are employed later on in the production stage, to prepare the bulk metal for manufacture into other useable forms, to treat the metal in such a way as to improve its properties with respect to the ability to work it, or to satisfy the needs of a particular application.

In reheat furnaces in the steel industry, for example, slabs and billets are heated up to about 1200°C prior to rolling. This may be carried out using electricity, including induction heating similar to that described earlier in this section. Alternatively oil or gas firing may be used, with the attendant exhaust gas waste heat, which is in part recoverable. In the USSR recuperative heat exchangers are widely used on gas-fired reheating furnaces (ref. 6.23), and in Japan Kawasaki Heavy Industries have developed a recuperative preheating system for use on oil-fired reheating furnaces. It is claimed that by recovering heat from the exhaust gases and channelling this hot air through an arrangement of nozzles and blowers onto slabs in a preheating chamber, energy savings of 20 per cent result. Installed on a 300 tonne per hour reheating furnace, it is estimated that this system would save $865 000 per annum.

Heat treatment 'per se' can be responsible for considerable energy savings in the overall route from the ore to the finished product, if correctly carried out. An interesting example of this is quoted in Metal Progress (ref. 6.24). Taking the amount of natural gas to manufacture a kilogram of steel from iron ore as 0.566 m^3, the example is based on a U bolt for locating large springs:

> "If the length of the bolt is 0.61 m and the loading is 62 142 kg, the bolt will weigh 4.81 kg using steel as received from the mill. With proper heat treatment the bolt need only weigh 2.45 kg. The energy invested in heat treatment would be 28 065 kJ per bolt, and the resulting saving would be 86 937 kJ per bolt due to the bulk savings made possible by heat treatment. The return on the heat treating energy investment is 381 per cent in system energy saved."

The shortage of natural gas in the United States has led to conservation in other heat treating processes, (ref. 6.25), the use of lower calorific value gaseous fuels is in some instances replacing natural gas. The Sunbeam Equipment Co. has constructed a case-hardening furnace with a horseshoe-shaped floor plan - the charge end of the hardening section faces the discharge end of the tempering section. Ductwork conveys exhaust gases at 900°C from the hardening section to the tempering section where the residual heat augments that being generated, saving 35 per cent on fuel bills. Idling furnaces are used for preheating in several factories, and heat treatment furnaces are being maintained when idle at substantially lower temperatures than normal (730°C) to conserve fuel.

6.6 Incinerators

The incinerator was until recently solely regarded as a piece of equipment to assist disposal of waste products which could not be used in a process or sold for re-use elsewhere. The material to be incinerated may be in the form of a gas, liquid or solid. Incinerators for disposal of gases, to prevent atmospheric pollution with obnoxious or poisonous matter, are commonly known as fume incinerators. Incineration of waste liquid can be used to remove organic pollutants and to recover inorganics. The disposal of solid waste material is the most widely applied use for incinerators, sizes ranging from compact units capable of dealing with a few kilograms per hour, up to very large schemes for disposal of all domestic and industrial waste in a large urban area.

The appreciation of the fact that the waste heat produced by any type of incinerator may be used to preheat the waste or combustion air, or can be fed to other parts of the plant, has led to an interest in heat recovery. Many incinerator manufacturers now offer optional heat recovery equipment. Alternatively the user may select his own recovery system for a special purpose. This section discusses some of the layouts used to date, and reviews some of the incinerators commercially available.

6.6.1 Fume incinerators
Fume incinerators are now used, in common with other types of equipment, to remove pollution from gaseous waste before it is discharged to the atmosphere. Most of the air pollutants, identifiable either by their odour, or visually, are combustible and many are in the form of hydrocarbons. The two basic types of fume incinerator used to deal with these gases are the thermal incinerator and the catalytic incinerator.

A thermal incinerator, one of which is shown in Fig. 6.31, consists of three major sections. These are the fume inlet plenum, the burner and the

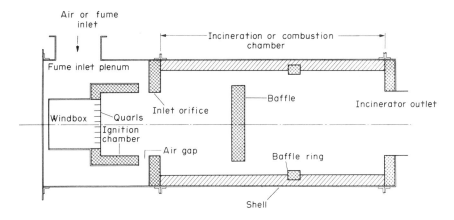

Fig. 6.31 Layout of a Hirt-Hygrotherm fume incinerator
(With acknowledgement to Hygrotherm Engineering)

furnace chamber. The unit illustrated, developed in California by Hirt/Hygrotherm, has several attractive features to ensure satisfactory fume incineration and low operating costs (ref. 6.26). A proportion of the fume input is internally by-passed through the burner for use as combustion air, and the elimination of the need for external combustion air can save up to 30 per cent on fuel expenditure. Burner design is an important feature of an incinerator, and Hirt use a short, wide flame having a flat profile. By directing the fumes across the flame front as they enter the furnace, rapid heating is achieved, guaranteeing efficient oxidation. Another desirable feature of fume incinerators is the baffle arrangement, which promotes high turbulence in the furnace, leading to effective mixing.

By relating the fuel gas input to the solvent concentration, efficient use can be made of the calorific value of the gases in the effluent being treated.

The recovery of heat from this type of incinerator can reduce fuel requirements by very substantial amounts. The system illustrated in Fig. 6.32, using a Hirt/Hygrotherm incinerator, was added to aluminium foil painting ovens where previously direct gas firing had been applied and the solvent-laden air was being discharged directly to atmosphere. In the new system the exhaust fume is taken from the fans to a common header duct via pressure controllers which ensure that the system operates satisfactorily under various loadings. The exhaust duct is taken to the incinerator forced draught fan which blows the fume through the preheater (2) where it is raised in temperature, before entering the incinerator inlet plenum (5). After passing through the incinerator combustion chamber (1) the flue gases re-enter the fume preheater (2) and then the two fresh air heaters (3) and (4) before entering the chimney. Fresh air to the ovens is blown by forced draught fans through the air heaters to the oven burner boxes (7) and (8).

Control of the incinerator temperature is achieved by a modulating split-phase system normally operating on the fuel valve, but using the fume preheater by-pass if the calorific value of the fume is high. Control of the oven temperature is individually attained also by means of modulation of the by-pass damper around the heat exchanger and the oven burner fuel valve.

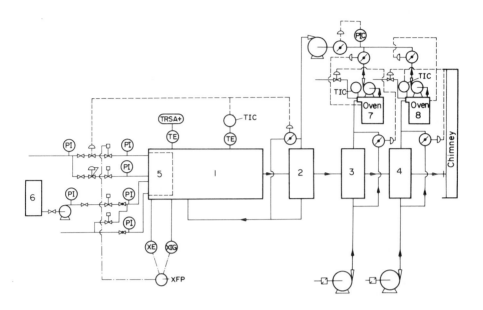

Fig. 6.32 Flow circuit of a Hirt-Hygrotherm fume incinerator
used as a heat source for ovens
(Installed VAW Germany by Hygrotherm Engineering Ltd.)

Normally all the heat required is provided by the heat exchangers and the burners are required to operate only during the warm-up period.

With total heat demands of 5.2 Gcal per hour to raise the temperature of the air entering the ovens, the incinerator was able to supply 4.7 Gcal per hour, or 90 per cent of the total heat requirement, while maintaining exhaust to atmosphere at an acceptable contamination level. The incineration temperature in this instance was 700°C.

In even more efficient installations Hygrotherm have used waste heat for fume preheating and heat supply to heat transfer oil which is in turn used to heat the oven by oil/air heat exchange, thereby eliminating large ducts and high fan heads, and giving greater flexibility of operation.

Not all fume incinerators burn organic waste. Sulphur dioxide (SO_2) is the most common gaseous atmospheric pollutant, because it is a by-product of fossil fuel fired boilers, furnaces, kilns, etc. As one step towards minimising this form of pollution, fuel oils are treated to remove most of their sulphur content, and some of these desulphurisation processes produce hydrogen sulphide (H_2S), itself a toxic and malodorous gas (ref. 6.27).

Treatment of the H_2S by an incineration process, followed by controlled cooling which permits a catalytic reaction to take place, can lead to the recovery of pure sulphur. The controlled cooling may be conveniently carried out by locating a waste heat boiler downstream of the incinerator combustion chamber. In an oil refinery in the Eastern United States producing 170 000 barrels per day, the desulphurisation process exhaust gases are cooled in four stages from 1393°C to 191°C, giving up 6.09 MW and generating 2.65 kg per second of saturated steam using feedwater at 118°C. This in itself is worth over $200 000 per annum in oil savings, to which must be added the value of the recovered sulphur. Csathy notes in reference 6.27 that the use of desulphurised oil in its own right leads to increased potential for heat recovery. This is due to the fact that a lower stack temperature can be used, as the cold-end corrosion threshold is reduced if this type of oil replaces that with a higher sulphur content.

Unless a conventional thermal fume incinerator can be equipped with some form of heat recovery equipment, it may be preferable to use a catalytic incinerator. A catalyst is a substance which can increase the rate of a chemical reaction without being changed itself during the process, and their use in fume incinerators can accelerate the oxidation reaction necessary to convert organics to carbon dioxide and water vapour. Because catalytic incineration can take place at much lower temperatures than thermal incineration (typically 250 to 350°C as opposed to 500 to 1000°C in thermal incineration), the quantity of heat needed to raise the fumes to treatment

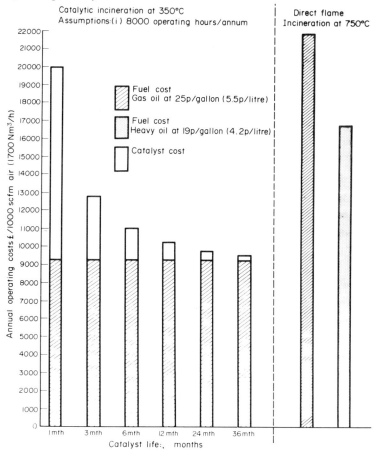

Fig. 6.33 The Engelhard catalytic DEODO pollution control system - operating costs. (Courtesy Engelhard Industries Ltd.)

temperature is much reduced. Heat recovery systems can of course still be incorporated downstream of the catalytic incinerator, as in thermal incineration plant.

A comparison of the operating costs of catalytic and thermal incinerators, prepared by Engelhard to illustrate their DEODO air pollution control system, is given in Fig. 6.33. The catalyst operates at 350°C, whereas the equivalent thermal incinerator would have to achieve 850°C for satisfactory fume combustion. (Catalyst life has been taken into account, but it is claimed that a life of only three months is not typical, even though some economy is still achieved over conventional system running costs).

Engelhard are finding considerable interest in heat recovery systems associated with this system, particularly when these units may be used to preheat incoming fumes. A reduction of 50 per cent in fuel costs, with a pay back period of twelve months, is considered usual when an efficient heat recovery system is used.

6.6.2 Liquid waste incinerator Arguments for heat recovery in liquid waste incinerators are identical to those for conventional thermal fume incinerator units. The material being treated may be a spent solvent or liquids containing both organic and inorganic matter, and potential also exists for recovery of inorganic salts.

A liquid incinerator with a large heat recovery installation is shown in Fig. 6.34. In this incinerator, the exhaust gas flowing at a rate of 22.9 kg per second is cooled from 1200°C (liquid incinerators often operate at higher temperatures than their fume counterparts) to 168°C, generating 10.4 kg per second of superheated steam at 1930 kN/m^2 and 400°C, (ref. 6.27). The financial savings, in terms of reduced oil consumption, amounted to £500 000 per annum.

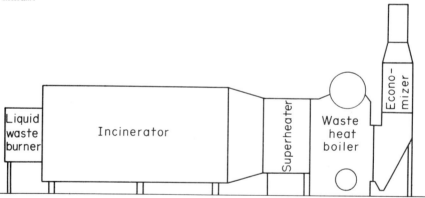

Fig. 6.34 A liquid waste incinerator with
a waste heat boiler installation
(Courtesy Deltak Corporation, Minneapolis, Minnesota, USA)

Originally developed by Nittetu Chemical Engineering of Japan, and available in the United Kingdom through Henry Balfour & Co. Ltd., the NICE liquid incinerator has a burner located at the top of the main chamber, above liquid nozzles which produce an atomised spray to aid combustion. Products of combustion and reaction pass down into a quench vessel, where the gases are cooled to about 90°C. The contents of the quench vessel may be treated in crystallisation, filtration and drying systems if it is required to recover materials. Heat recovery may be simply restricted to preheating of combustion air, or an evaporator can be used to vaporise incoming liquid, (ref. 6.28).

6.6.3 Solid waste incinerators
Most of us are familiar with solid waste incinerators used by local authorities for the disposal of municipal waste, and their use, including their application in waste heat and power generation systems is well documented, (refs. 6.29, 6.30). As shown in Table 6.4, the energy content of much solid waste is high, and if the waste is in a form which cannot be recycled for process application, an incinerator with waste heat recovery may be the answer to the disposal problem.

TABLE 6.4 Heating Values of Various Materials Compared with Oil Fuel

Constituent	Oil Fuel Equivalent (litres per tonne)
Dust	86
Paper (15% moisture)	286
Wood (20% moisture)	323
P.V.C.	524
Coal (CV 27 900 kJ/kg)	656
Polystyrene	870
Rubber	955

A modern incinerator for solid waste processing equipped with a variety of optional ancillary equipment, is illustrated in Fig. 6.35. This machine, known as the Consumat, and manufactured by Robert Jenkins Systems Ltd., comprises a refractory lined combustion chamber into which waste is loaded and air introduced. Initially the waste is heated by small auxiliary burners and undergoes a pyrolysis-type process at a temperature of up to 800°C. Compared to combustion brought about by the introduction of large quantities of air, this procedure allows the waste to decompose under quiescent conditions, thus minimising the carry-over of particles which would normally be emitted to the atmosphere via the stack.

As a further pollution control measure, smoke particles are oxidised above the main chamber by being heated to 1000 to 1200°C using a small burner. The gases are subsequently cooled to 800°C by mixing with cold air before being exhausted to atmosphere.

A waste heat recovery system for providing hot water or steam may be added to the flue, as shown in Fig. 6.35 and the Consumat system is also able to cope with liquid waste with little modification. The waste heat recovery unit is capable of producing 3.5 tonnes of steam from each tonne of waste processed. The Company finds that energy recovery becomes economically viable when disposal rates in excess of 150 kg per hour are used. Usually 50 to 60 per cent of the total combustion heat is recoverable.

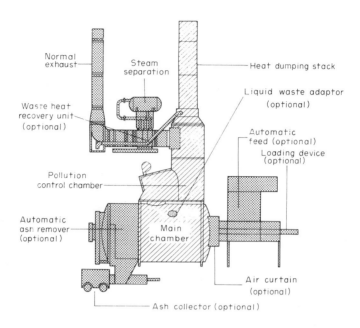

Fig. 6.35 The Consumat solid waste incinerator with heat recovery
(Courtesy Robert Jenkins Systems Ltd.)

It is more common for solid waste incinerators to use large quantities of combustion air, and because of the carry over of solid matter, a separator must be installed between the combustion chamber and any heat recovery equipment. This is illustrated in Fig. 6.36, where a cyclone separator fulfills the cleaning function. In this instance (ref. 6.27) it was necessary to use

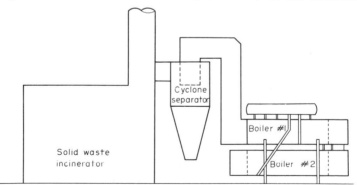

Fig. 6.36 Use of a cyclone separator to protect an incinerator waste heat recovery unit from contamination
(Courtesy Deltak Corporation, Minneapolis, Minnesota, USA)

a firetube-type waste heat boiler with high gas velocities to prevent settlement of any solid particles which were not captured by the separator.

A detailed analysis of contamination of a waste heat boiler installed downstream of an incinerator, in this case burning a variety of pumpable liquids, was recently published in the United States, (ref. 6.31). Flue gas temperatures at the exit to the secondary combustion chamber varied from 800°C to 1150°C because of the varying nature of the waste, a proportion of which was a chlorinated solvent. This produced hydrogen chloride when burnt, the quantity in the exhaust gas varying from 0.1 to 5 per cent by weight. This, together with fly ash, were the main contaminants.

After approximately 600 hours of testing, it was found that the carbon steel boiler tubes were satisfactory provided that purging of the boiler was regularly carried out and the tube surface was kept clean. It was recommended that the minimum tube diameter be 75 mm, as smaller tubes were more susceptible to fouling. From the fly ash collected in the inlet chamber of the test boiler, it was demonstrated that a cyclone installed upstream of the waste heat boiler, as in the solid incinerator plant illustrated in Fig. 6.36, would greatly reduce the possibility of plugging and the degree of tube fouling. It was also suggested that fairly high gas velocities (20 to 30 m per second) be maintained in the boiler to minimise ash deposit on the surfaces.

6.7 Thermal Fluid Heaters

As well as using steam or high pressure hot water for transferring heat around a plant to the various consumers, this heat may be generated in and supplied by a thermal fluid heater. These heaters operate in a similar manner to a domestic hot water central heating system, but instead of circulating water, a heat transfer fluid, commonly an organic with a boiling point above that of water, is used. This allows the fluid to be circulated at high temperature (organic heat transfer fluids can be used at temperatures between -50°C and 400°C) without the need for a pressurised system, as would be the case with a steam system. Typical of these fluids is Thermex (ICI trade name), a diphenyl-diphenyl oxide eutectic having a boiling point at NTP of approximately 260°C. The main disadvantage of these fluids is the fact that they tend to degrade or 'crack' at high temperatures, generally well in excess of their boiling point. They are also susceptible to oxidation.

There are three aspects of thermal fluid heaters which are of interest to the energy-concious plant engineer. These are their ability to use the combustion of waste products for firing the burners, the use of these systems to heat and cool plant in the same building, and their application in heat recovery and heat storage media, (the latter also being discussed in Chapter 8).

A comprehensive system using waste products for heating the thermal fluid, applied in a fibreboard plant, is illustrated in Fig. 6.37. The heater supplies three production presses running at 250°C (600 000 kcal per hour), a dryer (1 300 000 kcal per hour), melting tank (100 000 kcal per hour), space heating, steam raising (3 tonnes per hour at 18 bar), and hot water. During the production process, wood waste is available at an hourly rate of 450 kg. It would be expensive to remove this waste for disposal elsewhere, but with a heat content of about 1 200 000 kcal per hour, it would prove a valuable fuel, and was therefore used in the Konus-Kessel fluid heater, (ref. 6.32).

If waste heat is available and could usefully be applied in a liquid

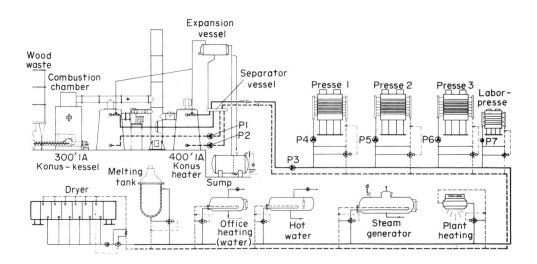

Fig. 6.37 Diagram of a comprehensive thermal fluid heating centre

thermal fluid loop, this heat could be collected at a high temperature (up to 400°C) without the need for an expensive high pressure system, as would be the case with steam. Similarly, the heat could be easily stored in the fluid at high temperature until required. Although not described in data available from the manufacturers of thermal fluid heater systems (see Appendix 4), a heat transfer system using the run-around coil for heat recovery could be used in conjunction with a high temperature fluid as part of a system.

The organic working fluids used in this type of heat transfer apparatus are somewhat limited in their operating temperature range and also in their heat transfer capability insofar as the film coefficients tend to be rather low. Both of these limitations may be overcome by using heat transfer salts, which have film coefficients two to three times those of organics, and can be stable at temperatures up to 550°C with very low vapour pressures. These heat transfer salts are typically a mixture of nitrates and nitrites, and while originally they were somewhat unpopular because of their high melting point (>100°C), water dilution techniques have overcome this. At higher temperatures liquid metals may be used, as shown in Table 6.5 (ref. 6.34).

Although the discussion has concentrated on liquid phase distribution systems, the versatility may be increased by using two-phase heat transfer. The use of these fluids has probably been under-rated in the past, and their application in high temperature heat recovery systems would be one area where their potential could be more fully realised.

TABLE 6.5 Operating Temperatures of Heat Transfer Media

Material	Temperature Range °C
Water and steam	0 to 200
Mineral oils	-10 to 315
Diphenyl/diphenyl oxides	20 to 400
Monoethylene glycol	-40 to 100
Isomeric benzyl benzenes	-50 to 330
Isomeric dibenzyl benzenes	-15 to 350
Terphenyls	200 to 400
Heat transfer salts (HTS)	155 to 540
Sodium and molten metals (& NaK eutectic)	200 to 700

6.8 'Non-Destructive' Waste Recovery Techniques

The recovery of waste has already been discussed in other Chapters in the context of water recovery. A previous section in this chapter details methods of directly incinerating waste products, but it is often possible to recover waste for re-use in a non-destructive manner, be it recycling in a manufacturing process or transformation into another product for use elsewhere. Waste can also be converted into a fuel which will be easier to burn and have a higher calorific value than untreated waste, be it solid, liquid or gaseous, used in incinerators.

Waste recycling need not be confined to material originating within the plant where the recycling is carried out. One manufacturer may profit from the treatment of the waste resulting from processes carried out in a different plant, or from waste which has its origins outside industry altogether. Waste recovery cannot be treated remote from economic considerations, however. It can involve considerable capital expenditure in recovery process plant, and the process itself may necessitate even more investment in effluent treatment plant as it is probable that a considerable proportion of the waste being processed may have to be disposed of as 'unrecoverable'. Obviously the cost of recovery must be sufficiently attractive to warrant investment in the recovery plant, and from the energy conservation point of view there are several factors to be taken into account, including the following:

(i)   The energy expenditure in recovering the waste, compared with that necessary to obtain the required material from natural reserves.

(ii) The benefits which can result from recycling - for example recycled glass can improve the glass melting process (see Chapter 4).

(iii) Benefits to industry must be weighed against longer term benefits which may occur in, for example, reutilisation of biodegradable matter.

There is a large and increasing number of private and government-supported organisations involved in waste recovery to provide fuel or to enable material to be recycled, and some of these, and the services they offer, are listed in Appendix 3. Recovery of oil for re-use, and the extraction of solvents from waste are among the most common facilities available to industry. Equipment may be purchased to enable a company to undertake its own recovery processes, or expert services carried out by companies specialising in distillation and other recovery methods may be used.

One of the more interesting recovery services offered is associated with heat transfer fluids such as Thermex and Dowtherm (see also Section 6.7). These can often become contaminated in use with polymeric material, carbon deposits, volatile decomposition products, or water. As these fluids are expensive and periodic refilling of a system with fresh heat transfer fluids is a necessary part of a maintenance programme, fluid recovery can offer substantial cost benefits. One process for their recovery, operated by Crewe Chemicals Ltd. (see Appendix 3), takes place in three stages. First of all, low boiling point contaminants such as water are removed, then a high vacuum distillation is carried out. Finally the residues are stripped by nitrogen blowing. Normally more than 99 per cent of the available high quality heat transfer fluid can be recovered from the crude.

6.8.1 The UK Waste Materials Exchange

One method of ensuring that industrial waste is wherever possible used to the best advantage of industry and the environment is that initiated by the Department of Industry in the United Kingdom. Known as the Waste Materials Exchange, it provides up-to-date information on waste originating from manufacturing processes throughout the country, and circulates lists of the waste, quantities involved, and location. The service excludes wastes for which there is an adequate commercial market, for example scrap metals, but includes typically acids and alkalis, catalysts, inorganic chemicals, rubbers and plastics.

Organised through the Warren Spring Laboratory at Stevenage, the Waste Materials Exchange was established in October, 1974. Each copy of the bulletin listing waste available includes reader's reply forms for indicating his wish to obtain or supply any of the materials listed as available or wanted, or to notify any materials for inclusion in the next issue. Requests for a particular material are forwarded to the company originating the waste so that they can make their own commercial arrangements and full confidentiality can be maintained.

The Production Engineering Research Association (PERA) is also offering a similar service to industry, in this case concentrating on surplus materials, including metals and plastics, rather than waste products.

A second study initiated at the Warren Spring Laboratory concerns the establishment of Waste Recovery Centres, and Fig. 6.38 shows the processes which could be included. The aim is to offer facilities for waste processing, with the emphasis on recovery, while maintaining costs such that they are comparable to those charged for current refuse disposal services.

Plant - Energy Recovery

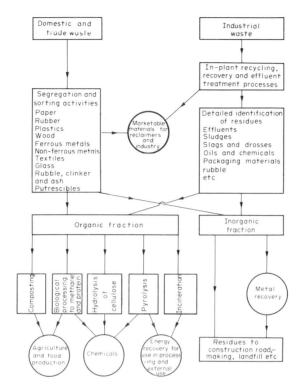

Fig. 6.38 Potential functions of 'Waste Recovery Centres'

6.8.2 **Conversion of waste into fuel** A study carried out recently in the United States (ref. 6.33) compared the routes and products of incineration (Fig. 6.39) and pyrolysis (Fig. 6.40) for waste treatment. While accepting that the newer incineration processes are not so destructive as earlier waste disposal methods, in which little or no product recovery was attempted, pollution can still be a problem, and it was suggested that pyrolysis would be a more beneficial waste disposal/recovery technique, providing fuel, including liquid hydrocarbons, which could supply energy remote from the waste disposal unit. (Incineration provides heat energy at the incinerator, but in the normal course of events no fuel is recovered which can be transported for use elsewhere).

Pyrolysis is defined as thermal decomposition without complete combustion. The fact that it can produce at least three different kinds of fuel - gas, oil and char, for instance - is seen by some as a disadvantage in that it increases recovery and marketing problems, but this drawback is probably over-rated. Systems such as that developed by Garrett Research and Development Company in California are directed at fuel recovery from urban refuse. By heating organic waste in a heat exchanger to about 500°C in an oxygen-free atmosphere, each tonne of refuse yields approximately one barrel of oil, 72.5 kg of char, and quantities of low energy gas. Recycling of the gas provides heat for the pyrolysis process.

Full commercial exploitation of pyrolysis and other similar processes is unlikely for some years, but it serves to illustrate the usefulness of urban waste, which, together with industrial waste, should be treated to obtain maximum benefits to industry.

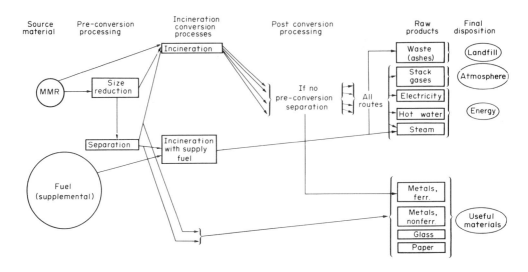

Fig. 6.39 Conventional incineration route for
waste disposal, with optimum recovery
(Courtesy National Aeronautics & Space Administration)

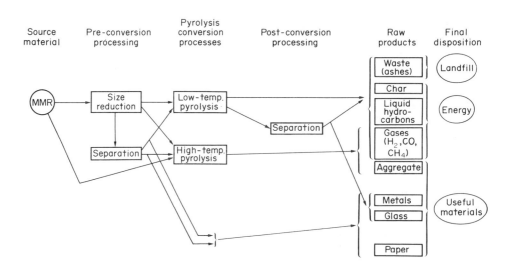

Fig. 6.40 Pyrolysis route for waste disposal
(Courtesy National Aeronautics & Space Administration)

REFERENCES

6.1 Anon. Heat recovery in air conditioned buildings. The Electricity Council, Publication EC2893, London, 1974.

6.2 Anon. Design concepts for optimum energy use in HVAC systems. The Electric Energy Association, Publication EEA-209, New York (1975)

6.3 Friedlander, G.D. Energy conservation by redesign. IEEE Spectrum, 10, 11, 36-43 (1973)

6.4 Anon. Heat transfer lighting fittings for integrated environmental design (IED). Osram-GEC Hirst Research Centre Report (1973)

6.5 Patel, K.N. and Murphy, W.J. Storage water application and service water supply systems to heat recovery. ASHRAE Journal, 16, 52 - 55, (1974)

6.6 Drummond, W.A. Economics of waste heat recovery from boiler flue gases The Plant Engineer pp 11 - 15, Feb 1975.

6.7 Glenn, R.D. Energy conservation opportunities in steam use. US National Bureau of Standards. Special Publication No. 403, Vol. II. Washington, D.C. (1973).

6.8 Urbani, A.C. Boiler blowdown and the economic viability of heat recovery. Proc. I. Mech. E. Conference on Energy Recovery in Process Plant, London, Paper C.20/75, 29-31, Jan 1975.

6.9 Danilov, B. Utilisation of steam boiler blow down. Processing May, 1975.

6.10 Bryan, D.J. et al. Applications of recuperative burners in gas-fired furnaces. The Institution of Gas Engineers, Communication 952, London, Nov. 1974.

6.11 Harrison, W.P. et al. The design of self-recuperative burners. J. Inst. Gas Engineers, 10, 538 - 563 (1970)

6.12 Berry, M.R. et al. A dry materials heat exchanger with regeneration. Trans ASME. J. Heat Transfer, Paper 73-WA/HT-14, pp 518 - 523, Nov.1974

6.13 Mills, L. Recover waste heat systematically. Power, pp 36 - 37, Dec. 1972.

6.14 Dickson, P.F. Heating and drying of aggregate. US National Asphalt Pavement Association, April, 1971.

6.15 Kolbusz, P. Industrial applications of heat pumps. The Electricity Council Research Centre, Capenhurst. Report ECRS/N845, Sept. 1975.

6.16 White, A. Batch type fluid bed dryers. Chemical Processing. Aug.1974

6.17 Horsley, M.E. Principles of waste heat recovery. Proc. Waste Heat Recovery Conference, The Institution of Plant Engineers, London Sept. 1974.

6.18　Williams, P.N. Two independently fired cupola blast heaters. Foundry Trade Journal, Feb 29th, 1968.

6.19　Leyshon, H.J. and Selby, M.J. Improved cupola performance by correct distribution of the blast supply between two rows of tuyeres. The British Foundryman, pp 43 - 55, Feb. 1972.

6.20　Anon. Big energy savings in iron foundries from BCIRA development. Press Release, British Cast Iron Research Association, July, 1975.

6.21　Anon. Electric melting for iron and steel foundries. The Electricity Council, Publication EC3128, London, Nov. 1973.

6.22　Ban, T.E. Effective energy utilisation from direct electric ironmaking Proc. Conf. on Energy - Use and Conservation in the Metals Industry. The Metallurgical Society of AIME, New York (1975)

6.23　Grigorev, V.N. and Gusovskii, V.L. Utilising fuel more efficiently in reheating and heat treatment furnaces. USSR Scientific and Technical Society for Ferrous Metallurgy, Stal. Vol. 3, pp 274 - 276. BLL Translation M-21957, (1974)

6.24　Clarke, B.N. Self-help conservation via effective heat treatment. Metal Progress (US) 108 95 - 96 (1975).

6.25　Frick, E.T. Guidelines for energy survival. Metal Progress (US), 108, 62 - 67 (1975)

6.26　Roots, D.C. The elimination of atmospheric pollution from coating and drying ovens. Hygrotherm Engineering Ltd. Publication (1975).

6.27　Csathy, D. Energy conservation by heat recovery. Paper 749086, 9th Intersoc Energy Conversion Engineering Conf., ASME, San Fransisco (1974)

6.28　Anon. Benefits of liquid waste incinerators. Processing, p. 41, Feb. 1976.

6.29　Rowe, W.G.E. and Huxford, R.C. Waste heat utilisation by 'co-operative' group schemes. Proc. Waste Heat Recovery Conference. Institution of Plant Engineers, London, pp 36 - 42, Sept. 1974.

6.30　Anon. The Incineration of Municipal and Industrial waste. Proceedings of Institute of Fuel, Vol. 1 & 2, London (1969)

6.31　Hung, W. Results of a firetube test boiler in flue gas with hydrogen chloride and fly ash. Trans. ASME, Proc. Winter Annual Meeting, Paper 75-WA/HT-39, Dec. 1975

6.32　Fuchshuber, P. Modern heat supply of an industrial plant. Betriebstechnik, 13,(1972)

6.33　Huang, C.J. and Dalton, C. Energy recovery from solid waste. Vol. 1 Summary report, University of Houston, Texas, NASA Report CR-2525, April, 1975.

6.34　Thorne, J.G.M. Closed circuit heat transfer systems and fluids, Processing, pp 49 - 52, March 1976.

Waste Heat Recovery Techniques

A large number of different techniques for recovering waste heat have been cited in previous chapters, and many applications have been discussed where waste heat recovery can reduce process operating cost and conserve significant quantities of fuel. In fact, after the implementation of 'good housekeeping' procedures, waste heat recovery is likely to be the major conservation method to be adopted in a wide range of industries, and can involve a substantial outlay of capital.

The discussion below describes the main methods of heat recovery from air, gases and liquids, including systems such as heat pumps which are only just beginning to be serious contenders in this field. (The heat pump is the only commercially available system for upgrading waste heat). Heat pipe heat exchangers which have been used extensively in the United States for gas-to-gas heat recovery are now available in the United Kingdom and Europe, and a section is devoted to these units. In addition a range of recuperators and regenerators capable of operating at very high exhaust temperatures are described.

7.1 Heat Pipe Heat Exchangers

The heat pipe heat exchanger used for gas-to-gas heat recovery is basically a bundle of finned tubular heat pipes assembled like a conventional air-cooled heat exchanger.

A heat pipe (ref. 7.1) is a sealed container having a wick, in which capillary action can be generated, lining the inside wall. The wick contains the heat pipe working fluid, which is the heat transport medium. If heat is applied to one end of the heat pipe, the liquid in the wick at that end evaporates, and the vapour flows to the cooler regions of the heat pipe, where condensation occurs, and the latent heat of condensation is rejected. On re-entering the wick, the condensate is then pumped back to the evaporator section in the wick, as illustrated in Fig. 7.1.

Because of the reliance on capillary action to return the liquid to the evaporator section, the performance of a heat pipe is a strong function of its inclination from the horizontal, the wick pore size, and the surface tension of the working fluid. The amount of heat transported is also determined by the latent heat of the working fluid, this being preferably as high as possible.

7.1.1 General description of unit In a gas-to-gas heat exchanger, as shown in Figs. 7.2 and 7.3 the evaporator sections of the heat pipes span the duct carrying the hot gas, and the condensers are located in the duct where heat transported from the hot gas is given up to cold incoming air. The flow through the heat exchanger should preferably be counter-current for maximum efficiency.

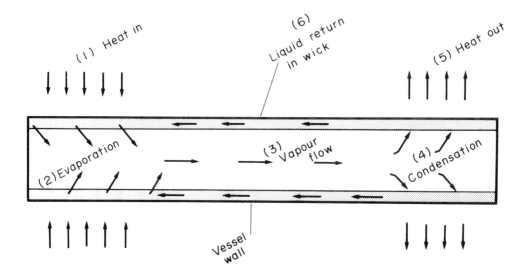

Fig. 7.1 Operation of a heat pipe

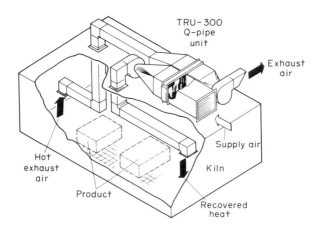

Fig. 7.2 Layout of a heat pipe heat exchanger
in a recovery application
(Courtesy Q-Dot Corporation)

Normally the heat pipes are mounted in a horizontal position, and the two ducts must be adjacent to one another at the point where the heat pipe heat exchanger is situated. One American company manufacturing heat pipe heat exchangers has patented a control system whereby the variation in heat pipe performance with its angle to the horizontal can be used to modulate the heat transfer from the hot to the cold duct. By tilting the complete unit a few degrees, so that the evaporators are above the condensers, the heat transfer can be gradually reduced to zero. This allows one to control the temperature of the gas being heated in instances where the ambient air or exhaust air may vary in temperature.

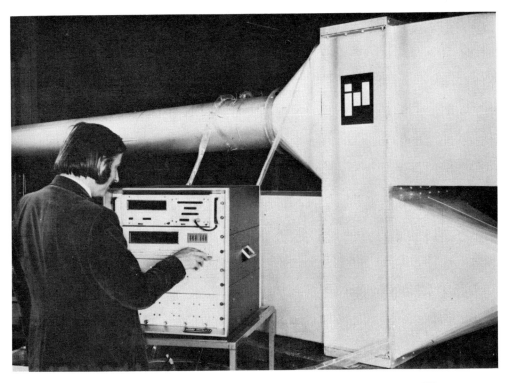

Fig. 7.3 A small heat pipe heat exchanger under test at IRD
(Courtesy IRD Co. Ltd.)

The materials and working fluids used in the heat pipe heat recovery unit depend to a large extent on the operating temperature range and, as far as external tube surface and fins are concerned, on the contamination in the environment in which the unit is to operate. The working fluids for air conditioning and other applications where operating temperatures are unlikely to exceed 40°C include the Freons and acetone. Moving up the temperature range, water is the best fluid to use. For very hot exhausts in furnaces and direct gas-fired hot air circuits, higher temperature organics can be used.

In most instances the tube material is copper or aluminium, with the same material being used for the extended surfaces. Where contamination in the gas is likely to be acidic, or at higher temperature where a more durable material may be required, stainless steel is generally selected.

The tube bundle may be made up using commercially available helically wound finned tubes, or may be constructed like a refrigeration coil, the tubes being expanded into plates forming a complete rectangular 'fin' running the depth of the heat exchanger. The latter technique is preferable from a cost point of view.

Unit size varies with the air flow, a velocity of about 2 to 4 m per second being generally acceptable to keep the pressure drop through the bundle to a reasonable level.

Small units having a face size of 0.3 m (height) by 0.6 m (length) are available. The largest units are about 5 m in length and 1.5 m high. The number of rows of tubes in the direction of the gas flow rarely exceeds 8, and is most commonly between 4 and 6, although several units may be used in series.

7.1.2 Performance

Heat pipe recuperators tend to be slightly less efficient than some other types because the heat pipe is in effect an additional thermal resistance when compared with units where a single wall separates the hot and cold gas streams. Also, the units generally transfer only sensible heat, except in instances when condensation occurs in the evaporator duct.

The performance of heat pipe heat exchangers may be expressed as an efficiency, or a 'recovery factor'. The recovery factor, based on the assumption that only sensible heat is transferred, is the percentage of energy recovered from the hot duct and given up to the cold air in the adjacent duct. The recovery factor is a function of the number of fins on the heat pipes, the number of tube rows, the gas velocity and density, and the ratio of the mass flows of the two gas streams. Fig. 7.4 shows the recovery factors for a 6 row

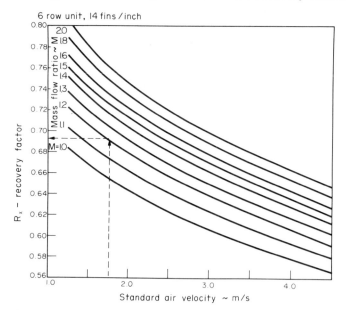

Fig. 7.4   The performance of a heat pipe heat exchanger over a range of operating conditions (Courtesy Q-Dot Corporation)

unit, for a number of values of M, the mass flow ratio. In most circumstances the face areas of the heat exchanger are equal, with the splitter plate separating the two gas streams located in the centre. However in cases where M is large, the face area of the evaporator and condenser sections may be adjusted, by moving the partition, so that the velocity, and hence the pressure drop through the duct carrying the highest mass flow is reduced.

At present heat pipe heat exchangers can cope with exhaust gas temperatures of up to 350°C, and 'specials' are available at high cost where temperatures in excess of this are likely to be encountered. (The use of these units in systems where the maximum temperatures can exceed 400°C can necessitate adoption of liquid metal working fluids in the heat pipe, adding considerably to the cost).

7.1.3 Applications

There is a wide range of applications for heat pipe recuperators in industry, commercial buildings and general heating and ventilating. Based on information supplied by Q-Dot Corporation, units have been installed to recover heat in the following industrial processes:

>   Air dryer regeneration
>   Automotive paint drying ovens
>   Bleaching ovens
>   Brick kiln regeneration
>   Cereal dryers
>   Chain line drying ovens (continuous process)
>   Industrial drying process regeneration
>   Industrial process (welding, plating & brazing)
>   Model shops
>   Paint spray booths
>   Pharmaceuticals
>   Phosphatizing plant
>   Pollution control
>   Refrigeration reheat
>   Solvent boil-off oven regeneration
>   Spray dryers
>   Tanning
>   Textile oven regeneration
>   Waste steam reclamation

In addition, there is a wide variety of applications in the heating and ventilating field, including:

>   Industrial process heat used for comfort conditioning
>   Hotels
>   Department Stores
>   Food Stores
>   Hospitals
>   Kitchens
>   Laboratories
>   Office buildings
>   Power plants
>   Schools
>   Ski Lodges
>   Swimming pools
>   Theatres
>   Universities
>   Utilities (US Power Companies)

The most significant difference between the industrial process applications of the heat pipe heat recovery unit and its use in most heating and ventilating applications is the temperature difference between the two gas streams.

It is obvious from the above list that the lower temperature differences in air conditioning applications have not precluded the use of this type of heat recovery unit here. However in the United States climatic extremes are greater than in the United Kingdom, and the use of the unit in the summer to cool incoming air is also possible; (the heat pipe heat exchanger is fully reversible). Thus utilisation at reasonable efficiency is probably possible for eight months per year. In the United Kingdom however, preheating of make-up air is likely to be the only requirement, and the quantities of heat involved suggest that in many installations the payback time would be much more than two to three years. (An example is given in the next section).

It is probable that industrial process applications will be the main market for these units in the United Kingdom. Initial calculations suggest that payback periods of one year are possible in some cases, and the added advantages of high temperature differences and year-round utilisation contribute to this.

Summarising, the three broad areas of application of heat pipe exchangers are process heat recovery, where the heat recovered from a plant exhaust may be used to preheat air or gas used in the same process; heat recovery from a process in which the recovered heat may be used for space heating, and heat recovery in air conditioning systems, where heat pipes may be used to preheat incoming air in winter or precool it in summer (refs. 7.2, 7.3).

The limitations on the application of the heat pipe heat exchanger, in addition to those already discussed, are mainly associated with the need to bring the hot and cold gas streams adjacent to one another at the heat exchanger location. In some cases this may involve an excessive amount of reorganisation of flow paths, coupled with the installation of lengths of new ductwork. It may be possible to overcome this limitation by mounting the tube bundle so that the heat pipes are vertical, with the evaporators (hot gas duct) below the condensers.

In many applications the limitations are a small price to pay for the advantage of no carryover of contamination between the ducts. The splitter plate also enables the two gas streams to operate at substantially different pressures, which is not possible with the rotating regenerator.

7.1.4 Cost data

Heat pipe recovery units have been in production in the United States for several years, and the approximate cost of units manufactured by the Q-Dot Corporation, the largest company in this field, (see Appendix 4 for manufacturers) is as follows:

Unit size	Material	Cost/m^3/hr flow *† (pence)
Small	Aluminium	16
Large	Aluminium	12
Small	Copper	21
Large	Copper	15
Small	Stainless Steel	32
Large	Stainless Steel	24

*The above prices do not include installation costs

†Costs are for 1975, and rate of exchange then existing

Q-Dot estimate that in industrial applications, where utilisation is high and fairly large temperature differences are involved, payback periods range from one to three years. In air conditioning and space heating applications, the payback period tends to be longer, possibly in the less attractive cases exceeding five years. (All figures are based on 1975 costs).

An example of their application and the cost savings involved, in this case the heat pipe units being manufactured by a United States company, Isothermics, is illustrated in Fig. 7.5 and described below.

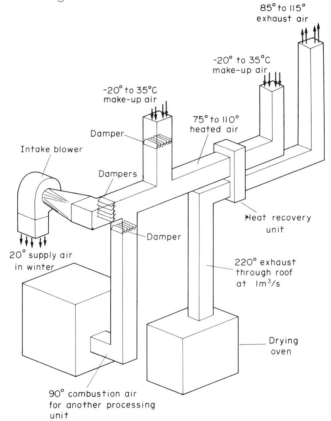

Fig. 7.5 A heat recovery installation, using heat pipes in a drying oven
(Used with the permission of Isothermics Inc. Augusta, NJ, USA)

Application: An oil fired drying oven exhausts 64 m^3 per minute air at 200°C. Heat input is 500 kW. It is required to heat make-up air to 20°C in winter and to heat process air to 95°C in summer. Oven operating twenty one hours per day. Allowable pressure drop 2.2 mm Hg.

System design: Heat pipe heat exchanger with 1.0 by 1.3 m face area, 4 tube rows deep. Tubes 19 mm o.d. with fin height 9.5 mm. All copper with nickel plate protective coating. Working fluids methanol and water. Weight 300 kg; total installation cost $10 000.

Results: Effective recovery 56%

Annual fuel saving $8 000.

7.1.5 Future developments

Based on experience in the United States, the heat pipe heat exchanger is well established in a wide variety of applications covering a range of temperatures. Its capital cost is competitive, and all evidence points to its reliability in use, provided that normal maintenance procedures are observed.

Two areas where development work could extend the range of application of heat pipe heat exchangers are the operating temperature range and the control of heat transfer. As mentioned earlier, the upper limit of the operating temperature range is currently, on commercially available systems, about 350°C. Future work on units employing liquid metal working fluids should concentrate on mass-production methods so that realistic costs can be achieved. (In the higher temperature spectrum, ceramic rotary regenerators are currently available which can fulfil a similar function).

Improvements in control are being investigated, (ref. 7.4). Currently control of the heat transfer rate between the hot and cold gas streams is effected by tilting the whole heat exchanger unit, which in many cases may be inconvenient. It is probable that much more precise control of heat flow could be obtained using the gas-buffered heat pipe, with only a moderate increase in capital cost. This would make these units useful in processes where the inlet air temperature has to be accurately controlled, and in air conditioning applications, where significant variations in the inlet air temperature occur throughout the day. (A gas-buffered heat pipe has an inert gas in the condenser section. As the temperature of the evaporator rises, the vapour pressure in the heat pipe increases and the surface area for heat rejection is increased because the vapour pressure reduces the volume occupied by the gas. Conversely a reduction in heat flow into the evaporator reduces vapour pressure, resulting in less surface for rejection and a higher heat pipe temperature as shown in Fig. 7.6).

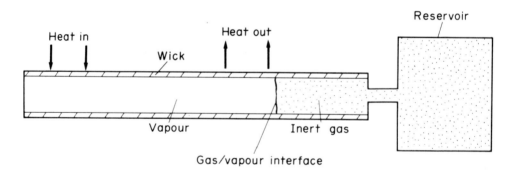

Fig. 7.6 The gas-buffered heat pipe; one method for temperature control

The practical application of gas-controlled heat pipes in a heat exchanger has been discussed by Basiulis and Plost (ref. 7.5). In a methanation process heat pipes can be used to maintain the methanation temperature, at the same time using recovered waste heat to generate steam, as illustrated in Fig. 7.7. Waste heat originates from the exothermic reaction between the hydrogen and carbon monoxide in the presence of a nickel catalyst. The heat pipe system eliminates the need for complex control systems.

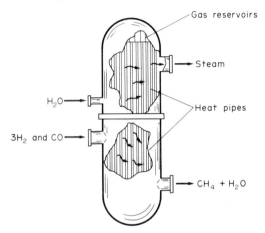

Fig. 7.7 Use of a heat pipe heat exchanger in a methanation process

Other work is being sponsored in the United States by the National Science Foundation (ref. 7.6) with a view to obtaining more basic data on the factors which affect performance. One of the aims of this work is to enable even cheaper units to be constructed. Optimisation of heat exchanger design using a computer program, (ref. 7.7) and a comparison of predicted and actual results obtained using commercially available heat pipe heat exchangers has been successfully made (ref. 7.8).

7.2 Liquid-Coupled Indirect Heat Exchangers

Liquid-coupled indirect heat exchangers sometimes known as run-around coils by the manufacturers, consist of two heat exchanger units connected to one another by a circulation system containing a liquid heat transfer medium. The operating principal is shown in Fig. 7.8. The circulating fluid, which is selected on the basis of the operating temperature level, is heated as it passes through a heat exchanger located in a hot gas stream. The liquid is then pumped through a heat exchanger located in a remote duct where heating is required, the heat removed, and the liquid returned to the heat exchanger in the hot gas duct. In applications such as air conditioning, where extremes of temperature are not met, the liquid used is often a water/glycol solution.

The theory of liquid-coupled indirect heat exchangers is given in ref. 7.9, and it is necessary here only to give a review of the parameters which determine the efficiency of this system. These are described in detail in a number of papers, which the reader may obtain from the manufacturers, (refs. 7.10, 7.11). This type of heat recovery unit has been developed as a commercial product by AB Svenska Flaktfabriken, of Stockholm, and the three most important components of the system, the cooling coil, heating coil and pump, are of standard well-proven design.

Compared to the heat pipe heat exchanger, the primary advantage of the

ECOTERM system, as it is called by the manufacturers, is the fact that it permits the remote location of the supply and exhaust ducts. (This is not unique to the ECOTERM system. It is possible to operate an air-coupled system, as discussed in the next section). It possesses the other advantages of the heat pipe heat exchanger when compared with regenerative heat exchangers.

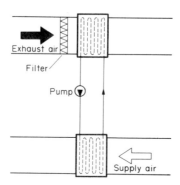

Fig. 7.8 Operating principal of a liquid coupled indirect heat exchanger, or 'run-around' coil
(Figure by kind permission of SF Air Treatment Ltd.)

The use of a pump in the run-around system introduces an additional unreliability factor, as well as requiring a power supply, unlike heat pipe exchangers. The manufacturers estimate that the pump consumption is approximately 1 per cent of the recovered energy, but the pump power is in part dissipated as additional heat (see design economic assessment below).

7.2.1 Thermal design The number of variables to be taken into account when design a run-around system is high, particularly where more than one supply and exhaust duct may be served by a single liquid loop. For this reason potential users are advised to contact the manufacturers, who are able to assess applications using computer optimisation programs, (ref. 7.12).

For a given total heat transfer area, the heat capacity flow rate of the medium (this being the product of mass flow and specific heat), as well as the area ratio of the supply and exhaust coils, should be selected so that the 'temperature efficiency' on the air supply side will be as high as possible.

The temperature efficiency of this type of heat exchanger is defined as:

$$T.E. = \frac{\text{Supply outlet temp.} - \text{Supply inlet temp.}}{\text{Exhaust inlet temp.} - \text{Supply inlet temp.}}$$

Temperature efficiencies are typically in the range 45 per cent to 65 per cent, and are also functions of the moisture content of the exhaust air and the ratio of the supply/exhaust air flows.

7.2.2 Applications The applications of run-around coils in industry and in air conditioning fields should be similar to those of the heat pipe heat exchanger, listed in Section 7.1.3. Fig. 7.9 shows a layout in a sausage skin factory in Sweden. Five identical systems were used to recover heat from the process exhaust, the heat being subsequently transferred to the process air supply, which was preheated. The total cost of this unit was £70 000, and was recovered in less than one year of operation.

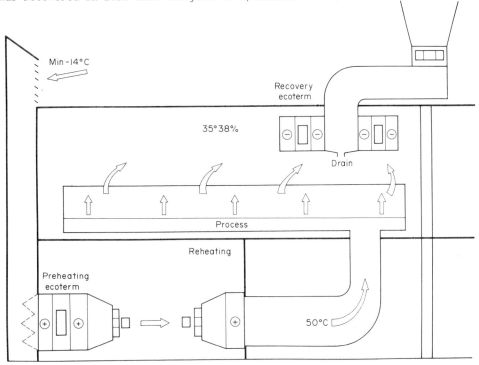

Fig. 7.9 A 'run-around' coil system used to recover heat in a sausage skin factory
(Figure by kind permission of SF Air Treatment Ltd.)

In many industrial processes the operating temperature may be too high to permit water to be used as the liquid transport medium. In these instances there is no reason why a heat transfer fluid such as Dowtherm or Thermex (a diphenyl/diphenyl oxide eutectic) should not be used as a working fluid, although no reference to the use of higher temperature working fluids has been found to date.

The versatility offered by run-around systems is such that one can envisage heat transport over considerable distances within a large industrial plant. The use of liquid to carry the heat, with the resulting small diameter pipes needed to connect the exhaust and supply heat exchangers, would make this more economical than a ducted air interconnecting system, both in terms of duct material and lagging.

7.2.3 Economics of run-around coils

An economic assessment of the use of this system in the air conditioning plant of a London hospital is given in the literature (ref. 7.10).

In order to assess the economics, it is assumed that heating is provided using an oil fired boiler, with oil costing £36.7/m^3 (August 1974). This corresponds to 0.46p/kW hr. The energy supplied to the pump and to the supply air fan is considered as additional heat during the heating season, but fan power is classed as a loss during the remainder of the year. Capital cost calculation is based on an annuity factor of 0.163 (ten years at an interest rate of 10 per cent) for all plant except the buildings, where a factor of 0.8 is taken.

The ventilation system has the following specification:

Supply air flow	: 11 m^3/sec
Exhaust air flow	: 11 m^3/sec
Exhaust relative humidity	: 55%
Supply air temperature required	: 18°C
Operating time	: 8760 hr/year

Using an Ecoterm coil, the reduced heat demand amounts to 57 050 degree-hours which at the appropriate air flow and oil price represents a reduction of the fuel cost of £3456 per year. When the heat recovery system is installed the boiler output can be reduced by 183 kW. Depending on the size of the boiler installation the resulting reduction in investment costs varies substantially from case to case. In the installation being considered, the reduction in investment costs is assumed to amount to £10/kW, i.e. a total of £1830.

The investment and operating costs for installing the heat recovery system are summarized in Table 7.1. According to the results given in this table, a net saving of £2621 is obtained for an increased investment of £3350, which means that the payback period (increased investment/net saving) is 1.28 years. If similar calculations are carried out for systems with six and ten row coils one finds that according to Table 7.1 the payback periods are 1.41 and 1.54 years, respectively. Accordingly, eight row coils seem to be the optimum size in this case.

TABLE 7.1 Cost Exercise on an Ecoterm Coil

Investments	Installation costs for the recovery system (coils, pump, pipework, filter)	£ 4 180
	Costs of increased building volume (£25/m^3)	£ 1 000
	Reduction in investment for the boiler plant	£ 1 830
	Increase in investment	£ 3 350
Capital and Operating Costs	Capital costs for the increased investment	£ 464/year
	Energy and power costs (pump and fans)	£ 271/year
	Service and maintenance	£ 100/year
	Reduced fuel costs	£ 3 456/year
	NET SAVING	£ 2 621/year

...Cont'd.

Number of Rows	System Efficiency (%)	Increase in Investment (£)	Net Saving (£)	Payback Period (years)
6	47	2910	2064	1.41
8	60	3350	2621	1.28
10	65	4150	2702	1.54

7.3 Gas-Coupled Indirect Heat Exchangers

The liquid coupled heat recovery system is only one of the techniques available for transferring heat from one location to another. Heat pumps, although they also upgrade the heat, are another type of heat exchange system which operates between remote heat sinks and sources, (see Section 7.9).

One of the several gas-to-gas recuperators which can be used to recover heat from flues for space heating and similar functions can be used to provide preheating of process air by linking it via a duct to the process air inlet, where optionally a second recuperator is located. While such a system, illustrated in Fig. 7.10 may not be as efficient as the liquid loop arrangement, the cost of the supply and exhaust heat exchangers is low, and the combination is worth considering where the two gas ducts are not too remote from one another.

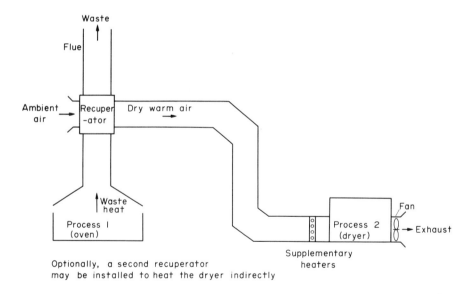

Fig. 7.10 The use of gas-to-gas recuperators to transfer heat between remote sources and sinks

7.4 Rotating Regenerators

The rotating regenerator (otherwise known as the Munter wheel, or the Ljungstrom heat exchanger, after its Danish inventor), has been used over a period of about fifty years for heat recovery in large power plant combustion processes, (ref. 7.13).

The operation of the rotating regenerator is shown in Fig. 7.11. Like the heat pipe heat exchanger (Section 7.1) the regenerator wheel spans two adjacent ducts, one carrying exhaust gas and the other containing the gas flow which it is required to heat. The gas flows are counter-current. As the wheel rotates, it absorbs heat from the hot gas passing through it and transfers the heat to the cooler gas flow. A later development, the hygroscopic wheel, is able to transfer moisture, as well as sensible heat, between the two ducts.

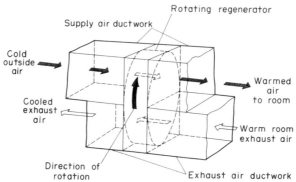

Fig. 7.11 Operation of the rotating regenerator
(Courtesy Wing International Ltd.)

Rotating regenerators, in common with many other heat recovery systems, can be used in hot climates for pre-cooling air used for conditioning large buildings, and the wheel works effectively in applications where the temperature differences between hot and cold airstreams are too low for effective use of the heat pipe heat exchanger.

The Wing Company (see Appendix 4), produce three different types of rotating regenerator. The most common form, also available from several other manufacturers, utilizes a wheel made up from a knitted aluminium or stainless steel wire matrix. This matrix is cheap, and the heat transfer efficiency is high as the airflow is exposed to a large amount of surface as it passes through the wheel. However, the pressure drop through this type of matrix can be relatively high, and the fouling of the matrix tends to be more severe than on other types.

Development of laminar flow wheels, in which the matrix is corrugated, resembling a small pore honeycomb, has alleviated the pressure drop and fouling problems of the mesh matrix, and this type of wheel is easier to clean. In terms of thermal efficiency the performance of a metallic corrugated matrix wheel is similar to that of the mesh type.

A third form of matrix used in rotating regenerators is non-metallic. Known as the hygroscopic wheel, this type can transfer moisture as well as sensible heat, and is particularly useful in heating and ventilating applications. The structure is similar to that of the metallic laminar flow wheel. While the hygroscopic wheel is likely to be up to 35 per cent more expensive than the metallic type, Wing claim that the increased capital cost is generally more than offset by the increased heat transfer attributable to latent heat recovery. The latent heat content can vary considerably from one application to another, and should be carefully assessed before settling for a particular unit.

The use of a stainless steel matrix permits operation of rotating regenerators at up to 820°C. In some processes regeneration is required at higher temperatures, and two instances have already been mentioned (gas turbines, discussed in Chapter 2, and steel industry preheating, detailed in Chapter 3). Corning manufacture regenerator wheels based on cellular glass-ceramic structures, which have demonstrated long life at 900°C, and rotating regenerators developed for automotive gas turbines, using a silicon nitride matrix, have been tested at temperatures in excess of 1000°C (ref. 7.14).

7.4.1 Performance and operation   In determining the performance and operating characteristics of a rotating regenerator, a number of factors must be taken into account, and these include the following:

Operating temperature: as discussed above, the type of matrix used, be it metallic or ceramic, depends upon the operating temperature range likely to be encountered in the application.

Operating pressure: in general it is desirable to operate this type of regenerator in situations where the exhaust and supply gas streams have similar pressures. If a knitted mesh is used for the matrix, there is a potentially large flow path available through the mesh if pressures are not equalised. In wheels of the laminar flow type, where the gas is restricted to movement in the axial direction, some pressure differential can be supported if suitable seals are fitted to the unit. To minimise carry-over of contamination, it is preferable to operate the exhaust stream at a marginally lower pressure than the supply stream.

Cross contamination: the possibility that contamination of clean supply gas could occur with this type of regenerator has led to the incorporation by manufacturers of purge sections, the operation of which is shown in Fig. 7.12.

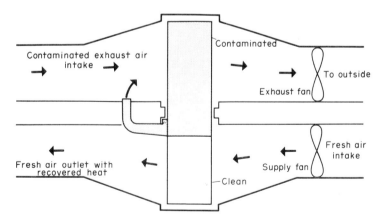

Fig. 7.12   Reduction of cross-contamination in a rotating regenerator by means of a purge section
(Courtesy Curwen and Newbery Ltd.)

Here a proportion of the supply air is used to scavenge the matrix section leaving the exhaust duct, and the contamination or residual exhaust gas is blown back into the exhaust duct. Proper purge section operation depends on correct fan location so that the pressure on the supply side is higher than on the exhaust side, and in cases where this requirement can be met, cross-contamination can be as low as 0.04 per cent by volume, and particle carry-over less than 2 per cent (ref. 7.15). Some manufacturers specify a carry-over of less than 0.1 per cent by volume in their literature, and where this factor is critical, their advice should be sought.

Purging is not obtained without sacrificing efficiency. The sizing of system fans should be increased by 10 per cent of rated volume to handle the greater gas flow requirements, and if the correct fan arrangement cannot be installed because of practical difficulties, seals can be incorporated at each radial partition, but these are not nearly as effective, cross-contamination rates possibly then approaching 8 per cent by volume.

On some of the higher temperature metallic regenerators, a separate fan specifically for purging may be fitted. These may use up to 4 kW in electrical energy, and, together with the much lower motor power for rotating the heat exchanger, should be accounted for in any cost analysis.

Very high temperature ceramic regenerators also suffer from cross-contamination, but in the case of gas turbine units, for a 4:1 engine compression ratio transverse flow accounts for only about 0.3 per cent of engine air flow at full speed conditions.

Unequal flow rates: Potential users will find that most manufacturers quote performance figures based on equal supply and exhaust flow rates. If the flows are unequal, correction factors must be applied. It is normal practice to size the generator for the maximum flow rate and to select the efficiency on the basis of the lower flow.

In situations where the inlet gas flow rate is much larger than the exhaust flow, say by a factor of 4, the rotating regenerator may be selected on the basis of the exhaust flow, and used to heat only a proportion of the supply air, say 25 per cent. The arrangement for this, shown in Fig. 7.13, permits the balance of the supply air to bypass the regenerator, after which it is mixed with the preheated air.

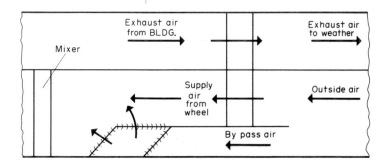

Fig. 7.13 Compensation for unequal flow rates in a rotating regenerator
(Courtesy Wing International Ltd.)

Waste Heat Recovery Techniques

Pressure drop: The pressure drop through the heat exchanger is a function of gas velocity and matrix design. Some manufacturers offer two types of matrix, one designed for optimum heat transfer capability which, because of its greater amount of surface, has a pressure drop which may be unacceptable in some applications. By compromising on thermal efficiency, a lower pressure drop unit having larger pores may be adopted. Typically a high efficiency unit (81 per cent at a velocity of 4 m per second) will have a pressure drop of approximately 35 N/m^2, whereas the corresponding low pressure drop system will operate at 76 per cent efficiency at the same velocity, resulting in a pressure loss of only 17 N/m^2, (Wing figures).

Control: Most rotating regenerators (excluding high temperature ceramic types) are driven by electric motors, the power of which on the largest wheels (4 m diameter) approaches 0.5 kW. For optimum heat transfer, rotational speed is typically 20 rev/min. As this speed reduces, the thermal efficiency is lowered, and hence speed control can be used to meet reduced heat load duties. Alternatively the supply air may be regulated using dampers and/or a bypass system. In most industrial processes where a constant duty is required, any form of modulation will be unnecessary, and a constant speed motor is recommended.

Details of the performance of a typical rotating regenerator are given in the tables below. The unit selected is the Wing CORMED metallic regenerator, designed for operation at temperatures between 150°C and 400°C. It is available in a variety of sizes, wheel diameter varying between 1.1 and 3.2 m. Maximum flow rate through the largest size of unit is about 70 000 m^3 per hour, corresponding to a velocity of 7 m per second. The performance data presented is for the model WCM-1400, representing the middle of the range of this type.

TABLE 7.2 Performance of Wing WCM-1400*

Efficiency - (based on average wheel temperature of 150°C)												
0.83	0.82	0.79	0.76	0.73	0.70	0.68	0.65	0.63	0.61	0.60	0.58	0.57
Velocity - m per second												
1.0	1.5	2.0	2.5	3.0	3.5	4.0	4.5	5.0	5.5	6.0	6.5	7.0
Pressure drop - N/m^2												
21	30	41	51	61	71	81	91	102	112	122	132	142
Flow rate - m^3 per hour												
4399	6599	8798	10998	13198	15397	17597	19796	21996	24196	26395	28595	30794

Similar tables are presented by manufacturers giving correction factors for unequal flow rates and the effect of temperature on pressure drop. Thus the selection procedure is quite simple and the user has no need to enter into elaborate thermal calculations.

The following table* is used to correct for efficiencies calculated at temperatures other than 150°C.

---

*All data is reproduced by permission of Wing International Ltd.

Industrial Energy

TABLE 7.3 Temperature Correction Factors

AVG. Temp. °C	Base efficiencies (AVG. Temp. = 150°C)						
	0.55	0.60	0.65	0.70	0.75	0.80	0.85	0.90
	Temperature corrected efficiencies						
65	0.51	0.56	0.61	0.66	0.72	0.77	0.82	0.88
95	0.52	0.57	0.62	0.67	0.73	0.78	0.83	0.89
120	0.54	0.59	0.64	0.69	0.74	0.79	0.84	0.89
150	0.55	0.60	0.65	0.70	0.75	0.80	0.85	0.90
175	0.56	0.61	0.66	0.71	0.76	0.81	0.86	0.91
205	0.58	0.63	0.67	0.72	0.78	0.81	0.86	0.91
230	0.59	0.64	0.68	0.73	0.78	0.82	0.87	0.92

Average Wheel Temperature = Exhaust Temperature Plus Supply Temperature Divided by 2.

The selection procedure is as follows:

Select a regenerator to handle 35 000 m^3 per hour of dry exhaust gas at 260°C with an equal supply airflow of outside air at 10°C for return to the process.

(i) Determine exhaust airflow at NTP

$$\frac{273 + 20}{273 + 260} = \frac{293}{533} = 0.54 \times 35\ 000$$

Flow at NTP = 18 900 m^3 per hour

(ii) Determine average wheel temperature

$$\frac{260 + 20}{2} = 140°C$$

(iii) Select wheel size and base efficiency from table, WCM-1400 has efficiency of 0.65 for nearest flow (19 796 m^3 per hour)

(iv) Correct base efficiency for average wheel temperature, Table 7.3. Negligible correction required.

(v) Determine supply air leaving temperature:

20 + (260 - 20) x 0.65 = 176°C

(vi) Determine heat recovered.

Flow x Sensible heat factor x Temp. rise

18 900 x 0.288 x 176 = 958 000 K cal per hour

7.4.2 Applications and economic assessments Within the context of the limitations discussed above, rotating regenerators can be used in similar applications to those listed in Section 7.1.3 for heat pipe heat exchangers, namely process preheating, the use of process waste heat for space heating and air conditioning heat recovery. They can operate at higher temperatures than existing commercially available heat pipe units.

Howden, who produce a range of rotating regenerators for power station boiler preheating, also manufacture smaller units for typical industrial applications (ref. 7.16) and the fuel savings which can be effected by recovering heat from a boiler, economiser or process plant are illustrated graphically in Fig. 7.14.

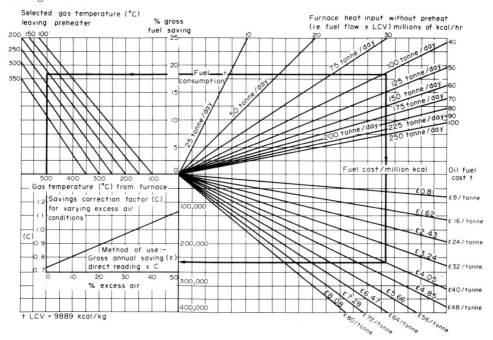

Fig. 7.14 Prediction of fuel savings using a graph based on rotating regenerator performance (Courtesy Howden Group Ltd.)

With the temperature of the gas leaving the boiler, economiser or process as a base, plot the desired gas temperatures through the air preheater on graph (A). Follow the grid lines through graph (B) which relates temperature drop to gas flow - a measure of the heat saved. A scale for evaporation rate is included, permitting either gas quantity or boiler capacity to be used. The final graph (C) converts the heat saved into annual money savings based on 6 000 hours operation. To obtain the most realistic figure, the fuel cost used should be the mean value during the economical life of the plant.

Another area of application, where a hygroscopic wheel can be used is in the printing industry, (ref. 7.18). In this instance a hygroscopic rotating regenerator was installed in the exhaust from coating and gravure machines at Harrison & Sons Ltd. These machines use large quantities of hot air for evaporating off solvents in the inks and coating materials. Heat is frequently wasted by directly exhausting the air to atmosphere following passage through the printing machine, although in some cases solvent recovery is practised.

In this particular factory, the air supply to the coating and gravure installation is heated by circulating heat transfer oil through finned tubes over which the air flows. The maximum heater output is 12 700 MJ per hour, at which rate consumption is 0.39 m^3 per hour of 3500 sec. fuel oil. By recovering 70 per cent of the heat in the exhaust air, the fuel bill can be reduced by £600 per week, based on current fuel oil prices. Two regenerators are used in the plant, and the location in one machine is shown in Fig. 7.15. They have diameters of 2 m and are 280 mm thick. The matrix is asbestos fibre sheet impregnated with lithium chloride. The hygroscopic properties attract the moisture, which is transferred into the preheating duct, and a purge section is fitted.

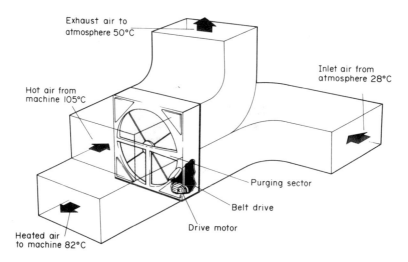

Fig. 7.15 Rotating regenerators used to recover heat from a coating and gravure plant in a printing works
(Courtesy The Engineer)

The total installation cost was £42 000, of which £11 000 is attributable to thermal wheel costs. Much of the ductwork used, which is included in the above cost, would be necessary to remove exhaust even if heat recovery was not used. In this case the payback period on the investment is substantially less than two years.

Independent detailed assessments of the performance of rotating regenerators are difficult to find, but one such study has been carried out at the University of Manitoba as part of the ASHRAE programme to develop standard procedures for rating air-to-air heat exchangers, (ref. 7.17). The work is directed at air conditioning applications, particularly in hospitals where the carry-over between air streams is of particular interest. No detailed results have yet been published, but trends indicate that peak efficiency (70 to 75 per cent) is achieved with rotational speeds in excess of 16 rev/min, and efficiency increases as the ratio of exhaust to supply air flow decreases from 1.0 to 0.5. The results of the study will probably be incorporated in future ASHRAE Standards, and it is expected that a more precise expression for the efficiency of this type of heat exchanger will be one outcome of the work.

7.5 Plate Heat Exchangers

Plate heat exchangers are becoming increasingly popular in a wide range of industries, in particular the food processing industry. The operation of this type of heat exchanger is illustrated in Fig. 7.16. The primary and secondary liquid flows are separated by a thin metallic membrane, which is corrugated to achieve maximum heat transfer by creating turbulence, as shown in Fig. 7.17.

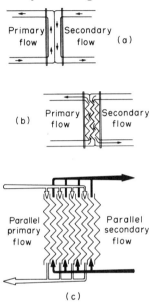

Fig. 7.16 Operation of a plate heat exchanger
(Courtesy Marine and Industrial Heat Ltd.)

Each plate is provided with a gasket which effectively seals in the fluid being heated or cooled. A complete installation involves a number of these plates, suspended on a frame, and compressed against a fixed rigid end plate by a moveable plate at the other end of the heat exchanger.

The main competitor of the plate heat exchanger is the conventional shell-and-tube type. While the latter type is well-established and has been improved considerably in terms of heat transfer efficiency in recent years, it has suffered from disadvantages associated with its size and poor liquid distribution. Cleaning of shell-and-tube heat exchangers can also be difficult, and because of the reversals in flow direction in the passes in the shell-and-tube unit pressure drops can be quite high. Undoubtedly the shell-and-tube heat exchanger is ideal for high pressures and arduous duties, but plate heat exchangers, once assembled in a supporting framework, can withstand working pressures of the order of 20 bar. A comparison of the heat transfer surfaces and the pressure drops through these two types of heat exchanger is given in Table 7.4, (ref. 7.19). These results, based on case studies made in the United States, show the significant reductions in surface possible with plate units, and with one or two exceptions, the lower pressure drop obtained with this type of heat exchanger.

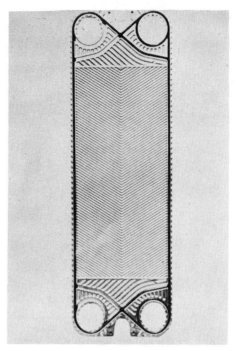

Fig. 7.17 Corrugation on a plate, used to promote turbulence
for good heat transfer
(Courtesy Robert Jenkins Systems Ltd.)

The size of plate heat exchangers is almost infinitely variable, and the unit can be increased or reduced in capability by changing the number of plates, which can amount to several hundred. A single frame can accommodate liquid throughputs of the order of 1000 m^3 per hour maximum at operating temperatures approaching 300°C. Stainless steel is the most common plate

TABLE 7.4 Comparison of Plate and Tubular Heat Exchanger Surfaces and Pressure Drops

Fluid A kg/s	Fluid B kg/s	Duty °C	Tubular Pressure Drop (kN/m^2)			Plate Pressure Drop (kN/m^2)		
			m^2	A	B	m^2	A	B
1.69 (hydrocarbon)	6.72 (water)	160A49 49B33	28	2.8	20.6	8.4	2.1	15.2
109 (water)	9.84 (seawater)	42A28 37B22	1130	131	131	330	74	61
7.42 (petrol)	2.52 (water)	105A35 35B26	84	22	31	26	9.7	31
2.2 (solvent)	24.3 (water)	60A40 35B26	170	20	31	41	21	24
18.8 (effluent)	59.5 (salt water)	106A38 41B20	140	72	55	36	31	69

material, but titanium, hastelloys, aluminium and tantalum are also available for special applications. While the plates used for primary and secondary fluids are identical, unequal primary and secondary flow rates can be accommodated.

One heat recovery application where plate heat exchangers have been successfully used is illustrated in Fig. 7.18. The APV Company supplied the plate heat exchanger for this installation (ref. 7.20), applied to large scale evaporation plant at the Marchon Division of Albright and Wilson Ltd., at Whitehaven in Cumberland.

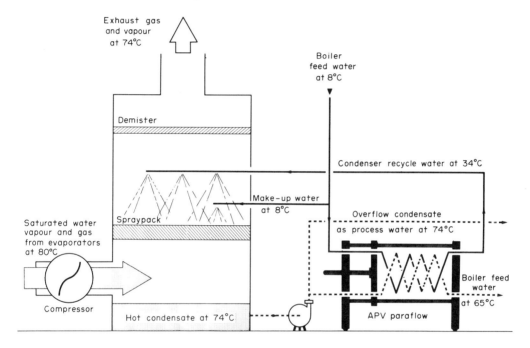

Fig. 7.18 An APV plate heat exchanger using waste heat to preheat boiler feedwater. (APV Company Diagram)

In 1964 a Paraflow type R55 plate heat exchanger was supplied for the purpose of providing hot water for the boiler feed. The heat source is exhaust vapour from two forced circulation evaporators which is ducted to a blower unit feeding a condenser. Hot condensate at 74°C is cycled through the Paraflow to heat the boiler feed water from 8°C to 65°C before returning to the condenser at 34°C. Overflow from the hot condensate is employed as hot process water.

The Paraflow has 106 stainless steel flow plates that give a total heat transfer area of 55 m^2, the whole unit occupying a space of only 3 x 0.75 x 2 metres. Nevertheless the division's services engineer estimated that it could supply approximately 750 000 litres per day of hot water, or the equivalent of half a boiler. The Paraflow has never needed to be opened up.

The original outlay on the installation, which included the Paraflow, its associated piping, the blower unit, the spray pack and tower, was £9200.

A more detailed quantification of the performance of plate heat exchangers as an aid to heat recovery can be made with reference to the dairy industry (ref. 7.21). Their use is widespread in a number of dairy processes, but it is in pasteurisation of milk where they have brought the greatest benefits. Fig. 7.19 shows the flow path in the pasteuriser. As the milk has to be cooled prior to being bottled, it is possible to preheat the milk entering the pasteuriser using heat from milk leaving the process for storage prior to bottling. In this way heat which would normally be rejected to cooling water is recovered and used effectively to reduce the energy expended in the pasteurisation process by almost 90 per cent. In the example shown, the quantity of regeneration is 86 per cent.

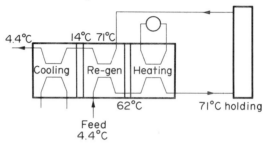

$$\% \text{ Regeneration} = \frac{62-4.4}{71-4.4} \times 100 \simeq 86\%$$

Fig. 7.19 The flow path in a dairy pasteurising plant
(APV Company diagram)

Obviously, by increasing the number of plates in a heat exchanger, one could theoretically approach a recovery efficiency of 100 per cent, but this would require an infinitely large heat exchanger owing to the lack of any temperature driving force available. This is illustrated in Fig. 7.20, which shows that while recovery of slightly over 90 per cent of the heat is possible while retaining a heat exchanger of sensible proportions, the number of plates needed for any further improvement makes further regeneration unrealistic. In order to assess the correct amount of regeneration in any particular process, account must be taken of the costs of energy, the capital cost of the plates, and pumping power. The effect of increasing energy costs in pasteurisation illustrates how this works. Until recently it has been standard practice to use regenerators having efficiencies of 85 per cent, but now it is worth the increased capital expenditure to install additional plates to give regenerator efficiencies in excess of 90 per cent. Fig. 7.21 shows how the operating costs of a pasteurisation plant capable of treating 5.7 litres per second of milk varies with the quantity of heat recovered. The degree of regeneration for minimum operating cost, as indicated on Fig. 7.20 is not achievable in practice for reasons of increased pressure drop etc., but a value of 92 per cent is feasible.

The heat transfer coefficients obtained with plate heat exchangers vary between 2500 k cal/m^2/hr/$^\circ$C and 6000 k cal/m^2/hr/$^\circ$C, and plates have typical surface areas of 1 m^2. Heat losses are negligible, as only the edge of the plate is exposed to the air, and no thermal insulation is necessary.

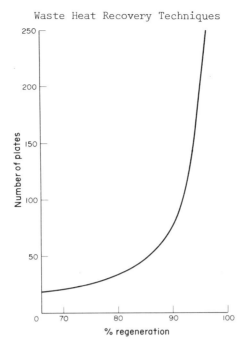

Fig. 7.20 Relationship between regeneration efficiency and number of plates. (APV Company diagram)

Wherever heat can be usefully recovered from a liquid process stream and used without being upgraded (i.e. if the liquid temperature is higher than about 45°C) some form of heat exchange is worth considering, and the plate type appears attractive, certainly from a first cost point of view.

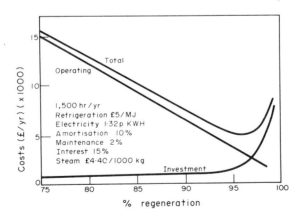

Fig. 7.21 Effect of heat recovery rate on the costs of a pasteurising plant. (APV Company diagram)

In comparing plate heat exchangers with equivalent shell and tube units, it must be appreciated that materials of construction should be taken into account - plate heat exchangers generally are of stainless steel or more expensive materials, dictated by the requirements of the application. A selection of plate heat exchanger manufacturers is listed in Appendix 4.

7.6 Economisers

An economiser is commonly regarded as a waste heat recovery unit unique to boilers, although now its application has spread to other processes where hot flue gas could usefully be harnessed to heat water or raise steam.

The majority of economisers are constructed for boiler applications, where they are used to preheat the boiler feedwater using the flue gases. In a boiler the heat transfer from the hot gas to the boiler tubes is comparatively poor, and as a result there is a very large temperature difference between the boiler shell and tube sides. This necessitates high flue gas temperatures, and by installing an economiser in the flue, some of the heat in this gas can be recovered. Where one economiser is insufficient to recover all the useful heat, an additional unit, or a sub-economiser, may be incorporated downstream of the main unit.

Economisers are normally constructed of steel, although originally cast iron was used as tube material. Most economiser tubes are finned, the steel fins being welded to the tubes. Their construction is much more substantial than on many other types of recuperative heat exchanger; tubes having outside diameters of 50 mm with approximately 1 m^2 of extended surface per metre length weighing over 10 kg per metre. Economisers used on large power stations can contain 100 000 m of tube, and are able to operate in conditions where flue gas temperatures approach $900^\circ C$, although in large power plant the temperatures are considerably lower than this.

The two main points to be borne in mind when using economisers are associated with the highest temperatures met on the tube side, and the lowest encountered on the gas side, (refs. 7.22, 7.23). 'Water hammer' can occur at the hot end of the economiser, created by the formation of steam pockets. Alternatively, if too much heat is removed from the flue gas, and condensation on the gas side occurs, this can lead to severe corrosion, which is more detrimental to the steel units than older cast iron economisers. It can lead to uneconomic operation if these restrictions are permitted to dictate the operating conditions of the economiser, and it is worth remembering that a $6^\circ C$ rise in boiler feedwater temperature can save 1 per cent on fuel consumption for the same boiler output. Alternatively, it may be stated that an increase in water temperature of $1^\circ C$ via the economiser lowers the flue gas temperature by $2^\circ C$.

Techniques for improving economiser performance, and at the same time overcoming the above problems, are illustrated in Figs. 7.22 and 7.23, (ref. 7.22). When the economiser was raising the feedwater to too high a temperature, the feedpump was kept at full capacity and some of the hot water from the economiser was allowed to flash into a 34 kN/m^2 process main, (Fig. 7.22). A relief valve at the economiser outlet controlled the flow to the flash tank, where the high temperature water was immediately flashed down to a lower temperature, resulting in about 10 per cent evaporation. Approximately 5 per cent of the total steam requirement was generated in this way, and additional economies of 15 per cent resulted from return of extra condensate.

Where condensation was occurring on the economiser tubes, a proportion of the output from the economiser was returned to the inlet to preheat the make-

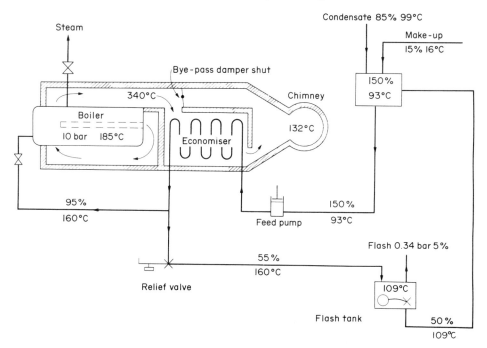

Fig. 7.22 An economiser installation with control of maximum feedwater temperature. (Courtesy HMSO)

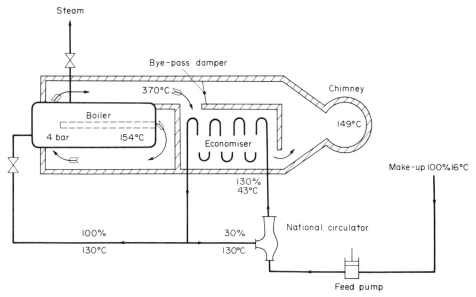

Fig. 7.23 Prevention of condensation in an economiser by raising the operating temperature. (Courtesy HMSO)

up before it entered the economiser. With a faster circulation rate the stack temperature was reduced sufficiently to effect an additional 3.5 per cent recovery without a return to conditions leading to condensation.

In large power generating plant, economisers operate at high temperatures and pressures, heating feedwater to temperatures of 200 to 300°C. However, in smaller boiler plant, where normal feedwater temperatures of only 40°C or thereabouts may be common, substantial economies can still be obtained using these heat exchangers, (ref. 7.24). Take as an example a boiler operating at 690 kN/m^2 without a superheater, using feedwater at an inlet temperature of 49°C. The total heat in the steam above that of the feedwater will be 2558 kJ/kg. An economiser installed in this boiler, heating the feedwater from 49°C to 127°C, would add approximately 424 kJ/kg, resulting in a saving in fuel of the order of 12.5 per cent.

As mentioned in the introduction to this section, the economiser, while normally associated with boiler heat recovery, can be used in any suitable exhaust gas stream as a process water heater, super-heater, or heater for liquids other than water, including heat transfer fluids. Economisers are being increasingly used in waste incineration plant, and furnace flue gases are used to generate process heat via economisers.

In gas turbines, the injection of water leads to a very significant reduction in the amount of air to be compressed, with a resulting reduction in its sensitivity to pressure ratio changes. The water also controls the maximum temperature, and is in effect part of the working fluid, (ref. 7.25). By using an economiser on a gas turbine, as shown in Fig. 7.24, this injection water can be preheated, and the economiser remains effective at lower temperatures than the gas-to-gas regenerator.

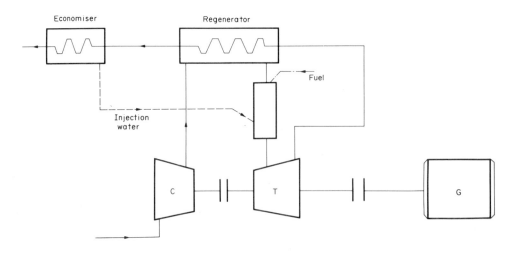

Fig. 7.24 Use of an economiser on a gas turbine to preheat injection water

The scope of application for the economiser is so wide that its association with boilers can be over-emphasised to the detriment of its other applications as a waste heat recovery unit. Several manufacturers are able to illustrate widely diverse successful applications, and some of the companies skilled in assembling these units are listed in Appendix 4.

7.7 Waste Heat Boilers

In a direct fired boiler the primary objective is to produce steam or hot water by using heat produced within the boiler by the combustion of a solid, liquid or gaseous fuel. A waste heat boiler, however, does not possess its own combustion system, but is supplied with hot gases from reactions taking place elsewhere, for example in furnaces. As with the economiser, described in Section 7.6, the term 'waste heat boiler' has evolved to cover a device which is capable of functions other than steam raising. Many waste heat boilers may not involve phase change, and in this respect the unit has much in common with an economiser. In this section, however, it is intended to concentrate on the steam-raising function of waste heat boilers.

The use of waste heat in exhaust gas streams to generate steam has a number of advantages compared to, say, gas-to-gas heat recover, and these are listed below.

(i) As the boiling process involves very high rates of heat transfer in terms of heat flux per unit area, the waste heat boiler is one of the most compact types of heat recovery unit.

(ii) Installations are in most instances lower on capital cost than other heat recovery systems of comparable duty.

(iii) Waste heat boilers can in general withstand high temperature exhausts without incurring problems of materials selection and life because the high heat transfer coefficients maintain the tubes at a comparatively low temperature, close to that of the fluid being boiled.

(iv) Response rate is high, and the duty may be conveniently varied by adjusting operating pressure on the steam side.

(v) Precise control of gas and water flow rates is not demanded.

The advantages above must be weighed against some of the drawbacks of steam raising. Assuming that a use exists in the plant for high grade steam, the need for good quality boiler feedwater implies the use of some water treatment process, which must be accounted for in capital and operating costs. It will also be found that the waste heat boiler alone will not recover sufficient heat to reduce the exhaust gas temperature as much as desired, and additional heat recovery equipment may be needed downstream of the boiler.

The relative merits of the various types of waste heat boilers are discussed below. In Appendix 4, a selection of manufacturers is given, covering most applications of these units.

7.7.1 Classification of waste heat boilers

Waste heat boilers may be classified in a similar way to direct fired conventional boilers, the three basic types being:

(i) Firetube boilers

(ii) Watertube boilers

(iii) Packaged units.

The firetube waste heat boiler raises steam by passing the exhaust gases which are used as the heat source through the boiler tubes. The water in the boiler is located outside the tubes. Firetube boilers, by their nature, have properties which indicate the most suitable applications. Their construction limits steam pressures to approximately 7000 kN/m^2, although developments should enable the operating pressure to be almost doubled (ref. 7.26), and gas-side pressure drops are higher than in watertube types. Where exhaust gas temperatures are very high, the tubeplate at the gas inlet end tends to be very vulnerable, and may limit life between major overhauls.

On the credit side, the firetube boiler is generally the least expensive of the two systems, and because it is less susceptible to fouling on the gas-side, can handle contaminated exhaust streams with relative ease. It is also much easier to clean the inside of the tubes should excessive fouling occur, than to clean external surfaces on a closely packed tube bundle. In an application where the exhaust gas pressure is very high (50 000 kN/m^2), the firetube boiler scores because the gas is contained within the tubes. However its efficiency can be low unless the exhaust gas being cooled has an acceptable film heat transfer coefficient.

A watertube waste heat boiler differs from the firetube design by the fact that the water is contained within the tubes and the exhaust gas flows over the external tube surface. As a result the range of application of the watertube is considerably greater, particularly at the large end of the scale, where very high duties and steam pressures are required. Its efficiency is substantially higher than the firetube type, particularly as extended surfaces can be added to the outside of the tubes to aid heat transfer from the gas in instances where the gas has poor heat transfer capabilities, or where the allowable pressure drop may be sufficiently low to restrict gas velocities.

The reliability of watertube boilers is commonly higher than that of their firetube counterparts, but liability to external fouling of the tube surfaces, and the increased difficulty of cleaning, must be taken into account when assessing the type of unit required in any application. This should also be considered should supplementary firing be necessary. (In this respect, the firetube type does not lend itself so well to supplementary firing as the watertube boiler).

As the watertube boiler flow paths are identical to those in superheaters, generators and economisers, coils fulfilling these latter functions can be incorporated in this type of waste heat boiler with little difficulty, extending its heat recovery capability substantially.

In the context of waste heat recovery, the packaged boiler is at present unique to heat recovery from water-cooled prime movers such as diesel and natural gas reciprocating engines. In the packaged boiler, heat from both the exhaust gas and the engine cooling jacket is combined to generate steam, the cooling jacket acting in effect as a feedwater heater. These may be used in conjunction with the total energy units described in Chapter 2, and their capabilities are discussed later in this chapter.

Waste Heat Recovery Techniques

7.7.2 Applications When selecting waste heat boilers, two major functional areas must be taken into account. These cover the uses to which the steam raised in the boiler will be put, and the function of the waste heat boiler with respect to the gas from which the heat is being extracted. The latter category involves three different kinds of processes, as follows, (ref. 7.27).

(i) The exhaust gas passes directly to waste. In this case the sole purpose of the waste heat boiler is to minimise heat losses in the exhaust before it is discharged to atmosphere. Boilers located behind many melting furnaces and prime movers such as gas turbines are examples of this area of application.

(ii) The exhaust gas requires cooling before it undergoes further processing. As well as raising steam, the boiler here performs a useful function in cooling the exhaust for use further downstream. Pyrite roasting plant and sulphur combustion plant are examples where the process stream requires the boiler to perform such a function.

(iii) As in (ii), but the stability of the processes is a function of the rate at which the exhaust gas is cooled. Here the waste heat boiler is also acting as an integral part of the overall process control system. Cracked gas coolers in ethylene plant are an important application satisfied by the use of waste heat boilers performing this function.

On the 'steam' side, the waste heat boiler has three principal uses:

(i) Provision of steam for process or space heating. In this application the steam raised will normally be at a comparatively low pressure.

(ii) Steam for power generation. This necessitates the raising of very high pressure steam, commonly superheated, and is a feature of combined steam and gas turbine cycles (Chapter 2).

(iii) Application as a stripping medium in a process, (little utilisation).

7.7.3 Waste heat boilers in gas turbine sets The need to make full use of heat recovery systems in order to economically justify 'in-plant' electricity generation or the use of prime movers to drive compressors and similar units has already been stressed. This and the subsequent section describe the performance of waste heat boilers on such installations.

As detailed by one of the waste heat boiler manufacturers, (Conseco - see Appendix 4), the efficiency of a gas turbine generator set by itself is only 25 per cent. Hot water heat recovery systems and low pressure steam systems recover up to 75 per cent of the available heat in the turbine exhaust and can produce an overall efficiency of 80 per cent. The application of a high pressure steam waste heat boiler results in recovery of up to 65 per cent of the available heat, giving an overall efficiency value of 72 per cent.

The potential heat recoverable from turbine exhaust gas is approximately 10 kW/kg/sec per degree C temperature drop in the exhaust gas. With an exhaust mass flow of 15.4 kg per second and a gas temperature drop through the heat exchanger from 554°C to 160°C, the recovery rate is approximately 6.1 MW. As a comparison, if the equivalent fuel is used in a fired boiler having an efficiency of 80 per cent, the heat of combustion is slightly greater than 7.9 MW. This heat release rate is equivalent to 14 litres per minute of fuel oil or 720 m³ per hour of natural gas.

This particular manufacturer estimates a payback period on the heat recovery equipment of about three years, taking into account operating and maintenance costs.

In the gas turbine waste heat boiler illustrated in Fig. 7.25, a supplementary firing capability is provided (ref. 7.23). This releases the particular process for which the steam is being raised from its dependence on the performance of the gas turbine, and also allows boiler efficiency to be raised. (Normally the waste heat boiler for a gas turbine is sized on the basis of the exhaust flow rather than the required steam output, making the unit relatively expensive. Thus any means whereby its output could be greatly increased with little additional expenditure is of interest). As considerable amounts of oxygen still exist in the exhaust gases, this can be used to assist supplementary firing, and in some cases it is possible to increase the steam raising capacity over and above that produced by the unfired boiler by 200 per cent. As a result a very large gain in boiler output is achieved at the expense of the installation of simple grid burners. (In the unit illustrated, the bypass stack is included to permit boiler operation at lower loads than those dictated by exhaust mass flow, if the need to reduce steam production arises).

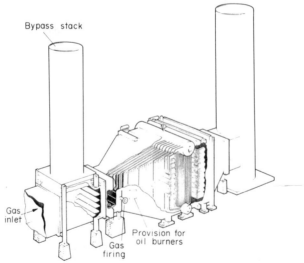

Fig. 7.25 A waste heat boiler used to recover heat from a gas turbine exhaust, aided by supplementary firing

7.7.4 Steam raising using heat from reciprocating internal combustion engines

The arguments in favour of using as much of the waste heat as possible from engines such as diesel and gas driven reciprocating units are similar to those applying to turbines, and some of the systems used for this are briefly

Waste Heat Recovery Techniques   231

described in Chapter 2. The Vaporphase system, manufactured in the United
States by Pott Industries Inc., is one of the most popular types of waste heat
recovery units for reciprocating engines, and their solutions are appropriate
to many installations of varying sizes. Vaporphase heat recovery systems
range in size from 100 kW units to 15 000 kW installations serving very large
16 cylinder multiple unit generating sets.

While the waste heat boiler and jacket heat recovery units can be separate,
the packaged system manufactured by Pott Industries combines both functions in
a single boiler, as shown in Figs. 7.26 and 7.27. In common with many other

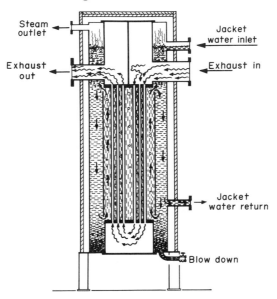

Fig. 7.26 A vertical 'Vaporphase' waste heat recovery system
           serving internal combustion engines

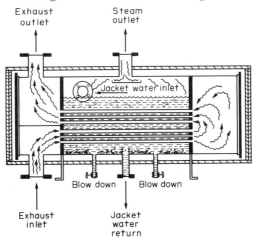

Fig. 7.27 The horizontal 'Vaporphase' waste heat boiler

waste heat boilers for engines, the unit also acts as an engine silencer.

Efficiencies of up to 75 per cent are claimed for the Vaporphase system, and the convenience associated with a single unit for all heat recovery functions is very attractive. A multiple installation serving three generating sets is illustrated in Fig. 7.28.

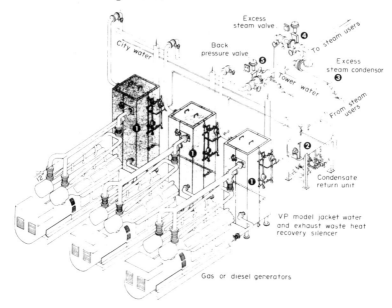

Fig. 7.28 Multiple installation for waste heat recovery from three generating sets

7.8 Recuperators

A recuperator is a gas to gas heat exchanger normally employed to recover waste heat from hot waste gases. Its widest application is as a preheater for air entering boilers and furnaces, and in this respect is a direct competitor of the rotating regenerator, described in Section 7.4, and, to a lesser extent, heat pipe exchangers.

The advantages claimed for the use of recuperators are several, and are listed as:

(i)   Reduction in fuel requirement

(ii)   Reduction in the quantity of excess air

(iii)   Higher flame temperatures are possible

(iv)   The quantity of unburnt fuel is reduced

(v)   Combustion proceeds more rapidly

(vi)   The oxidisation of stock is minimised.

There are two basic types of recuperator, those which rely on convective heat transfer, and units which transfer heat via radiation. The latter tend

Waste Heat Recovery Techniques

to be used at the high temperature end of the scale.

7.8.1 Convection recuperators Convection recuperators can be used where gas approach temperatures are less than 1000°C, although higher temperatures can be permitted if special materials and construction techniques are used. Originally all recuperators were made from ceramic materials, but these units suffered from serious leakage problems, and have today been largely superceded by metallic recuperators.

There are two basic types of convection recuperator, those which use cast tubes and those using drawn tubes which are assembled in bundles, in common with many other types of heat exchanger, (ref. 7.29). The use of cast tubes is normally recommended for low pressure applications, where leakage is unlikely to be a significant problem. It is most common to bolt each tube to the header boxes, and thus tube replacement is relatively easy.

The tubes used are available either plain or with a wide variety of extended surface configurations. Wide pitching of surface projections is used when the exhaust gas flow may be heavily contaminated. In some cases, the outside of the tubes may be left completely bare. However the overall heat transfer through a tube of this type may be maintained at an acceptable level by retaining the extended surfaces on the inside of the tube, through which the air to be heated is passed.

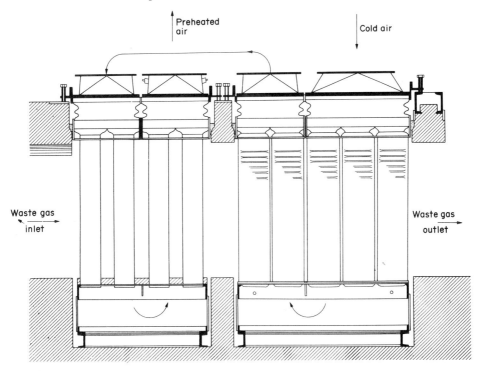

Fig. 7.29 A 4-pass cast composite tube recuperator installation (Courtesy Thermal Efficiency Ltd.)

An alternative method of location used on cast tube recuperators involves the use of end flanges with integrally cast steel expansion joints. This permits the tubes to be welded together to form tube banks of any size. A 4-pass horizontal flue cast composite tube recuperator designed by Thermal Efficiency Ltd. is illustrated in Fig. 7.29. It can be seen that both bare and externally finned tubes are used in this installation. These rows of plain spun alloy steel tubes, containing a percentage of chromium and nickel, are arranged in front of the composite tubes to safeguard the latter from localised heating, created by non-uniform and excessive radiation. The chromium provides a resistance to oxidation at high temperatures, and the nickel content improved ductility in areas where high thermal stresses are likely to be encountered. Typical uses of these units include soaking pit and re-heat furnace recuperation, where an additional requirement is for resistance to abrasive and sintered dust laden gases. Although tube replacement is not as conveniently carried out as with bolted units, each tube can be removed from the bundle once the weld beads have been ground off.

Composite tube recuperators are used exclusively as convection type heat exchangers with waste gas temperatures of up to $950^{\circ}C$. Using the plain spun tube system described above, the temperature range can be slightly extended. Drawn steel tube recuperators are available in many forms. Each tube bundle is attached to header boxes, and the construction technique used allows the tubes, and individual tube bundles, to expand relative to one another. In some systems the tubes are also bent at their mid point to minimise stresses arising from thermal expansion. These recuperators are often used where it is required to recover a considerable proportion of the radiation heat, and the tubes are generally not finned. However conduction through the wall is enhanced by the fact that whereas cast recuperator tubes have wall thicknesses of the order of 8 mm, drawn tubes may have a thickness of only 3 mm.

The discussion so far has concentrated on units in which the exhaust gas flows outside the tubes in a single pass, while the air passes inside. However, 'flue-tube' recuperators, which resemble shell and tube heat exchangers, have been used where dirty gases are involved. As it is easier to clean the inside of a tube, the dirty exhaust gas was passed through the inside of the tubes, while the air to be preheated was circulated across the tubes via a series of baffles.

It has been stated above that metallic recuperators have largely superceded the refractory type. However, high pressure/high temperature ceramic recuperators are being developed for some specialised applications, and the reader will find data on one of these units, developed by the British Steel Corporation, in Chapter 3. A number of manufacturers are able to supply ceramic recuperator tubes capable of operating at exhaust gas temperatures of up to $1800^{\circ}C$. It is claimed that the sealing problem on this type has been overcome using a ceramic fibre packing based on aluminium silicate. An example of such a recuperator, with tubes of 'Carbofrax' and 'Refrax' produced by the Carborundum Company, connected directly to the outlet of a high temperature kiln, is designed to accept 0.86 m^3 per second of gas at $1800^{\circ}C$, giving an air pre-heat temperature of $1200^{\circ}C$ and a heat transfer rate of 104 kW.

For much less arduous duties, some very simple recuperative heat exchangers are available. The Econovent EX static heat exchanger, illustrated in Fig. 7.30, has been developed by the Munters organisation of Sweden to recover heat from wet or saturated exhaust air and gases, which can contain large quantities of latent heat. This company also produce rotating regenerators,

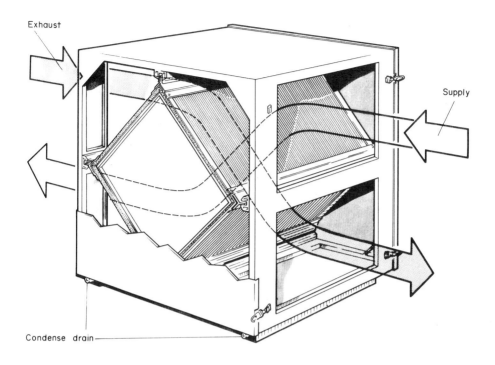

Fig. 7.30 The Econovent EX heat exchanger, typical of a number of simple, low cost recuperators available

and the static unit overcomes carry-over problems associated with this type of heat exchanger. The Econovent EX, it is claimed, will recover as much as 70 to 75 per cent of the sensible and latent heat in the exhaust gas stream, and this heat is given up to adjacent narrow ducts carrying the air to be heated, thus the unit in some ways resembles a plate heat exchanger used for gases. This unit is constructed for modular installation, each module handling 0.5 to 1.5 m^3 per second of gas. Collection and draining of condensate is also accomplished within the heat exchanger.

An equally simple recuperative heat exchanger which can be located in ducts and flues to recover heat for space heating or process use is illustrated in Fig. 7.31. It is based on the core of a gas fired heater manufactured by ITT-Reznor, and the core design is such that no cross-contamination can occur. The performance of the ITT unit is inferior to several other types of heat exchanger, (typically 35 to 55 per cent recovery) but the unit is cheap and simple to maintain.

7.8.2 Radiation recuperators

A radiation recuperator takes the form of two concentric cylinders, the air to be heated normally flowing through the outer annulus, while the exhaust gases flow through the central duct. Alternatively the unit may be built up with tubes between two headers. A radiation recuperator of this type is illustrated diagrammatically in Fig. 7.32.

Fig. 7.31 A recuperator manufactured by ITT-Reznor, used for space heating

Fig. 7.32 A tubular shell radiation recuperator, used in high temperature gas streams, (1400°C preheating to 700°C) (Courtesy Thermal Efficiency Ltd.)

Waste Heat Recovery Techniques 237

Compared with the convection recuperator, the radiation type offers very low resistance to gas flow and in most instances never needs cleaning. The dirtiest of exhaust gases can be permitted through it, and by its nature this type of recuperator can also act as part of the chimney or flue.

The size of a radiation recuperator can vary considerably, the largest units being about 50 m long and 3 m in diameter. Radiation recuperators can be constructed using tubes to separate the exhaust gas and air, rather than a single annulus. These are used in instances when preheat temperatures in excess of 600°C are required. It is preferable to use parallel flow of air and gas in these, as the tubes tend to be subjected to near equal temperatures along their whole length, thus keeping temperature-induced stresses to a minimum. A typical application of this type of unit is in glass-melting tank installations.

The radiation recuperator is generally regarded as being the most reliable of the two main types available, and has the longest life.

As well as applications involving heat recovery from boiler and furnace exhausts, the radiation recuperator may be used in conjunction with a radiant tube heater, forming a self-contained radiant tube heater and recuperator unit, as shown in Fig. 7.33. The recuperator replaces the normal exhaust stack on individual radiant tubes, absorbing heat which is then used to preheat combustion air for the burner. Typical savings on a gas-fired radiant tube can be

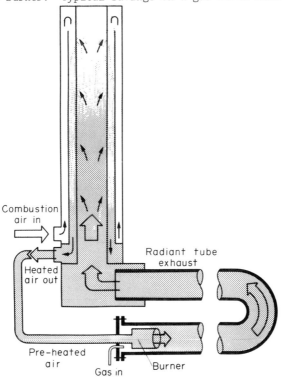

Fig. 7.33 A unit combining the functions of radiant tube heater and recuperator - the Gibbons-Holcroft Radiant Tube Recuperator.

over 20 per cent in terms of reduced gas volume requirement. Chrysler Corporation's New Process Gear Division have installed recuperators on their furnace radiant tube combustion system, and have achieved a fuel reduction of 27 per cent (ref. 7.30). At the same time productive capacity has been raised by one third and thermal pollution has been reduced. This plant is one of the largest heat treatment facilities in the United States. The investment expenditure on the recuperators represented only 4 per cent of the total cost of the four-zone furnace, and the saving in gas has been quantified by Chrysler as 17 m^3 per hour. Even under standby conditions, Chrysler estimate that a 22 per cent fuel saving is obtained.

7.9 The Heat Pump

All the heat recovery systems described so far in this chapter are used to recover waste heat which is at a sufficiently high temperature to be re-used directly for space heating, preheating of combustion air, or in another process. Many of the sources of waste heat are at a very high temperature, particularly exhaust gases from furnaces, boilers etc. In other cases heat recovery can be usefully applied even though the heat recovered may be low grade, because the heat sink may not require raising more than a few degrees in temperature. Typical applications of this type include air conditioning and drying processes, where temperature differences between inlet and exhaust flows are low.

There are many instances in industry where the temperature of the medium (be it gaseous or liquid) carrying the waste heat, is at too low a temperature to be usefully employed elsewhere if recovered. In most instances liquid effluents, rather than exhaust gases, fall into this category, although the heat pump described below may be applied in either case. Shell and tube and plate heat exchangers, particularly the latter, may be operated with low liquid temperature differences (for example a plate heat exchanger can recover heat from hot condensate at 74°C to raise the temperature of boiler feedwater from ambient to 65°C), and these units are comparatively cheap and, if well-maintained, reliable. However in many industrial effluent streams, satisfactory use of the heat in the stream, which may be at as low a temperature as 25 to 30°C, can only be accomplished if this heat can be upgraded (i.e. raised to a higher temperature), and then directly re-used in the same process, or applied in another part of the plant where higher grade heat is required. Alternatively such recovered heat could be used for factory and office space heating.

The heat pump system is able to fulfil this function.

7.9.1 The ideal heat pump cycle

In a power cycle, heat is received by a working fluid, which then proceeds to do work, rejecting the heat at a lower temperature. In a heat pump or refrigeration cycle the reverse happens; heat is received at a low temperature, and work is then done on the fluid, which proceeds to reject the heat at a higher temperature. A refrigerator is associated with the removal of heat, or cooling, whereas a heat pump is designed with the principle purpose of supplying heat at a high temperature. Heat pumps are able to provide both heating and cooling, and are widely applied in air conditioning applications for this purpose as detailed in the preceding chapter. In this context, however, we are primarily concerned with the heat pump operating on a heating-only cycle.

The basic circuit of a heat pump consists of an evaporator, condenser, compressor and expansion valve, connected as shown in Fig. 7.34. The

Waste Heat Recovery Techniques

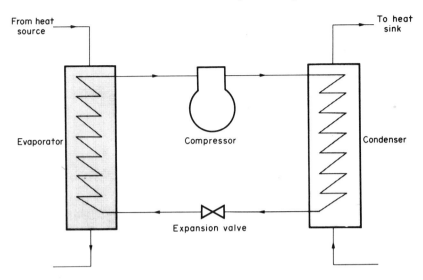

Fig. 7.34 The basic heat pump circuit

evaporator receives the low grade heat from the waste heat source, the heat being taken up into the heat pump circuit by evaporation of the working fluid being circulated. The resulting vapour is then passed through a compressor, where its pressure and temperature are increased, before being pumped to the condenser where it gives up the heat collected at the evaporator plus the heat equivalent of the work of compression. As this heat is rejected, the vapour condenses, and hot condensate then passes through an expansion valve, cooling in the process. The working fluid, commonly a fluorinated hydrocarbon, remains in a closed sealed circuit throughout the operation.

The ideal heat pump cycle is represented by a reversed heat engine operating on the Carnot cycle, illustrated in Fig. 7.35. The performance of the heat

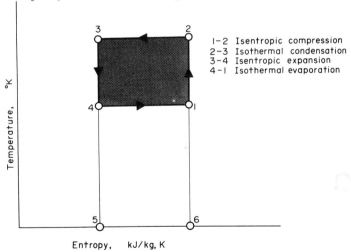

Fig. 7.35 The Carnot cycle - the ideal heat pump operating cycle

pump is commonly expressed as the ratio of the heat supplied at the condenser to the heat equivalent of the work put in at the compressor. Alternatively, the 'coefficient of performance' (COP), of the unit may be written thus:

$$COP = \frac{T_1}{T_1 - T_2}$$

where T_1 is the condensing temperature

and T_2 is the evaporating temperature,

both expressed in absolute degrees.

A typical temperature range over which a heat pump might operate is from 30°C to 80°C. The ideal coefficient of performance is therefore 353/(80 - 30), or 7.06. This means that 7.06 times the quantity of work supplied is delivered to the condenser as heat. In practice however, the heat pump cycle has a much lower coefficient of performance than that suggested by the reversed Carnot cycle, and is more accurately represented by the Rankine cycle, illustrated in Fig. 7.36. In an actual plant the work in expanding the work fluid would be dissipated as mechanical friction, and an expansion valve is used, resulting in the stage 3 - 4 lying on a constant enthalpy line. While no work is now obtained from the expansion process, this does not affect the heat extracted from the condenser, but reduces that taken up in the evaporator. The expansion valve makes the process irreversible. After expansion the working fluid is in the form of a saturated liquid, which then picks up heat in the evaporator, ideally being boiled off to a saturated vapour, absorbing the heat at constant temperature.

A practical heat pump cycle differs from the reversed Carnot cycle in two other respects. The compression is normally carried out in the superheat

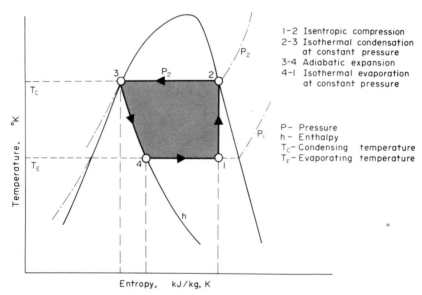

Fig. 7.36 The Rankine cycle, which more accurately represents practical heat pump operation

region, with desuperheating occurring in the condenser. Also, some additional liquid cooling (subcooling) occurs between the condenser exit and the expansion valve, prior to the adiabatic expansion between points 3 and 4, whence the cycle is repeated. In actual cycles the pressure of the fluid drops as it flows through the heat exchangers, but this is not shown on the cycle diagram in Fig. 7.37, (ref. 7.31).

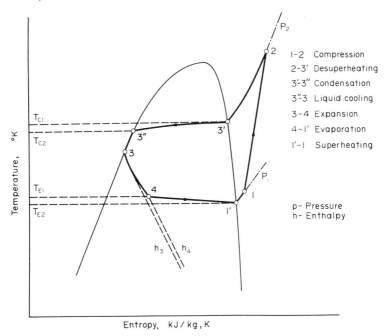

Fig. 7.37 Modified Rankine cycle, taking into account cycle inefficiencies

7.9.2 Heat pump sources The heat pump can operate successfully only if a suitable heat source is available to evaporate the fluid being circulated in the system. With a few notable exceptions, discussed later, the heat pump has to date been applied in situations where both heating and cooling are required, such as year-round air conditioning and space heating using waste heat rejected by refrigeration units.

It is in the United States where most of the major developments in heat pump technology have taken place, and while there have been periods in its evolution when the engineering of the concept has been called into question, particularly from the point of view of reliability, the use of heat pumps in the above applications is now commonplace. A substantial proportion of the market for these units is in domestic air conditioning, and Table 7.5 lists some of the manufacturers of these units, with data on duties.

We are primarily concerned with the heat pump working on a heating cycle, and the heat sources available are numerous, (ref. 7.32). In air conditioning applications, ambient outside air is the most common source of heat, being passed over the finned evaporator using a fan. The major drawback of such a heat source is its temperature variation. At the times when most heat is

TABLE 7.5 Heat Pumps for Domestic Air Conditioning

Manufacturer	ARI Standard Rating				C of P Heating
	Cooling		Heating		
	Capacity (W)	Power Input (W)	Capacity (W)	Power Input (W)	
Lennox Industries Inc.	7 000	3 800	7 000	3 200	2.18
(air source)	26 400	11 400	27 200	9 700	2.80
American Air Filter Company Inc.	2 780*	1 100	2 780	1 150	2.41
(water source)	14 600*	6 800	15 000	5 500	2.72
Carlyle (Carrier) Air Conditioning Ltd.	5 850	3 300	6 450	2 800	2.30
(air source)	34 300	14 700	34 300	12 100	2.84
General Electric	5 300	2 800	5 300	2 500	2.12
(air source)	35 000	15 000	35 000	12 600	2.78
Fedders Corporation	6 450	4 000	6 450	3 800	1.70
(air source)	16 400	8 800	17 500	7 200	2.43
York Division of Borg-Warner	5 600*	2 000	6 150	2 000	3.07
(air & water source)	14 600	7 900	15 800	6 600	2.40
Westinghouse Electric Corporation	5 300	2 900	5 000	2 400	2.08
(air source)	17 300	8 900	17 300	6 900	2.50
Low Impact † Technology	-	-	7 000	-	3.00
			38 000		3.00
Temperature Ltd.	1 460	850	2 280	900	2.53
(UK)	3 800	1 860	5 430	1 950	2.78

*Water source heat pump
†Figures are approximate

required, the ambient air will obviously be at a low temperature, and this often necessitates the use of a back-up heating system to provide a boost to the performance when the heat pump is unable to provide all the heating requirements. In an attempt to overcome this difficulty, a number of natural heat sources which do not succumb so markedly to seasonal or even daily variations have been tried. In the 1950's, it was popular to bury the evaporator coil a few feet below the ground surface, but the extensive amount of piping required to satisfy the heat requirements because of poor heat transfer coefficients between the soil and the evaporator was one of the reasons why interest in this system was not sustained. Solar energy is now increasingly of interest as a heat source for heat pumps in domestic applications, but as this is, like the ambient air, limited by the fact that it is available at times when heat is least needed, energy storage systems are normally also included in the 'package'. (A comprehensive list of references which will enable the reader to trace the evolution of the heat pump in more detail is given in Appendix 5).

Some heat pumps, as can be seen from Table 7.5, are designed with water as the heat source and/or heat sink. In a heating application water has proved more attractive than most in this respect as it is rarely subjected to the wide variations in temperature which restrict the use of ambient air heat sources. River water has been used in several well-documented applications of heat pumps in the United Kingdom, including the Royal Festival Hall, which used the river Thames as the heat source, (ref. 7.33). Of course, only a small proportion of potential heat pump applications could make use of naturally occurring water heat sources, but, as stressed in the introduction to this section, such heat sources are commonplace in industry, with important differences: they are normally at a somewhat higher temperature than river water and are even less subjected to seasonal changes in conditions. Also, the duty of any heat pump applied in such a situation will in general remain essentially constant insofar as industrial process requirements are concerned. (This would not of course be true of a heat pump used to provide space heating in an industrial context). It must be stressed that here we are primarily interested in waste heat from processes other than electricity generation, which is considered separately in Chapter 2, (ref. 7.34).

7.9.3 Selection of heat pump working fluid

Normally the working fluid used to transport the heat around the secondary circuit of a heat pump is called the refrigerant. In many cases the fluid is identical to that used in a conventional refrigerator, and in higher temperature heat pumps derivatives of these fluids, commonly denoted using the prefix 'R', e.g. R12, R21, and manufactured by companies such as ICI, Dupont and Imperial Smelting Corporation, are used.

A comparison of the vapour pressures of the various refrigerants, at evaporator and condenser temperatures, for a variety of applications ranging from low temperature food freezing to high temperature heat pumps is given in Table 7.6, (ref. 7.35). The only fluid in the listing not belonging to the classification described above is ammonia, which has too high a vapour pressure at heat pump temperatures to be of use in such systems. Pressure is not the only criterion which is used in determining the correct working fluid to be used in a particular application. The amount of heat which can be transported by the refrigerant is a strong function of its latent heat, or enthalpy, values of which are listed in Table 7.7. The specific volume of the working fluid determines the displacement of the compressor needed to deal with the flow of fluid necessary to meet the duty. Critical pressure and temperature influence the limiting conditions beyond which the fluid cannot be successfully used, and the miscibility of the working fluid with the compressor lubricating oil is a significant factor in the selection of fluids.

As well as the above properties, such considerations as cost have to be taken into account, although the market for most of the commercially available refrigerants is sufficiently large and competitive to keep prices within reasonable bounds. In addition, as the secondary circuit is in theory a closed system, loss of refrigerant should be minimal, and the quantity required in even a large system will not be a significant proportion of the capital or running

TABLE 7.6 Saturation Pressures with Various Refrigerants at Conditions Varying from Applications in Low Temperature Refrigeration to High Temperature Heat Pumps

Typical Application	Temperatures Evap.	Temperatures Cond.	Refrigerant	Pressures Evap. $kg/cm^2 abs$	Pressures Cond. $kg/cm^2 abs$	Remarks
Food Freezing	−40°C	+35°C	R12	0.65	8.6	Usually two stage compression
			R22	1.07	13.8	
			NH3	0.73	13.8	
Food Storage	−20°C	+35°C	R11	0.16	1.5 ⎫ Low pressure refrigerants
			R21	0.28	2.6	
			R114	0.38	3.0 ⎭	
			R12	1.54	8.6 ⎫ High pressure refrigerants
			R22	2.50	13.8	
			NH3	1.94	13.8 ⎭	
	−10°C	+35°C	R11	0.26	1.5	
			R21	0.45	2.6	
			R114	0.60	3.0	
			R12	2.24	8.6	
			R22	3.60	13.8	
			NH3	2.96	13.8	
Water Chilling	+1°C	+35°C	R11	0.43	1.5	
			R21	0.75	2.6	
			R114	0.94	3.0	
			R12	3.26	8.6	
			R22	5.25	13.8	
			NH3	4.56	13.8	
	+1°C	+50°C	R11	0.43	2.4	High condensing temp. from use of air cooled condenser
			R21	0.75	4.0	
			R114	0.94	4.6	
			R12	3.26	12.4	
			R22	5.25	20.0	
			NH3	4.56	20.7	
Heat Pump	+25°C	+70°C	R11	1.05	4.2	
			R21	1.83	6.7	
			R114	2.18	7.4	
			R12	6.6	19.0	
			R22	10.5	30.5	
			NH3	10.2	35.0	
	+25°C	+80°C	R11	1.05	5.3	
			R21	1.83	8.4	
			R114	2.18	9.5	
			R12	6.6	23.2	
	+25°C	+90°C	R11	1.05	6.7	
			R21	1.83	11.3	
			R114	2.18	12.3	
			R12	6.6	29.0	
	+25°C	+100°C	R11	1.05	8.3	See also Table 7.7
			R21	1.83	14.0	

Waste Heat Recovery Techniques

costs. Safety is an important factor; the refrigerant should be non-toxic, and must be compatible with the components of the circuit into which it is likely to come into contact. It must also not be inflammable. At present one of the factors which is limiting the use of heat pumps at temperatures above 110 to 120°C is the thermal stability of the working fluid. Decomposition can occur at temperatures below the critical temperature.

In addition to the more common refrigerants listed in the tables, azeotropic mixtures of 31/114 and 12/21 have been developed and patented by Allied Chemicals. Their capacity is some 60 per cent higher than R21, and both have critical temperatures in excess of 110°C.

7.9.4 Heat pump compressors The compressor is the most important single item in a heat pump installation, and is available in a number of forms, the three most commonly used in refrigeration practice being the reciprocating, screw and centrifugal types.

It is the reciprocating compressor which has been most popular to date in heat pump applications. It can operate at high pressures, and retains a reasonable efficiency when not operating at full power. There are three distinct forms of construction of the reciprocating compressor. Hermetic and semi-hermetic compressors are constructed with the electric motor drive enclosed in the same sealed unit as the compressor itself. This type of assembly was originally developed for refrigeration duties, and is comparatively cheap to manufacture, the need for rotating seals being eliminated. The sealed units are somewhat limited as far as operation at the higher temperatures required by industrial heat pump applications are concerned, and they are also limited by the choice of prime mover, this being in all cases restricted to electric motors. This can be overcome by using an open reciprocating compressor, which can be driven by any type of prime mover. (It may be necessary to employ a belt drive with the open system instead of direct drive, depending upon the compressor model and prime mover type).

The Westinghouse Templifier heat pump, described later in this section, uses a centrifugal compressor driven by a hermetic electric motor. This unit is available in a range of sizes up to 300 kW compressor power. It is based on the Westinghouse Model CE centrifugal compressor range which are single stage units with an internal hydraulically operated modulating control system. In general, centrifugal compressors are more expensive than their reciprocating counterparts, particularly at the low duty end of the scale. However, it can accommodate much greater loads than most of the commercially available reciprocating units, and has lower vibration levels than this type.

Screw compressors have considerable potential in the larger industrial heat pump applications. (The Grasso Monoscrew MS10, with a maximum discharge temperature of 90°C, has a capacity range of up to 1.5×10^6 kcal per hour. A single stage unit operating between temperatures of -10°C to 25°C has an output of 1.436×10^6 kcal per hour with a power input to the compressor drive of 361 kW.)

7.9.5 Practical industrial heat pumps There is a number of options available to the potential user of heat pumps for heat recovery in industry. Because of the importance of all forms of heat pump in the future to the overall energy conservation programmes, the systems are dealt with in some detail. Five different systems are described, three of which are based on electric motors as the compressor prime mover. One system is driven by an internal combustion

TABLE 7.7 Theoretical Performance of Refrigerants
at Typical Heat Pump Temperature Conditions

Evaporator 50°C   ) Assume compressor swept volume of 28.32 m³/hr (1000 cu ft/hr)
Condenser 110°C

REFRIGERANT		11	21	113	114	12	31/114	12/31
Evaporator Pressure	kN/m²	220	385	103	425	1190	758	1200
Condenser Pressure	kN/m²	1000	1610	560	1700	3950	2920	4140
Critical Temperature	°C	198	178	214	146	112	142	118
Compression Ratio		4.45	4.2	5.3	3.95	3.35	3.86	3.4
Specific Volume	m³/kg @ 50°C	0.0801	0.0619	0.1298	0.0320	0.0146	0.0324	0.0170
Weight circulated	kg/sec	0.098	0.126	0.060	0.245	0.530	0.242	0.450
Net Refrigerating Effect	kJ/kg	114.5	145.5	86.4	48.5	36.6	118.8	65.0
Heat of Compression	kJ/kg	27.9	35.0	25.6	18.6	18.6	35.0	25.6
Capacity	kW	13.9	22.8	6.76	13.0	28.3	37.6	41.7
Coefficient of Performance		5.1	5.2	4.35	3.5	2.9	4.48	3.55

The net refrigerating effect is obtained by subtracting the enthalpy of the refrigerant at the condensing temperature from the enthalpy of the superheated vapour entering the compressor at the evaporator temperature. The heat of compression is the difference between vapour enthalpies upstream and downstream of the compressor.

reciprocating engine, and the final unit described - at present largely hypothetical - is based on the use of a gas turbine drive. The selection of the best system is obviously a strong function of the type of application, and these applications are used to illustrate the functioning of each type of heat pump.

The electric heat pumps are described first, these being:

(i)   Liquid-liquid heat recovery
(ii)   Air-air heat recovery
(iii)   Combined refrigeration and heating.

Liquid-liquid heat recovery (electric heat pump): The use of electric-driven heat pumps to recover heat from waste water or other liquids for re-use in heating process streams, or heating water for space heating, is exemplified by the Westinghouse Templifier heat pump, the compressor/motor assembly having been described above. Templifier units are either single or two-stage heat pumps, based on centrifugal compressors, covering a range of heating capacities up to approximately 1.3×10^6 kcal per hour with liquid leaving temperatures (on the primary side of the condenser) of up to 110°C.

Figure 7.38 shows the Templifier in an application where water is to be delivered at 82.2°C using heat available from waste water being rejected at 32°C. In this instance two stages of compression are used, and the use of a flash collector tank from which a proportion of the vapour is bled to the inlet of the second stage compressor, acts to improve the process efficiency, (ref. 7.36). The coefficient of performance of the Templifier varies with

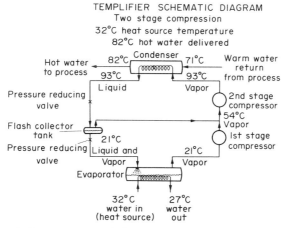

Fig. 7.38 The Templifier heat pump applied to recover heat from a waste water stream
(Courtesy of Westinghouse Electric Corporation)

conditions at the evaporator and condenser, and where the temperature differences between the heat source and heat sink are large, the two stage compression arrangement is normally recommended. The curves shown in Fig. 7.39 illustrate the COP's that can be obtained with the Templifier system when cooling water acting as the heat source by approximately 6°C and heating

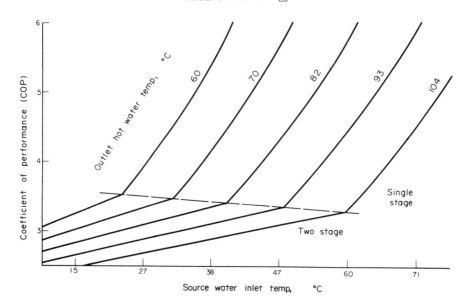

Fig. 7.39 Coefficients of Performance (COP) attainable with a Templifier unit in a water-water heat recovery situation (Courtesy of Westinghouse Electric Corporation)

process water through a range of approximately 6 to 20°C to the outlet temperatures shown. If a heat pump is required to provide process hot water at 82°C using returning water at 74°C, using waste water available at 43°C, from Fig. 7.39 a Templifier having a single stage compressor will meet this duty, and will have a COP of 3.7. The operating cost of the heat pump, assuming electricity is available at 1.5p / kW hr, can be compared to that of an oil fired process heater operating at a seasonal efficiency of 70 per cent.

Medium fuel oil cost	: 5.2p / litre
Electricity cost	: 1.5p / kW hr
Conversion efficiency of oil heater	: 70 per cent
Fuel oil heat content	: 7.56×10^3 kcal/litre
Heat pump COP	: 3.7
Total process heat required	: 500 kW hr

(i) <u>Direct electric heating</u>

Cost = 500 × 1.5 = £7.50

(ii) <u>Oil heating</u>

Cost = $5.2p \times \dfrac{500 \times 10^3 \times 0.86 \text{ kcal}}{7.56 \times 10^3 \text{ kcal/litre}}$ = £2.96

(iii) <u>Heat Pump</u>

Cost = $\dfrac{500 \times 1.5}{3.7 \text{ (COP)}}$ = £2.03

Thus in the particular example shown, the heat pump is more economical than oil heating as a means of providing the process heat. The comparison with direct electric heating shows that costs are reduced in direct proportion to the COP, but it is unlikely that many cases occur in industry where direct electric heating of a fluid can be replaced by a heat pump.

A further illustration of the Templifier performance may be given by relating heat outputs at the condenser to the power input required by the compressor motor. These are given in Table 7.8 for a single stage unit heating water from $54^\circ C$ to $65.5^\circ C$ using source water at $35^\circ C$. (In these instances the source would be cooled to $29.4^\circ C$). Four Templifier models, having motors ranging from 63 kW to 291 kW, are shown. The power required to overcome primary circuit pressure drops in the evaporator and condenser is not included in any of the above analyses.

TABLE 7.8 Example of Templifier Performance*

Templifier Model	Water Flow		Templifier Heating Capacity	Power Input	Approximate Operating Weight
	$35^\circ C$ Source	$65.5^\circ C$ Delivery			
	m^3/hr	m^3/hr	Thousands Kcal/hr	kW	kg
TP050	33	21	228	63	2300
TP063	71	44	486	132	5300
TP079	106	66	728	189	6700
TP100	169	105	1167	291	10100

*Nominal size single stage Templifier heating delivery water from $54^\circ C$ to $65.5^\circ C$ while cooling source water supplied at $35^\circ C$ to $29.4^\circ C$. (Courtesy Westinghouse Electric Corporation)

Westinghouse have studied a number of applications of the Templifier unit, and Table 7.9 sets out the type of industry and the processes in the industry where the Templifier could be applied. Temperatures given in the table are those required at the condenser outlet, (ref. 7.37).

Air-air heat recovery (electric heat pump): One of the successful heat pump applications arising out of work at the Electricity Council Research Centre at Capenhurst is in the drying of timber, (ref. 7.38), although the technique is equally applicable to other drying processes (ref. 7.31).

The efficiency of a dryer can be increased with the use of heat recovery systems (as discussed in Chapters 4 and 6) and other heat exchangers such as run-around coils and heat pipe bundles, described earlier in this chapter, are regularly used. These systems all rely on heat recovery from the exhaust, which is then vented to atmosphere, the heat being transferred into fresh incoming air. However, in theory the efficiency of a dryer in which the air is completely recirculated could approach 100 per cent, but this state is prevented from being achieved by the increasing humidity of the circulating air. Full recirculation can be implemented in practice if a dehumidifier is located in the drying kiln to remove the moisture from the exhaust air, as

TABLE 7.9 Potential Applications and Temperature Requirements of a Templifier Heat Pump*

Major Process Use (Temperature requirements in °C)	Meat Products	Dairy Products	Canned, Cured & Frozen Foods	Grain Mill Products	Bakery Products	Sugar	Confectionary Products	Beverages	Miscellaneous	Textile Mill Products	Apparel & Other Textile Products	Lumber & Wood Products	Pulp & Paper Mills	Chemical and Allied Products	Petroleum Refining	Fabricated Rubber Products	Fabricated Metal Products	Machinery, except Electrical	Electrical Equipment & Supplies	Transportation Equipment
Washing/sanitizing/cleanup (60)	X	X	X	X	X	X	X	X												
Cooking (100 – 115)	X	X					X													
Pasteurization (65)		X																		
Blanching (85)			X																	
Dye heating (88)										X										
Pressing (100)											X									
Log soaking (90)												X								
Metal cleaning and plating (60 – 90)																		X	X	X
Paint drying (80 – 120)																		X	X	X
Drying (105)										X										
Mold release solution (85)																X				
Vessel heating														X	X					
Heat tracing															X					
Evaporators		X												X						
Process heat (hot water)														X						
Paper machine													X							
Pulp heating													X							
Sterilization (110)		X	X																	
Scalding (50 – 60)	X																			

* Courtesy Westinghouse Electric Corporation

251
Waste Heat Recovery Techniques

shown in Fig. 7.40. The dehumidifier used is a heat pump, and the flow circuit is illustrated diagrammatically in Fig. 7.41. The humid air from the timber stack is passed across the heat pump evaporator, which cools this air below the dew-point, removing both sensible and latent heat. As a result of

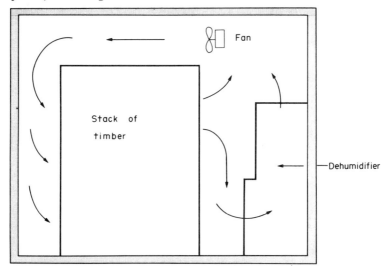

Fig. 7.40 A heat pump dehumidifier developed by the Electricity Council, and used in a drying kiln

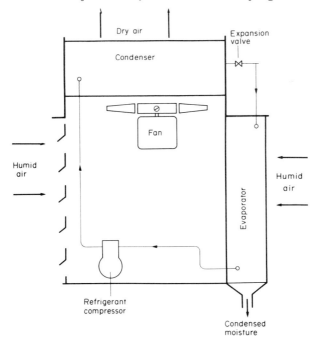

Fig. 7.41 The flow circuit in the dehumidifier (Courtesy The Electricity Council)

this cooling, condensation of a proportion of the water in the air occurs, and this is drained off, the condensate flow rate indicating the rate of drying. The heat removed from the air during dehumidification is then upgraded and given up in the condenser to the resulting dry air. This air is then passed through the timber stack, removing more moisture. In experimental kilns, coefficients of performance of about 3.6 have been obtained, implying fuel savings of approximately 50 per cent.

This system has also been employed to dry compressed air, preventing corrosion in the circuits and damage to the control equipment, and is being studied for brick kiln applications.

Combined refrigeration and heating using an electric heat pump: It is possible to make use of both the heat recovered, in the form of useful energy rejected at the condenser, and the cooling effect which results from the heat removed from the evaporator primary circuit. Obviously this is economically the most desirable state of affairs, and can frequently be used to good effect in plants where refrigeration, or even less demanding water chilling requirements exist.

Typical of such a requirement is an injection moulding plant, where chilled water is required to cool the moulds, hence solidifying the plastic in the component being moulded. In addition heating is required in the factory to maintain comfort conditions. A heat pump capable of fulfilling both of these functions has been installed in the factory of the Link 51 Plastic Division at Telford in the United Kingdom. Based on a Prestcold compressor, the unit replaces a conventional cooling tower, and supplies water at 7.2°C at a rate of up to 1140 litres per minute to the moulding machines (ref. 7.39). The heat extracted from the water is used for factory space heating. By using the controlled conditions afforded by the heat pump, the rate of production of the injection moulding machines could be maintained at a high rate, independent of ambient conditions which could affect the performance of the cooling tower.

With a condenser heat output of 325 kW, the system, illustrated in Fig. 7.42, is estimated to save about £15 000 per annum, and the heat pump, in conjunction

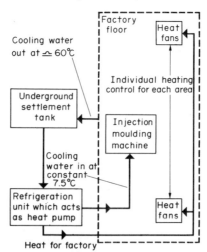

Fig. 7.42 A heat pump functioning as a water chiller and space heater (Courtesy The Engineer)

with a number of other less expensive heat and light conservation measures, is estimated to have a pay-back period of approximately three years.

Heat pumps driven by reciprocating internal combustion engines: The above examples of heat pumps driven by electric motors show that the system is able to show the most significant savings when used to replace direct electric heating. A heat pump with a COP of approximately 3.5 will, if seen from the point of view of conserving natural fossil fuel reserves, bring the conversion efficiency of electric power plant effectively up from 30 per cent to slightly over 100 per cent. Obviously this can be used as an argument for concentrating more resources on the development of the all-electric economy, and to some extent this has been used by Westinghouse in support of the Templifier, (ref. 7.36).

Cost differentials between electricity and other sources of energy, particularly at present natural gas, are working against the use of electric heat pumps for 'heating only' applications in the United Kingdom. In many cases the use of an electric heat pump to recover heat from a warm effluent stream would be more expensive, both in terms of capital and running costs, than the cost of producing the heat directly using a gas-fired packaged boiler

It is possible to identify three groups of applications for heat pumps in industry, only one of which is an obvious contender for electric heat pumps. These are as follows:

(i) Heat recovery from a process effluent stream where the original process heating is by gas or oil - electricity for a heat pump drive is probably too expensive.

(ii) Heat recovery from a coolant used to remove energy in heat form from, say, extruders, die casting moulds etc. - in this case electricity could be used, but alternative drives are cheaper to run, unless the refrigerating duty is also beneficial.

(iii) Heat recovery in a system where heating and/or cooling is currently done using electricity - an electric heat pump with a COP of 3 would cut running costs by over 60 per cent and would be fully acceptable.

The alternative drives for compressors include diesel engines, steam engines and gas engines, the latter being available either as a conversion of a diesel engine or as a gas unit designed as such from conception.

Of the above, a quick analysis of any engineering directory will show that the steam engine has almost disappeared from production. However diesel and gas engines are readily available and, certainly in the case of the diesel engine, are comparatively cheap owing to their mass production for automotive and other uses.

Diesel and gas engines are, of course, only about 30 per cent efficient, but by using the waste heat generated in the water cooling system and in the exhaust, (plus, on larger units, in the oil cooler), total efficiencies approaching 80 per cent may be obtained. If we add to this the coefficient of performance of the heat pump system driven by such an engine, the economics begin to look very attractive.

The use of these engines as static prime movers is of course not new - such units have been operating as electricity generators in industrial plant for decades, and the 'total energy' concept, where one attempts to find use for as much of the mechanical and thermal energy as possible available from the prime mover, has developed alongside the generator. Generating capacities ranging from 50 to 20 000 kW driven by gas engines (reciprocating or gas turbines) are common, particularly in the United States where very large gas-driven reciprocating internal combustion engines are available.

Much more interest is now being shown in using these engines for driving air and refrigerating compressors (see Chapter 2), particularly in applications where a use can be found for the waste heat. A heat balance for a 4 cycle naturally aspirated gas engine is given in Table 7.10, (ref. 7.40).

TABLE 7.10 Typical Heat Balance for a 4-Cycle Naturally Aspirated Gas Engine (kcal/kW hr)

Heat Output	Load (per cent)		
	40	70	100
Shaft power	345	605	860
Jacket water	280	640	1000
Exhaust - recoverable	185	365	535
- unrecoverable	285	415	535
Lubrication oil	60	80	100
Radiation and other	255	270	285
TOTAL	1410	2375	3315

As an alternative to the conventional water jacket cooling system on engines of this type, an 'ebullient' system may be used. This cooling system involves evaporative heat transfer, and operates with a low temperature differential, peak temperatures being in the range 100 to 125°C. As the coolant ascends in the engine jacket and riser leading to the steam separator, evaporation takes place and increasing amounts of steam are generated, and the system is aided by natural circulation brought about by the changes in coolant density. In the separator, steam at about 100 kN/m^2 is removed from the coolant and transported to the process where it will act as the energy source. The water from the separator is returned to the engine, supplemented by the condensate returning from the process heated by steam.

When an ebullient system is specified for engine cooling, a waste heat boiler utilising exhaust heat will provide additional steam. These can be obtained incorporating silencers. Should the exhaust heat be required to provide hot water, this can also be implemented.

It has been suggested, (ref. 7.40) that the waste heat available from the internal combustion engine could be used to directly boost the temperature of the primary fluid leaving the condenser. This would allow the heat pump to give, in effect, higher outlet temperatures than those possible with electric driven units, as the final temperature rise will not be dependent on either

Waste Heat Recovery Techniques   255

the compressor or refrigerant properties.

A unit driven by an internal combustion engine is less efficient when not running at full load, but in many industrial applications load variations are minimal, and this should not be a major factor in restricting development of the system.

A heat pump circuit, using the exhaust and water jacket heat from a gas engine driving the compressor, is illustrated in Fig. 7.43. This represents a water-to-water system. When the greatest demand is for cooling, ports a and b of the valves would be open; when heating is required, ports b and c would be opened. Figure 7.44 compares the energy utilisation of electric and internal combustion engine compressor drives.

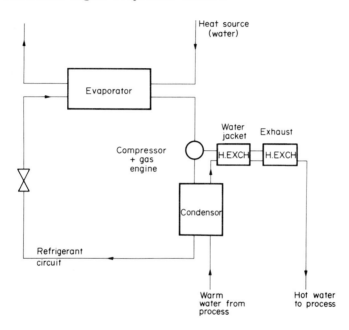

Fig. 7.43 Circuit of a heat pump driven by a natural gas reciprocating engine

7.9.6 Other heat pump types There is a number of other types of heat pump available, or in the process of development. Because their significance as far as industry is concerned is at present minimal, they are not described in detail. However, the reader will find adequate information on their perform- ance and status in the references listed.

The thermoelectric heat pump relies on the fact that when an electric current passes through a junction between two different conductors, heat is either absorbed or rejected at the junction, depending on the direction in which the current is flowing. A completed circuit must consist of two of these junctions, and the opposite effect will occur at the second junction. Thus, as a current is passed around the circuit, heat will be taken in at one junction and rejected at the other. This effect can be quite marked when

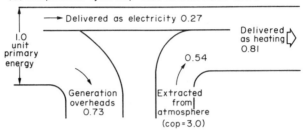

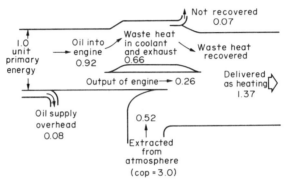

Fig. 7.44 A comparison of effective energy utilisation of heat pumps driven by (a) an electric motor and (b) an internal combustion engine

certain combinations of material are used.

At present these devices have very low power capabilities, and COP's are lower than with vapour compression types. Of potential significance is the fact that they can operate at temperatures well in excess of 100°C, (ref. 7.41).

Absorption cycle machines do not have a compressor. Mechanical processes are replaced by chemical reactions and the working fluid is used in conjunction with a second medium, known as the absorbant, in which it is highly soluble. The solution is heated in a generator, as shown in Fig. 7.45, by an external heat source. High temperature vapour given off in the generator passes to a condensing heat exchanger and the liquid returns to the absorber. The liquid working fluid is reduced in pressure and evaporates in a low temperature heat exchanger, which functions in the same way as a vapour compression cycle evaporator. The vapour is then redissolved in the absorbent.

The two most common types of absorption cycle machines, used as air conditioners, operate on lithium bromide and water, or ammonia and water. The lithium bromide system has the highest COP and may be attractive in cases where waste heat such as steam is available to heat the generator (ref. 7.42).

The United States Institute of Gas Technology is currently undertaking a major study of heat actuated heat pumps based on Rankine, Stirling and Brayton cycles. While the work is not yet at a stage when industry can benefit immediately from results, the progress should be monitored with interest by anyone considering heat actuated heat pumps, (ref. 7.43).

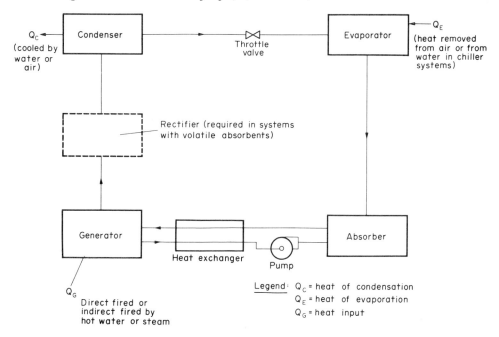

Fig. 7.45 Absorption cycle heat pump
(Courtesy American Gas Association)

7.10 Other Heat Exchanger Types

The majority of the heat exchangers described so far are designed specifically for heat recovery applications. However, there are several other types which, although most commonly used for cooling or heating duties which do not normally involve heat recovery, can equally as effectively be used in this area. Also, a number of heat exchangers are available in materials which open up the areas of application considerably.

Shell and tube heat exchangers, available at comparatively low cost if made using commonly available materials, can be used for heat recovery from process streams or waste water. These heat exchangers are also available with tubes manufactured in Teflon, primarily for handling acids, and also can be fabricated in graphite. The use of shell and tube heat exchangers is not confined to liquid-liquid systems. In some cases the amount of tube surface can be increased, as shown in the Lordan unit in Fig. 7.46, permitting its use for heat recovery from, for example, exhaust gases.

As mentioned in Section 7.5, heat exchangers of this type are in competition with plate heat exchangers in many areas of application. Lordan units can be made using low cost materials (instead of stainless steel) and there are advantages in retaining a stainless steel tube-side assembly with a carbon steel shell. Alternatively copper or cupro-nickel tubes may be used in combination with carbon steel tube plates, bonnets and shells. A number of considerations enable the choice of spiral tube type or plate heat exchanger to be made, as pointed out by D.J. Neil Ltd. (see Appendix 4):

Fig. 7.46 Interior of a Lordan shell and tube heat exchanger illustrating one technique for increasing surface area in a given volume (Courtesy D.J. Neil & Co. Ltd.)

(i)   Pressures - high pressures necessitate the use of spiral tube type heat exchangers.

(ii) Unequal primary and secondary flows - in many instances a spiral tube unit will be cheaper and smaller than the equivalent plate heat exchanger.

(iii) Temperature cross-over less than 10 per cent of the primary temperature difference - spiral tube type unit probably cheaper than a plate heat exchanger.

The following example, relating to the use of a spiral tube heat exchanger manufactured by Lordan, demonstrates heat recovery in a canteen dishwashing machine, as illustrated in Fig. 7.47.

Heat recovery scheme for canteen dishwasher using spiral tube heat exchangers
Background

Canteen for 600 people. Dishwasher in use 365 days x 7 hours = 2555 hours per year. Dishwasher water consumption 1.4 m^3 per hour, raised from 10 to 90°C by steam at 200 kPa. Steam assumed to have a fuel cost of £1.50 per 1000 lbs.

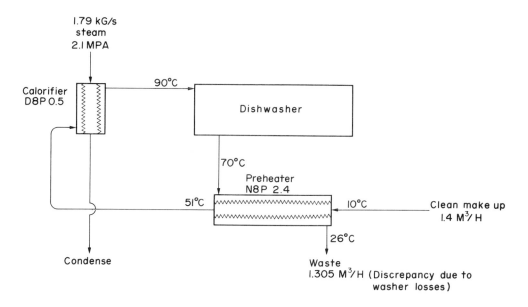

Fig. 7.47 Lordan spiral tube heat exchangers used as calorifier and heat recovery unit on a dishwasher (Courtesy D.J. Neil & Co. Ltd.)

With Calorifier Only

Q/hr = 1400 kg/hr x 80°C = 112 000 kcal/hr = 130 kW
Steam consumption 3.66 kg per second
Fuel cost £1854 per annum
Cost of calorifier type D8P1, £285

With Heat Recovery

(a) Pre-heat 1.4 m³/hr, 10/51°C by waste water 70/26°C
(NB Due to losses in the machine, waste water discharge is reduced to about 1.305 m³/hr)
Q/hr = 1400 kg/hr x 41°C = 57 400 kcal/hr = 67 kW = GAIN

(b) Heat 1.4 m³/hr, 51/90°C by steam at 200 kPa
Q/hr = 1400 kg/hr x 39°C = 54 600 kcal/hr = 63 kW
Steam consumption 1.79 kg per second
Fuel cost £907 per annum

	£
Cost of Preheater, type N8P2.4	580.00
Cost of Calorifier, type D8P0.5	210.00
Total	790.00
Less cost of original calorifier D8P1	285.00
Additional capital cost of heat recovery	505.00
Fuel cost, calorifier only	1854.00 p.a.
Fuel cost, heat recovery	907.00 p.a.
Saving in fuel cost	£ 947.00 p.a.

This comparison ignores installation costs, but it is clear that a gross payback period of well under one year can be expected.

There are other variations in the internal geometry of shell and tube heat exchangers, involving extended surfaces or improved layout directed at making the unit truely counter-flow.

REFERENCES

7.1   Dunn, P.D. and Reay, D.A. Heat Pipes. Pergamon Press, Oxford, (1976)

7.2   Behrens, C.W. Heat Pipes: Breakthrough in thermal economy ? Appliance Manufacturer, pp. 73 - 76, Nov. 1973.

7.3   Rogers, B.T. Passive heat recovery as an energy conservation measure. Building Systems Design, Pt. I, Feb. 1972, Pt. II, March 1972.

7.4   Vasiliev, L.L. et al. A study of heat and mass transfer in heat pipe based exchangers. II Int. Heat Pipe Conf., Bologna, Italy, (1976)

7.5   Basiulis, A. and Plost, M. Waste heat utilisation through the use of heat pipes. Trans. ASME, Paper 75-WA/HT-48, New York, Nov. 1975.

7.6   Feldman, K.T. Simplified design of heat pipe heat exchangers. Proc. II Int. Heat Pipe Conf., Bologna, Italy, March/April, 1976.

7.7   Amode, J.O. Analysis of a heat pipe heat exchanger. Ph.D. Dissertation, Univ. of New Mexico, Albuquerque, (1975)

7.8   Amode, J.O. and Feldman, K.T. Preliminary analysis of heat pipe heat exchangers for heat recovery. Trans. ASME, Paper 75-WA/HT-36, New York, Nov. 1975.

7.9   Holmberg, R.B. Heat transfer in liquid-coupled indirect heat exchange systems. Trans. ASME, J. Heat Transfer, Paper 76-HT-E, pp 499 - 503, Nov. 1975.

7.10 Fulton, R. Ecoterm - a method of waste heat recovery. Fläkt-SF Air Treatment Ltd., Publicity Literature, (1975).

7.11 Strindehag, O. and Astrom, L. Energy conservation by Ecoterm. Fläkt Review, Sweden, (1974)

7.12 Strindehag, O. A liquid coupled system for heat recovery from exhaust gases. Building Serv. Eng., 43, 52 - 56, (1975)

7.13 Applegate, G. Heating a factory for nothing. Heating & Vent. Eng., 44, 21 - 24, (1970)

7.14 Mortimer, J. Exchanging heat to save fuel. The Engineer, pp 34 - 37, 2/9 August, 1973.

7.15 Applegate, G. Heat regeneration by thermal wheel. Proc. Waste Heat Recovery Conference, Inst. Plant Engnrs., London, 25 - 26 Sept. 1974.

Waste Heat Recovery Techniques

7.16 Anon. Waste heat recovery systems. Processing, p. 46, May, 1975.

7.17 Fisher, D.R. et al. Performance testing of rotary air-to-air heat exchangers. Trans. ASHRAE, Pt. 1 pp 322 - 332, (1974)

7.18 Butler, P. Use the Munter wheel to recover your heat and make big savings. The Engineer, p. 24, 20 Nov. 1975.

7.19 Congram, G.E. Plate heat exchangers save fuel, space and downtime. Oil & Gas Journal (US), 22 Sept. 1975.

7.20 Anon. Regeneration and the energy crisis. APV Spearhead - Engineering in the Process Industries, No. 3, June 1974.

7.21 Starkie, G.L. Some aspects of energy conservation in dairy process plant. J. Soc. Dairy Technology. 28, 121 - 129, (1975)

7.22 Bunton, J.F. Recovery of low grade heat. Proc. Inst. Plant Engineers Waste Heat Recovery Conference, London, 25 - 26 Sept. 1974.

7.23 Lyall, O. The Efficient Use of Steam, HMSO, London, (1956)

7.24 Lock, A.E. Boiler Economics (Part 3) Industrial Process Heating, 12, 32 - 33 (1972)

7.25 Wood, B.D. Applications of Thermodynamics. Addison-Wesley, Reading, Mass. (1969)

7.26 Fanaritis, J.P. and Streich, H.J. Heat recovery in process plants. Chemical Engineering, 80 - 88, (1973)

7.27 Anon. Waste heat boilers and their uses in the chemical and metallurgical industries. Schmidt'sche Heissdampf - Gesellschaft GmbH Prospectus, 1969.

7.28 Gibson, T. Waste heat boiler possibilities. Proc. Inst. Plant Engineers. Waste Heat Recovery Conference, London, 25 - 26 Sept. 1974.

7.29 Kay, H. Recuperators - their use and abuse. Iron and Steel International, pp 231 - 240, June, 1973.

7.30 Anon. Recuperator saves natural gas in heat processing. Instruments and Control Systems (USA), pp 72, 74, Jan. 1976.

7.31 Kolbusz, P. Industrial applications of heat pumps. Electricity Council Research Centre, Report ECRC/N845, Sept. 1975.

7.32 Macadam, J.A. Heat Pumps - the British experience. Building Research Establishment Note N117/74, Dec. 1974.

7.33 Montagnon, P.E. and Ruckley, A.L. The Festival Hall heat pump. J. Institute of Fuel. pp 1 - 17, Jan 1954.

7.34 Kolbusz, P. The use of heat pumping in district heating. Electricity Council Research Centre, Report ECRC/M700 Feb, 1974.

7.35   Anon. Heat pumps. A new application for Howden screw compressors. Howden Journal, Howden Group Ltd., Glasgow, 14, 33 - 39, (1976)

7.36   Ross, P.N. The Templifier for process heat. Proc. EEI Conservation and Energy Management Division Conf., Atlanta, Georgia, March 16 - 18, 1975.

7.37   Anon. Westinghouse Templifier. Westinghouse Electric Corporation Bulletin TP, Feb. 1976.

7.38   Kolbusz, P. The improvement of drying efficiency. Electricity Council Research Centre, Report ECRC/R476, Jan. 1972

7.39   Williams, E. Keep the factory fires burning by extracting heat from cooling fluid. The Engineer, p. 18, 1/8 Jan. 1976.

7.40   Baker, M.L. Systems for extracting and utilising engine rejected heat. Trans. ASME, Paper 63-OGP-6, New York, ASME, 1963.

7.41   Farrell, T. Thermoelectric heat pumping. Electricity Council Research Centre, Report ECRC/R844, Sept. 1975.

7.42   Ellington, R.T. et al. The absorption cooling process. IGT Research Bulletin No. 14, Chicago, (1957)

7.43   Wurm, J. An assessment of selected heat pump systems. Annual Report, (Feb. 1973 - Feb. 1974). Institute of Gas Technology, Project HC-4-20, Chicago, March, 1974.

Energy Storage

The ability to store energy in a useful form for use at a later stage when demand increases, or to allow energy to be collected at a time when it is available at a lower than normal cost, has attracted an increasing amount of attention in recent years.

Most people are familiar with at least one form of energy storage, the use of special radiators in the home to store heat collected during 'off-peak' hours when electricity is available at a lower cost. The heat stored in the radiator 'fill' is released during the next few hours, and the rate of heat release can be modulated using forced convection or other means.

Industry has used thermal storage devices for many years, commonly known as heat accumulators. Some accumulators are employed to store steam at high pressure for release later at a lower pressure for process use. These can cope with large fluctuations in demand. The use of hot water as a storage medium is also popular, and a major advantage of hot water storage over other sensible (and latent) heat storage media is that the water can be used both as the heat storage medium and the heat transport medium, eliminating the necessity for a heat exchanger. The term 'heat accumulator' is not as familiar now, its function being taken over by thermal storage boilers, which, as their name implies, can be used for both heat generation and storage. The accumulator generally formed an additional piece of plant linked to the boiler proper.

Other solid and liquid heat storage media are being developed, particularly in association with solar energy exploitation, and some are known as phase change materials (PCM). A PCM storage medium takes in heat while it melts, and heat flows from the storage medium as it solidifies, thus the heat storage capacity of a PCM is a function of its latent heat of fusion. Salts such as sodium hydroxide are typical of this type of storage material.

So far the introductory remarks have all been concerned with the storage of thermal energy. There are several other energy forms which can be stored, and these are listed below.

(i) Potential energy storage, (pumping water to a higher elevation, use of compressed air, or the energy in springs).

(ii) Chemical energy storage, (use of hydrogen, either in liquid or gaseous form, or energy storage in secondary batteries).

(iii) Kinetic energy storage, (flywheels)

(iv) Energy storage in electromagnetic fields, (capacitors or superconducting magnets)

While the potential of some of these non-thermal storage systems is unlikely to be of any immediate interest to industry, they are all being considered for use in primary energy generation processes. The reader is referred to the paper by Ramakumar et al (ref. 8.1) for further data on these non-thermal techniques.

8.1 Thermal Storage Boilers and Accumulators

The main function of an accumulator or a thermal storage boiler is to improve the flexibility of a conventional boiler. Thermal storage can be used to even out the peaks and troughs in the demand for steam, or to perform an identical function with respect to the steam supply. Its importance can be best illustrated with respect to the use of waste heat boilers to raise steam for electricity generating plant. If fluctuations occur in the steam supply, which will be completely dependent on the operation of the process from which the heat is being recovered, difficulties may arise in guaranteeing satisfactory generator output performance, particularly if the demand for process steam is at times large in relation to the waste heat boiler supply capability (ref. 8.2).

In an electricity generating plant incorporating a Ruths accumulator and a steam turbine, designed to meet the above conditions, the accumulator may be situated upstream of the turbine, and the electrical output varies as the process steam demand. An alternative arrangement would be to site the accumulator downstream of the turbine, where, for a given pressure differential across it, the accumulator storage capacity would be greater. However, in the former case, variations in the steam supply rate are taken up by the accumulator before steam reaches the turbine, and the electricity generation output is higher because the loss to the turbine of pressure differential is at a high pressure level, where heat loss is less. This is in spite of the fact that some degree of superheat is lost in the accumulator circuit.

A Ruths accumulator (ref. 8.3) is a large cylindrical steel vessel almost completely filled with water, the vessel being mounted horizontally to expose the greatest water surface. The accumulator is charged by introducing the steam through nozzles located below the surface of the water, the nozzles being so designed to promote rapid water circulation. As steam is pumped into the water, the pressure in the accumulator rises, resulting in a rise in the boiling point of the water. This in turn permits more steam to condense and more heat to be stored.

When steam is required to be withdrawn from the accumulator, indicated by a reduction in pressure in the supply/exhaust line, the pressure in the body of the accumulator is reduced. This causes the surplus heat in the water to be rejected as flash. Flow restrictions are included in the circuit to prevent steam escaping too quickly.

Thermal storage boilers have been developed to enable the functions of the boiler and accumulator to be combined in a single unit. Typical of a modern thermal storage boiler is that produced by Babcock and Wilcox (Operations) Ltd. This boiler is available in a range of sizes, and combines the features of the Lancashire and Economic boilers with that of an accumulator to give high efficiency rapid steam raising, together with a large water capacity and the ability to even out steam pressures to within 8 kN/m^2 of the required operating pressure, attributable to a special feed control system.

The principle of this control system is that when a peak demand occurs the feed regulator automatically reduces or stops the flow of feed water to the boiler. The water level is allowed to fall and the sensible heat stored in the boiler water is given up to increase the rate of evaporation. The firing rate remains constant, maintaining a high thermal efficiency. A drop in the steam demand causes the feed regulator automatically to increase the flow of water to the boiler, so the water level rises. With the constant firing rate maintained, sensible heat is again stored in the boiler water in anticipation of further peak demands. In addition to acting on normal thermal storage principles, the boiler can also be operated as a conventional unit with modulation plus override for bringing in the thermal storage facility when the change in steam demand cannot be met by modulating control. This function is illustrated in Fig. 8.1.

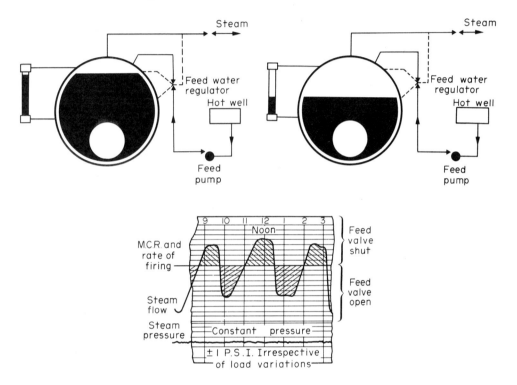

Fig. 8.1 The operation of the Babcock thermal storage boiler

The boiler can cater for peak steam demands up to 33 per cent above the boiler maximum continuous rating for periods of up to one hour, meeting lesser peaks for longer periods. As a constant firing rate can be maintained throughout operation, the flue gases are kept at design temperature so that condensation and subsequent corrosion cannot occur because of numerous shutdowns or lowering of the firing rate (which would also lower boiler efficiency).

A second thermal storage boiler system, which, it is claimed, overcomes the problems associated with intermittent firing, is marketed under the name 'Metro-Flex Thermal Storage System'. Its operation is based on a special firing technique and as well as offering a thermal storage capacity from the point of view of steam demand, the temperature within the boiler, being maintained at a fairly constant and high level, increases boiler life by minimising thermal shock. The energy conservation aspects of the system are also emphasised by the manufacturers, the Metro-Flex System offering the greatest economies when the load conditions are at their most difficult. The firing cycle of this system is compared with that of a conventional on-off burner and a modulating burner in Fig. 8.2. An on-off burner cuts out comparatively

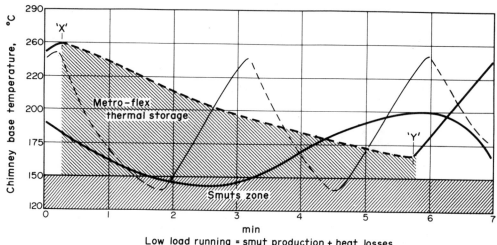

Fig. 8.2 The Metro-Flex thermal storage system compared with the characteristics of on-off and modulating burners

frequently, and when this occurs, cold air is drawn through the boiler, dissipating heat and necessitating reheating the whole system, thus wasting energy. A conventional modulating burner system may not supply a sufficiently high heat input to prevent cooldown of the boiler and flue gas, leading to pollution and corrosion. The Metro-Flex system of burner operation ensures firing at the most efficient rate with a high heat input, but when the burner cuts out, the flue is completely isolated from the boiler, eliminating the possibility of cold air being drawn through the system. In this way, heat loss from the boiler is reduced to a minimum, and it therefore acts as a heat store. Savings obtained are normally in the region of 5 to 6 per cent of the fuel bill, but on one oil-fired plant a saving of 9.8 per cent was achieved, directly attributable to the prevention of shut-down cooling losses.

8.2 Thermal Insulation

The use of thermal insulation as a means of reducing heat loss was frequently advised in Chapter 5, its application being appropriate in a large number of items of plant and ancillary equipment. Attention was also drawn to the fact that the cost of insulation of industrial buildings, (unfortunately plant contained within the building is not specifically included), may be subjected to tax relief in the United Kingdom, and to date this is the only direct financial inducement to energy conservation initiated by the United Kingdom government.

The popular concept of thermal insulation is a jacket of fibrous material surrounding a vessel or pipe carrying a hot fluid, and to a large extent this remains true today. However, glass may be regarded as a thermal insulator in buildings, and the latest concept in process thermal insulation involves covering the surface of a hot liquid with a layer of plastic balls.

Thermal insulation blanket: The commonest form of thermal insulation used in industry is the fibrous material applied as lagging on pipes and vessels. A wide range of materials is available for insulating features such as these, varying considerably in their cost, thermal conductivity and applicable temperature range. The application of thermal insulation has, until comparatively recently, been largely on an 'ad hoc' basis, proceeding in a plant as and when convenient. This has often meant that pipes requiring to be lagged have been found to be located too close to a wall, or one another, for the effective insulation thickness to be applied. It is important that the need to use thermal insulation is appreciated at the design stage, so that proper provision can be made for its application.

The most critical factor in selecting insulating material is the correct determination of its thickness. One United Kingdom manufacturer, Fibreglass Ltd., produce a guide to the calculation of the optimum thickness, relating heat savings to the cost of the material, according to the graph in Fig. 8.3. Obviously the optimum thickness in a particular application is also a function of the temperature of the fluid in the pipe or vessel to be lagged, a low temperature hot water pipe requiring less insulation than a high pressure steam main.

Depending upon the type of fuel used, it is possible to draw up a table relating the surface temperature of the pipe to the amount of fuel wasted per unit length of pipe. Table 8.1 shows typical results, and also includes losses from a flat surface such as a tank, (ref. 8.4).

The environment in which the insulation is situated can affect its performance, and thus influences the selection. Good resistance to water and water vapour if the use is outside a building or underground are important. If the insulation is not protected, the onset of frost could cause it to break up. A high water content will also have a detrimental effect on the thermal insulating properties. Allowances should be made for shrinkage, and the durability of the insulation with respect to vibration, thermal cycling and the possibility of mechanical damage. Application of a thermal blanket should take into account the fact that it may need to be removed at some future date, for maintenance to the pipework, flanges or valves which it is insulating; (it is important to cover all parts of a pipe, including flanges and valves). When insulation is removed to allow access to the pipe, it should be done with care so that the insulation can be re-used. Insulation with an outer metal shell as cladding can be conveniently located in two halves around a pipe using clips, and is readily removed and re-applied.

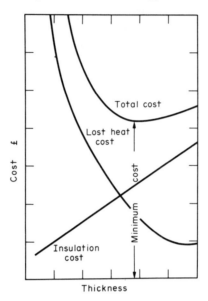

Fig. 8.3 The economic thickness of thermal insulation
(Reproduced from a Fibreglass Ltd. publication)

Most insulating materials are comparatively inert and safe in use. Where a particular fire hazard or potential for chemical attack exists, treatment of the insulation may be necessary, and the manufacturers will be able to advise on these aspects.

TABLE 8.1 Heat Loss From Hot Metal Surfaces Expressed in Fuel Quantities per Annum

Fuel	Surface Pipe N.B. mm	Units	Temperature of Hot Surface (°C)	
			90	425
Coal	25	tonnes/m	0.13	2.07
	50	"	0.26	3.12
	200	"	0.75	12.00
	flat	tonnes/m^2	1.18	19.69
Fuel oil	25	litres/m	101	1471
	50	"	164	2223
	200	"	537	8579
	flat	litres/m^2	881	14094
Gas	25	therms/m	37.7	552
	50	"	62.3	935
	200	"	201	3224
	flat	therms/m^2	331	5293

Energy Storage

A recent plant survey carried out by the Esso Petroleum Company at three factories owned by a large group revealed two major applications for thermal insulation where previously it had not been employed. The insulation of fuel storage tanks and boiler hotwell and condensate return lines led to the savings listed in Table 8.2 (ref. 8.5).

TABLE 8.2 Savings Using Insulation

Location	Current Heat Loss GJ/year	Saving by Insulation GJ/year	Equivalent fuel saving Litres
Fuel storage	1561	1108	33 600
Boiler hotwell & condensate return	2791	2216	54 500

The importance of glass as a thermal insulator in commercial buildings is discussed in Chapters 5 and 6.

Insulation of hot liquid surfaces: A technique developed by Capricorn Industrial Services of London (see Appendix 4), known as the Allplas system, is proving most effective in minimising heat loss (and evaporative loss) from the surfaces of hot liquids. The Allplas system is based on the use of hollow polypropylene balls, which are floated on the surface of liquids as a 'blanket' to provide almost complete cover without the disadvantages of solid lids. The balls are resistant to most acids and other industrial fluids, and can be used at temperatures considerably higher than the boiling point of many common fluids. The balls tend to soften at temperatures between $120^{\circ}C$ and $145^{\circ}C$, and can therefore be used on most process tanks. The following data on performance was obtained from the Allplas publicity literature.

"Many firms using Allplas balls have reported big fuel savings when heating open tanks. To establish the exact amount of savings possible under controlled conditions, the National Engineering Laboratory at East Kilbride carried out a series of heat loss tests (NEL report 70/14/6386/1), and the results are shown in Fig. 8.4. The tests were made on an open tank which was insulated with polyurethane foam on the sides and bottom, so that the minimum of heat was lost from these surfaces. The amount of heat needed to maintain the tank at a given temperature for a given period was noted with the surface unprotected (curve A), with one layer (curve B) and with two layers of balls (curve C), the results being shown in the diagram. With the water maintained at the constant temperature of $90^{\circ}C$, heat loss was reduced by 69.5 per cent by a single layer of balls, and by 75.5 per cent by a double layer.

In industry, insulation is rarely so efficient and usually much heat is 'exported' by dipping cold and withdrawing heated components from the tank. On the other hand, Allplas balls enable extraction equipment to be run at lower air flow rates, or to be dispensed with altogether. Since heat losses rise steeply with increasing air velocity, reduced ventilation increases savings in fuel (as well as capital costs). Fuel savings under industrial conditions with a single layer are usually of the order of 50 per cent and often higher. In addition to saving fuel, Allplas blankets help to maintain the solution at a much more uniform temperature, a fact which is of considerable value to many process industries.

Evaporation is also drastically reduced. In many industrial applications it is suggested that loss of solution by evaporation has been completely eliminated. The NEL report referred to above shows that this view is not far short of the truth at 80°C, evaporation of water is reduced by 88 per cent with one layer of balls. Two layers reduce evaporation by 89 per cent. These figures refer to still-air conditions. It is this drastic reduction in fume generation under an Allplas cover which often allows extraction fans to be switched off, thus eliminating a powerful cooling effect."

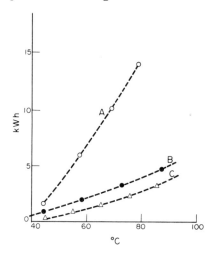

Fig. 8.4 The effect on heat losses of covering a liquid surface with plastic balls. (From "Technical Data on Allplas", reproduced by permission of Capricorn Industrial Services Ltd.)

Further data on the performance of this insulation system is given in reference 8.6.

8.3 Heat Storage Media

Thermal storage boilers and accumulators described in Section 8.1 are associated primarily with steam raising, and operate either in conjunction with a boiler, or in the case of the thermal storage boiler, combine both functions.

Heat storage for steam raising is probably the most widely applied storage application in industry. However, because of the increased awareness of energy costs, other storage media for conserving waste heat which can be collected at 'off-peak' tariff rates, or accumulating heat from alternative sources such as solar energy, are becoming of greater interest to industry and commerce. The types of storage media available are described below, and in later sections information is given on applications of heat storage.

8.3.1 Sensible heat storage
The hot water tank used in most houses as a heat storage device, can also be applied in commercial and industrial

buildings, both for storing water for services required during occupation and to meet 'off-peak' heating demands (see Chapter 6).

A major advantage of thermal energy storage in water is the fact that no heat exchanger is required, (except in cases where it is necessary to give up the heat to a warm air circulating system). Thus the heat storage and heat transport media are identical. Storage tanks for liquid are easily obtainable in a range of sizes, and the system is comparatively simple to install.

The storage tank comes into its own when it is possible to use waste heat as the heat source. Steam may be used to heat an industrial storage vessel, as opposed to being stored itself in an accumulator. However when using a water storage tank, the control arrangement is important. It is undesirable to immediately allow cold water to flow into a storage tank as soon as the hot water is drawn off in any quantity because the cold water immediately tends to flow towards the bottom of the tank, where the temperature sensor actuates the steam inlet valve. A more efficient system incorporates a set temperature control, which allows the water to rise to this temperature before any more cold water is permitted to enter the tank. Control is required to ensure that the water supply is arrested when the tank is full, and also that the steam supply is terminated when a maximum temperature is achieved, (ref. 8.7). A good steam-heated storage tank can rapidly achieve a uniform temperature.

The material of the storage tank can be plastic, copper or galvanised mild steel. If the water (or other medium) being stored is not being re-used directly, but serves a heat exchanger, a concrete or rubber lined pit would be satisfactory, as in the example described in Chapter 4 for storing the warm effluent from a dyeing plant.

Sensible heat storage has been investigated by the Central Electricity Generating Board, as a means of allowing plant to be run under constant conditions, independent of electricity demand, and also to permit load fluctuations to be met swiftly (ref. 8.8). While this particular study concentrated on the use of water as the sensible heat storage medium, liquid sodium was also considered, in conjunction with the sodium-cooled fast reactor. It was concluded that while water was economically attractive for power station heat storage, sodium was ruled out because of its current high price.

On the domestic scene, the storage of heat in a solid medium has also become a common feature of space heating systems. A recent study carried out by the Electricity Council (ref. 8.9), included investigation of both solid and liquid sensible heat storage media, for use at higher temperatures than a water storage system.

Molten salts and aluminium would be used in their molten states, but the top end of the operating temperature range (750°C) would be too high for safe operation of a salt such as sodium hydroxide, which is limited to 600°C, where its storage capacity is 494 kW h/m^3. The Electricity Council had considerable reservations concerning the use of high temperature molten materials, in spite of their low weight, and safety and cost considerations ruled them out at an early stage.

Of the solid media available, ceramics or alloys based on cast iron are the best materials from the capacity point of view. However, the maximum operating temperature is determined by other factors, such as the performance of the

TABLE 8.3 Properties of Sensible Heat Storage Media (ref. 8.9)

Materials	Temperature range for active storage calculation (°C)	Capacity (kWh/m^3)
Olivene } Forsterite	250 - 750	395
Magnesite	250 - 750	437
Feolite	250 - 750	512
Cast Iron Alloy	250 - 750	565
Molten Salt (e.g. sodium hydroxide)	250 - 750	640
Aluminium	250 - 750	664

insulation and the heating element life, (this particular system was designed for storing energy supplied by 'off-peak' electricity). The extraction of heat is also a function of the thermal diffusivity of the core material. In the prototype unit, cast iron was selected as the storage media, although the Electricity Council imply that if Feolite had been available when the prototype storage system was constructed, this may have been used instead. (Feolite, a sintered iron oxide in brick form, was developed by the Electricity Council specifically for thermal storage applications. It is cheaper than cast iron and can be used at temperatures of up to 1000°C.)

The first full-scale application of this system in industry resulted from the Electricity Council work, and the storage system has been operating for several years in a plant producing alkyd resin. This plant is described below.

"Alkyd resin, the basic ingredient of varnishes, is produced in closed stainless steel kettles and the resin plant constructed to demonstrate the commercial use of the thermal storage principle has a working capacity of 1.6 m^3 and a maximum processing temperature of 260°C. Heating starts immediately after charging the kettle with reactants and continues until the required processing temperature is reached. The desired rate of rise of resin temperature is usually 1°C per minute and this requires an initial rate of heating of about 190 kW, falling steadily as the temperature of the store falls and that of the resin rises. The kettle is then maintained at a constant temperature to allow the necessary reactions and refluxing to take place. The heat required for the process is stored in about 7000 kg of special, high-temperature resisting cast iron blocks and this store has an internal heat transfer area of 32 m^2 and a voidage of about 30 per cent. The store (Fig. 8.5) is heated at the rate of 80 kW during the night for about 7½ hours to provide a total heat charge of about 2 GJ (approx. 600 kW h). During the day-time processing period the heat is transferred by circulating air between the thermal store and the heating surface of the process vessel in a closed duct system. The temperature of the resin is readily controlled within ± 1.5°C of the required processing temperature by switching the circulating fan on or off. The whole process takes eight to ten hours to complete."

Energy Storage

It has been found that the capital cost of the thermal storage system in the resin plant is comparable to that of an oil-fired installation, and the running cost per litre of product is equal to, or marginally less than that for oil. Although a gas-fired system would have a lower initial cost (70 per cent of that of the thermal storage system), its running cost would be up to 50 per cent greater than that of the new system. Other advantages claimed for electric thermal storage heating include simplicity, ease of maintenance and clean operation. Future proposals for the use of this method in the glass industry are described in Chapter 4.

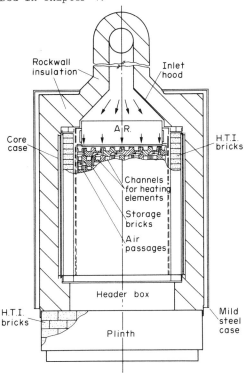

Fig. 8.5 Detailed view of a thermal store in a resin plant
(Courtesy The Electricity Council)

8.3.2 Heat of fusion storage media As with the sensible heat storage methods discussed in the previous section, the thermal capacity of a heat of fusion, or latent heat, storage medium depends upon its volume. In the case of the latter type, the volume needed is a function of the latent heat of fusion of the material used. Unlike a sensible heat storage system, the heat of fusion storage medium gives up or absorbs heat at a constant, or near constant temperature. In order to do this, however, it must change its phase. When receiving heat, the material melts, and it rejects this heat when it resolidifies. Latent heat of fusion materials in general offer a greater storage capacity per unit volume than sensible heat storage media. Typical performances are given in Table 8.4.

In spite of the high storage capacity of this type of material, there are several drawbacks associated with their use which makes them less attractive, at least in the short term, for industrial use on a large scale. The materials listed in Table 8.4, which are the most popular for latent heat storage, have a rather limiting operating temperature, being more appropriate for domestic heating; (in fact most of the development in this area has been associated with domestic solar energy systems) although their use in conjunction with heat pumps is possible. With a eutectic temperature of 590°C, germanium sulphide has been suggested as a material suitable for industrial use (ref. 8.12). Unfortunately, while sulphur is cheap, germanium is at present an expensive commodity, although its recovery from fly-ash on a large scale could make such a system more competitive.

TABLE 8.4 Heat of Fusion Storage Media (refs. 8.10, 8.11)

Material	Melting Temp. °C	Density kg/m^3	Heat of fusion J/g	Heat of fusion J/cm^3
$NaC_2H_3O_2, 3H_2O$	58	1297	265	340
$Na_2S_2O_3 \cdot 5H_2O$	48	1650	209	344
$Ca(NO_3)_2, 4H_2O$	47	1858	154	283
P116 (wax)	47	785	209	163
$FeCl_3, 6H_2O$	36	1617	223	359
$Na_2CO_3, 12H_2O$	36	1522	265	400
$LiNO_3, 3H_2O$	30	-	307	440

The attraction of germanium as part of a storage compound, and one which it has in common with other semiconductor materials, is that it contracts when changing from the solid into the liquid phase. One of the most difficult problems to overcome in the design of a storage system of this type is integration with the heat exchanger. Materials such as those listed in Table 8.4 expand on melting, and have a positive coefficient of expansion in the liquid phase. It is difficult using a material of this type to transfer heat into the reservoir, as the liquid lies on top of the solid phase, with the hottest liquid above the remainder. Placing the heat exchanger on the top would be unsatisfactory because convection would not assist heat transfer, and conduction downwards into the solid phase would be a slow process. Placing the heat exchanger in the solid itself could create difficulties associated with stresses brought about by the volume changes on melting.

It is claimed by Van Vechten (ref. 8.12) that the 'contract on melting' characteristic of many semiconductor materials would overcome this problem. The solid phase would be above the liquid phase, and with the heat exchanger located at the bottom of the storage vessel, convection would assist both heat input and heat removal.

REFERENCES

8.1 Ramakumar, R. et al. Solar energy conversion and storage systems for the future. IEEE Trans. Power Apparatus and Systems, PAS-94, 6, Nov./Dec. 1975.

8.2 Ungoed, W.P.C. and Sayer, C.E. Waste heat steam for electricity generation. Waste Heat Recovery, Chapman & Hall, London (1963)

8.3 Lyle, O. The Efficient Use of Steam. HMSO, London (1956)

8.4 Harris, J. The economic value of insulation. Maintenance Engineering April 1975.

8.5 Anon. Energy saving: the fuel industries and some large firms. Energy Paper No. 5, Department of Energy, HMSO, London, (1975)

8.6 Anon. Ball blanket prevents boiler corrosion. The Heating and Ventilating Engineer, Nov. 1968.

8.7 Gillies, J. Heat storage and waste heat recovery. The Plant Engineer, 14, 257 - 262, (1970)

8.8 Gardner, G.C. et al. Storing electrical energy on a large scale. CEGB Research, 2, 12 - 20 (1975)

8.9 Gibbs, M.G. et al. Thermal storage for industrial process heating. Report ERC/M504, Electricity Council Research Centre, Capenhurst, Sept. 1972.

8.10 Altman, M. Conservation and better utilisation of electric power by means of thermal energy storage and solar heating. National Science Foundation, University of Pennsylvania, Report UPTES-71-1, Oct. 1971.

8.11 Lorsch, H.G. Thermal energy storage devices suitable for solar heating. 9th Intersoc Energy Conversion Engineering Conf.(1974)

8.12 Van Vechten, J.A. Latent heat energy storage is feasible. Electrical World. Aug. 15, 1974.

New or Specialised Processes and Plant Having Energy - Saving Potential

There are a considerable number of industrial processes which to date have been commonly regarded as being appropriate only in high technology industries such as aerospace, or in specialised production methods in the industry where they were developed. These processes can be used for example, for heating, welding, cutting or shaping with less energy consumption than some of the more commonly accepted techniques.

While initially the capital cost of the equipment necessary to perform these tasks was, and in some cases still is, comparatively high, they can offer advantages other than lower energy consumption. Electron beam welding, for example, minimises the use of weld preparations, hence reducing machining time, and is a sufficiently repeatable process to allow higher quality and consistent welding, thus reducing wastage. Similar arguments may be made in favour of laser welding.

This chapter briefly describes some of the techniques which are available, or are under development, which can offer saving such as the above, and Appendix 4 lists manufacturers of the equipment described. A number of Laboratories and Research and Development establishments, both in the commercial sector of industry and run by government agencies, can give advice on the most effective utilisation of these techniques and the reader is referred to Appendix 3 for the addresses of these organisations.

Some of the processes described will have been mentioned in previous chapters, more particularly in sections devoted to energy conservation in the various industries considered in Chapters 3 and 4. Typical applications will also be given below in the appropriate sections.

It is also hoped that this chapter will, in showing that new and advanced technology can be employed beneficially in a wide variety of industries, help towards fostering a more receptive attitude in industry to new types of tools etc. One of the most unfortunate aspects of new and advanced technologies is the lack of expertise devoted to easing the transition from the laboratory to industry, and this failing is particularly evident when one experiences contact with the more conservative and older industries. Both parties, the laboratory and industry, are often at fault; the former in part through premature over-selling of a technology, and the latter by a reluctance to invest and a lack of appreciation partially fostered by a certain amount of disbelief in the benefits accrueing to the use of new technology. The fact that some of the equipment described in this chapter is anything but new, and in at least one instance extremely unsophisticated, speaks for itself.

9.1 Dielectric and Microwave Heating

If a material is a poor thermal conductor, the application of heat by conventional methods such as radiation or convection will have only a limited effect on the inside of the material, and unless considerable care is exercised, the outer surface of the material is likely to be excessively heated before the centre reaches the required temperature. Obviously the lower the thermal conductivity of the material being heated, the more inefficient will be the heating process, in that losses will be greater, (ref. 9.1).

Dielectric and microwave heating offer considerable advantages in situations such as this because heat is generated throughout the body, independent of the thermal conductivity. Thus heating can be both uniform and rapid. Of particular significance in drying processes is the fact that some materials, including water, react much more favourably to the application of these forms of heating than others. Thus water can be readily evaporated from many media using these methods.

Dielectric and microwave heating units consist of a generator capable of producing high frequency energy, (in the case of dielectric heating the frequency is normally below 200 MHz, in microwave heating the 897 and 2450 MHz bands are used). The normal method of applying dielectric heating is to put the material to be heated between two electrodes, connected to the h.f. generator. The electrodes may be in contact with the material, or separated to allow a conveyor to pass between them, if a continuous drying or heat treatment process is envisaged. Alternatively, only one part of the material can be heated, as in adhesive curing or plastic film welding. These systems are illustrated in Fig. 9.1.

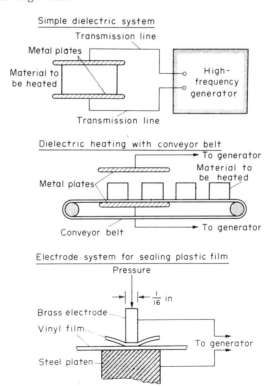

Fig. 9.1 Dielectric heating methods (Courtesy The Electricity Council)

In microwave heating the generator transmits the energy to the surface of the material to be heated, and the very high frequency of this energy allows it to penetrate deeply into the material. Unlike dielectric heating, the means of application can be more varied. In paper drying, for example, as shown in Fig. 9.2, the power is developed along a waveguide through which the paper sheet passes. Any energy not absorbed by the paper is taken up by the water load at the end of the waveguide. A second type of microwave heating, shown in Fig. 9.3, takes the form of a metal walled cavity, or oven, in which the microwave energy is directed at the load from all angles, (ref. 9.2).

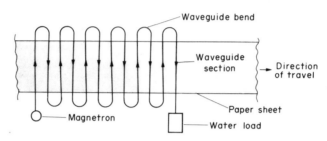

Fig. 9.2 Microwave heating in a paper drying process
(Courtesy The Electricity Council)

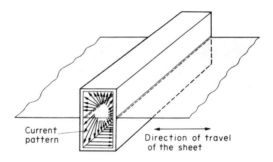

Fig. 9.3 A microwave oven configuration
(Courtesy The Electricity Council)

One of the most detailed technical and economic assessments of the application of microwave heating has been carried out by Dunlop Limited, during development of a method for preheating conveyor belting prior to vulcanisation, (ref. 9.3). (This report also gives a first-class review of these heating techniques, including advantages and disadvantages). Ways of improving the production efficiency of conveyor belt presses and continuous vulcanisers are constantly reviewed because of the very high capital cost of the equipment. Preheating of the belting would lead to a reduction in curing time, but the main problem was centred around the technique used for this heating. Circulation of hot air and infra-red heating proved unacceptable because of undesirable skin heating effects. Dielectric heating was less attractive than microwave heating because of a lack of flexibility concerning product shape, and the fact that the compound had a low electrical breakdown threshold. Also the dielectric should be homogeneous, otherwise some parts will have a different heating rate than others. Microwave heating overcame all of these factors, however, and was selected for the preheating process.

New or Specialised Processes

Using a preheater, reductions in curing time of one third have been obtained. In terms of savings in capital expenditure, this is very significant. With an increase in output of 40 per cent using a flat bed press with a charge length of 12 m, the cost of the preheater necessary to do this was £90 000. The flat bed press cost £800 000 and thus a 40 per cent increase in throughput was obtained with an additional capital expenditure of only slightly over 10 per cent. Based on an electricity charge of 2p/kW hr, the running cost of the preheater was 1.3p per metre length of belt. No additional labour was required, and similar savings were demonstrated using microwave heating for continuous vulcanising. Data on these processes is available from the Electricity Council, (ref. 9.4).

9.2 Electron Beam Welding

Electron beam welding differs fundamentally from other welding processes in that it uses a focussed stream of rapidly moving electrons to generate the heat energy necessary to weld the metal joint being treated.

The basic components of an electron beam welding machine are shown in Fig. 9.4. A heated filament emits electrons which bombard a tantalum cathode button. Electrons are emitted from this bombarded button, accelerated by high voltage, and then focussed so as to impinge on the workpiece to be welded, normally in a vacuum chamber. The focussing arrangement ensures that the beam is concentrated on to a very small area of the workpiece, and as a result the energy concentration which can be achieved is many times greater than that obtained using conventional welding procedures. The heat generated at the focal point of the beam results from the extremely high kinetic energy of the electrons. The conversion efficiency of this energy into heat is very high, and this, coupled with their high velocity, gives deep local penetration of heat into the weld joint. Thus weld depth to width ratios of 20:1 can be achieved, compared with 2:1 which is typical of tungsten inert gas welding. This may be illustrated by the fact that 10 kW can be concentrated onto a 0.25 mm diameter spot during welding.

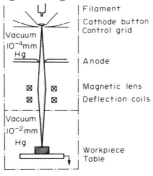

Schematic of electron beam welding

Fig. 9.4 Layout of an electron beam welding machine (Courtesy IRD Co. Ltd.)

A high voltage 10 kW electron beam welding machine is highly versatile with respect to weld depth, and can be used to weld 0.025 mm foils or 65 mm depth welds in steel plate in a single pass.

9.2.1 Advantages of electron beam welding

Electron beam welding, by its nature, has a number of advantages over other welding methods. Obviously by concentrating all the energy directly at the weld joint, and ensuring that the energy in the electrons is efficiently converted into heat, the process is inherently an efficient user of energy. A number of other features which ensure minimum materials wastage also indirectly contribute towards conservation.

This welding method produces a weld with uniformly distributed stress by virtue of the low overall heat input, (in spite of the highly concentrated energy, the input is low compared to conventional welding, which heats up much of the surrounding metal), high depth to width ratios and the parallel sided welds characteristic of the process. Fully machined components can be welded, and because of the very small molten pool, weld shrinkage is much less than that obtained with other techniques, and is uniform throughout the joint.

The beam, because it is narrow, can be directed into restricted spaces where welding would normally be impossible. For example a gear assembly with a gap between the gears of only 3 mm could be welded, as shown in Fig. 9.5. In addition, this figure illustrates the ability of electron beam welding to join dissimilar metals. A vast range of materials can be welded using this technique, including many which are unweldable using other methods. Some materials can be welded in the fully heat treated condition without a reduction in mechanical properties. Tool steel can be readily welded to mild steel, and copper can be welded to aluminum.

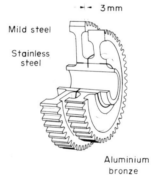

Fig. 9.5 Electron beam welding used in applications where access to conventional welding techniques is impossible (Courtesy IRD Co. Ltd.)

As mentioned above, the process can weld material of various thicknesses. Foils can be welded to thick plates, and a welded joint having a varying thickness (for example an aerofoil shape) can be made in a single pass, and welding speeds are high.

Filler materials are not normally employed in electron beam welding, and in general the weld preparations are less complex than those required for conventional welding methods. The mating surfaces are normally machined, and special weld preparations may be necessary for welding bellows and diaphragms, as shown in Fig. 9.6.

New or Specialised Processes

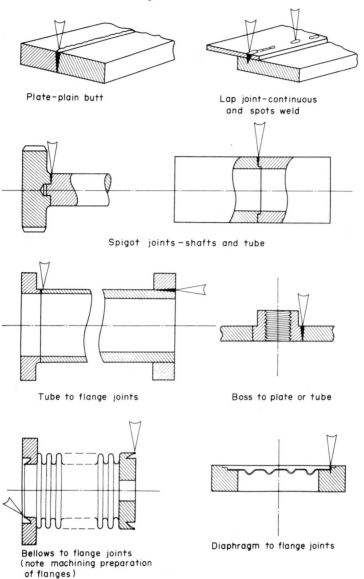

Fig. 9.6 Some weld preparations unique to electron beam welding
(Courtesy of Electron Beam Processes Ltd.)

Electron beam welding in air: High power (greater than 10 kW) electron beam welding, and in many installations most lower power welding, has to date been carried out under vacuum. While this has limited the application somewhat in mass production industries, it offers advantages such as minimal contamination and obviates the need for shielding gases. However work at the Westinghouse Research Laboratories (ref. 9.5) using a non-vacuum electron beam welder of 60 kW power has resulted in a major extension of the capabilities of this welding tool.

To date, only low power electron beam welds have been used outside a vacuum chamber. In air the beam power density is degraded by scattering of the electrons by gas molecules. This has led to the general conclusion that electron beam welding in air requires the gun and workpiece to be located very close to one another, because if the power density is reduced, the rapid melting conditions and narrow weld zones cannot be maintained. With beams of high power and high power density, however, the gas in the beam path is heated to temperatures in excess of 1000°C, reducing its density and hence its ability to scatter the beam. Thus the high power density of the beam can be retained in air over considerable distances, (several cm). Westinghouse have produced single-pass butt welds with a depth to width ratio of 4:1 in 3.8 cm thick steel at a speed of 0.77 cm/sec. Seam welds of two 3 mm thick hot-rolled steel sheets can be produced at speeds of up to 15 cm/sec. It is claimed that these high power machines need not be significantly larger or more costly than those of much lower powers, and with the energy efficiency of the process exceeding 50 per cent, the energy utilisation is approaching the theoretical optimum. (This is based on the ratio of the theoretical energy to melt the metal in the weld area to that expended by the welder).

60 kW is not the maximum power attainable, but even at this level it is believed that the cost justification for electron beam welding in air is now sufficiently great to allow a considerable broadening of its application.

It is possible to obtain samples of electron beam welding on jobs which may be useful transferred to this process with the assistance of a number of companies, some of whom are listed in Appendix 4.

9.3 Fluidised Bed Technology

As with many other techniques which tend to arouse much more interest at times when their attractiveness becomes evident as a result of a new emphasis in a particular field, in this case energy conservation, fluidised bed technology is anything but new. Leva (ref. 9.6) cites fifteen years of development in his study of fluidisation, written in 1959, and the technique was used by the Romans as a means for purifying ore.

The phenomenon of fluidisation relates to a particular mode of contacting granular solids with fluids, either liquid or gaseous. Consider a bed of sand particles resting in a vessel with a porous bottom, as shown in Fig. 9.7 (a). If air is then passed upwards through the porous base, there will be a

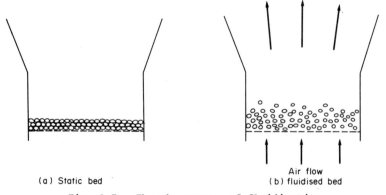

Fig. 9.7   The phenomenon of fluidisation

New or Specialised Processes

particular airflow rate at which the sand particles will be moved slightly away from one another and become suspended in the flow. The bed of sand particles will then resemble a high viscosity liquid, and the particles can be moved around with the expenditure of much less energy than when they were tightly packed together. The sand is then said to form a 'fluidised' bed. There are a number of states of fluidisation. A bed in which the onset of fluidisation has just occurred is known as a quiescent fluidised bed, whereas at much higher air flows the particle movement is aptly described as a turbulent fluidised bed. Then the flow becomes sufficiently to entrain the particles and carry them upwards by some distance, the bed becomes dispersed. Most fluidised bed applications require operation in the quiescent or turbulent region.

Solids in a fluidised bed are perfectly mixed because they can move in any direction, horizontally or vertically, within the confines of the bed. Gas passing through the bed experiences a pressure drop which is a hydrostatic head, independent of gas flow, once fluidisation occurs. The gas flow needed to create fluidisation is a function of particle size, weight and shape, the size being the most important. Typically sizes of particles range from 0.05 mm up to several mm in diameter. In most cases it is necessary for a fluidised bed to accommodate a wide variety of particle sizes, and the airflow necessary to fluidise the largest particles may cause carry over, dispersing the smaller particles. In dryers using this technique, filters will be provided downstream of the bed to collect these particles.

9.3.1 Fluidised bed combustion The major research effort associated with fluidised bed technology has been aimed at the development of high intensity efficient combustion techniques. Of these, the work on fluidised bed boilers is perhaps the most significant. A fluidised bed boiler is basically a box containing boiler tubes, at the bottom of which is an air distributor which takes the form of the porous supporting grid, as shown in Fig. 9.8, (ref. 9.7). In some units a bed of inert granular particles makes up the layers directly above the grid, these particles being heated up to a temperature which will

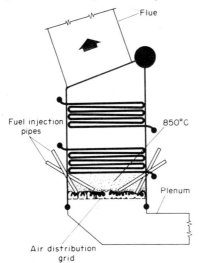

Fig. 9.8 A fluidised bed boiler

support combustion of, for example, coal. Preheated coal particles are then fed into the bed, and the heat is absorbed by the inert particles making up the bed, and act as the medium for heat transfer to the boilers. Liquid and gaseous fuels can also be used.

Fluidised bed boilers can operate under pressure, and it has been suggested (ref. 9.8) that the operation of a bed at 16 atm pressure could produce a most effective compact unit suitable for raising steam for very large power stations, as illustrated in Fig. 9.9. (Combustion intensities are up to 50 times those obtained in conventional pulverised coal furnaces, even though the temperature of the bed is kept at between 750°C and 950°C).

It is now possible to convert conventional boilers to fluidised bed combustion, permitting simple changes in fuel as and when required, be it coal, oil or gas (see Appendix 4), and such a system can use poor quality fuels, (ref. 9.9). Fluidised beds can also be used for incineration and coal gasification (ref. 9.10).

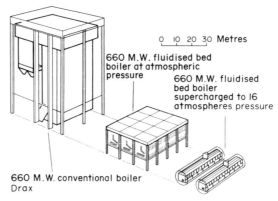

Fig. 9.9 The effect of pressurisation on the size of fluidised boilers having an output sufficient for a large power station

9.3.2 Fluidised bed dryers and coolers Fluidised bed dryers have been described in Chapter 5, in the particular context of heat recovery, and little need be added here as far as the simple dryers are concerned.

This technique can also be used to cool particulate matter, cooling being effected both by the airflow through the bed and by tubes carrying cold water which are 'immersed' in the bed. Most of the heat is removed by the water, the air functioning largely to maintain fluidisation.

9.3.3 Component cleaning using fluidisation The latest application of fluidised bed technology is in the area of thermal cleaning. Known as the Thermal Cleaning Bath (TCB), the unit will, it is claimed, clean dyes, workpieces, and even complex sintered alloy components at temperatures of up to 450°C. Success has been achieved in the removal of polymers from filters, a normally very difficult cleaning task.

The TCB relies on the large heat sink or reservoir effect of the bed to maintain a constant uniform temperature and very high heat transfer rate to the component being cleaned.

New or Specialised Processes

9.4 Electric Foil Heating Elements

The variety of electric heating elements now available is very large. In addition to standard units marketed by a number of companies, the user has access to tubular and strip sheathed heating elements which he may form himself to fit a particular application. Mineral insulated heating cable, generally having a diameter of less than 10 mm, is also available for application by the user, normally being supplied in a range of power ratings on the basis of linear Watts per unit length. Heating elements with finned exterior surfaces are offered for air and gas heating in process lines, and more sophisticated band heaters for tank temperature control are typical of the range of units available.

Flexible surface heating elements, formed using thin heating wires supported by a carrier which can be made in a variety of materials, including glass fibre, rubber or PVC, are a convenient method of heating pipelines and are particularly useful for laboratory work, where a heater may, because of its infinitely variable shape, be applied in many different applications.

The above types of heating elements have been in use for many years, and although their applications, and the number of forms in which they are available, have increased substantially in scope over this period, little change in their basic form has occurred. However, the potential of electric heating has been widened by the development of etched foil elements. Similarly the cost, already very competitive, should be much reduced.

Etched foil elements are similar in appearance and manufacture to printed circuit boards, (ref. 9.11). A thin layer of metal foil is bonded to a substrate, a circuit pattern is printed on its surface with a resist, and the uncoated metal is then etched away, leaving the desired conductive path. Unlike conducting copper circuits, the function of the etched heater is to produce a high I^2R loss in order to generate heat. Metals such as Inconel 600 and stainless steel are used. The element in a typical foil heater is shown in Fig. 9.10. This is backed by the insulating material, which is selected on the basis of temperature, electrical and other environmental considerations.

Fig. 9.10 An element in a typical foil heater

Etched foil heating elements probably have the greatest potential in the appliance manufacturing industry, where they offer an extremely safe, reliable and cheap form of heater. It offers several major advantages over the conventional tubular or strip types which will also be of interest in a number of industrial contact heating applications. By its nature, the etched foil heater is essentially a surface heater. While tubular heating elements tend to concentrate their heat in an area limited by the diameter and length of the element, relying on conduction to transmit heat to regions lying between each element, a foil unit which covers the entire area is much more capable of providing uniform heating. Thus the foil element is able to produce the same heating effect as an array of tubular elements by operating at a lower temperature. This leads to increased life, and it is claimed that it is often possible to achieve the desired performance with a lower wattage than the conventional electric heater.

The versatility offered allows one to construct heaters which concentrate more heat in some areas than in others, as shown in Fig. 9.11. Obviously flexibility is no problem with this type of unit, and etched foils should offer similar characteristics, with more uniform heating, to the tape heaters mentioned in the introductory remarks to this section.

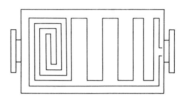

Fig. 9.11 An etched foil heater designed for uneven heat flux generation

Optimum performance is obtained if the heaters are bonded directly on to the area to be heated. These heaters are currently limited to temperatures of about $300^\circ C$, but this is not too restrictive as good contact and temperature uniformity permits operation at a lower temperature than with other electric heating elements. The total thickness of film insulated heaters is only about 0.2 mm, and thus the thermal mass of the heater is very low and response times are correspondingly short. In spite of the temperature limitation created largely by the restrictions enforced by the insulating medium, surface fluxes of up to 6 W/cm^2 are possible.

The etched foil heaters may exhibit different thermal expansion characteristics to those of the materials to which they are bonded, although these can often be closely matched. If this is not possible, resilient insulation can assist in overcoming any difficulties.

This type of heater appears to offer many advantages, and although its temperature limitations may preclude its use in some industrial applications, it can provide very uniform heating with economies in power consumption.

9.5 The Laser as a Welding Tool

The word LASER is an acronym for Light Amplification by the Stimulated Emission of Radiation. Although the process of stimulated radiation emission was defined by Einstein over fifty years ago, it was not until 1960 that it was employed to produce light amplification in the first laser. Subsequent progress has been very rapid and many different types of lasers, employing gaseous, liquid or solid state lasing media have been developed.

In the operation of a laser the absorption of energy provided by an external source causes a significant fraction of the atoms or molecules in the lasing medium to be, for a short period, excited into electronic, vibrational or rotational states having a higher-than-normal energy. These atoms or molecules then decay naturally back to their stable (lower energy) state with the emission of the excess energy in the form of electromagnetic radiation, including light. As a result of stimulated emission the quantum of radiation energy given out by each atom or molecule can cause further emission when it interacts with a second excited atom, thereby causing it to de-excite. When this process is carried out by a large number of atoms appreciable amplification of the initial radiation energy can result.

The techniques used to obtain a significant fraction of the atoms or molecules in an excited state - called a population inversion - depend on the type of laser. In a pulsed solid state laser the excitation is obtained by directing a pulse of light energy into the laser medium from a high power flash lamp. In a laser which uses gas mixtures, energy may be fed into one species by a d.c. discharge and transfer of energy to the second (lasing) species may take place by atomic collisions, so that atoms of the second species then emit laser radiation while decaying back to their ground (un-excited) state.

In high gain systems a single pass or transit of light of the appropriate wave-length through the laser may yield a sufficiently intense output beam. In situations where the gain is lower the output can be enhanced without the system being unduly long by making the volume containing the laser medium (generally a cylinder) into an optical cavity by the use of reflecting mirrors. In this way the light can be reflected many times through the laser medium before emerging through one mirror, which is made partially transmitting for this purpose.

The main features of the light obtained from lasers which make them different from other sources are that it is:

(a)   monochromatic, i.e. of a single colour or wavelength

(b)   coherent, i.e. all the wave trains that constitute the laser beam are in step

(c)   directional, being emitted as a well-defined narrow beam

(d)   capable of high intensities.

The output wavelength of the various lasers now available range from the ultra-violet part of the spectrum, through the visible to the far infra-red. Power levels of present continuous lasers range from milliwatts to hundreds of kilowatts, while pulsed lasers having output energies of less than 1 Joule have been built, with pulse lengths from 10^3 to 10^{12} sec and with power levels up to gigawatts (10^9 W).

The high energy output of the laser has been used for many years to enable the device to be used as a welding tool. While much of the early data acquired by laboratories investigating the laser in this and other industrial applications tended to concentrate on the scientific aspects of laser operation, rather than on the practical aspects of interest to industrial users, this is no longer the case.

Laser welding is closely related to electron beam welding as far as its operating advantages and metallurgical benefits are concerned. It was claimed that laser welding was preferable in many industrial applications because no vacuum chamber was needed. However, as described in Section 9.2 electron beam devices operating successfully in air at high powers have now been developed, and the above claim therefore loses much of its impact. The laser is a simple unit, and in comparatively low power welding applications, such as spot welding, it is compact and can be hand-held, as shown in Fig. 9.12. The simplicity of lasers extends throughout the range of weld powers, into the region where several kilowatts are used. The same optical system and shield gas can be used for welding most metals, ranging from steel to zirconium (ref. 9.12). As with electron beam welding, no fluxes or welding rods are needed, and there are also no electrodes.

Fig. 9.12 A hand-held laser used for spot welding
(Courtesy IRD Co. Ltd.)

It may be thought that because metals tend to be poor absorbers of infrared radiation at room temperature, very large amounts of power will be needed for laser welding. However good reflectors such as aluminium can be welded using 1 kW lasers, because the ability of a metal to absorb radiation increases rapidly once its melting point has been reached. (Similar amounts of power can be used to cut metal - in this case the laser is used to vaporise the metal, rather than just melting it).

9.5.1 Characteristics of laser welds In laser welding, the depth of weld penetration and the weld thickness, is, for a given laser power, a function of the point on which the laser beam is focussed. Fig. 9.13 shows that maximum penetration, and the corresponding minimum weld thickness, is obtained if the beam is focussed on to a point between 1.2 and 2.5 mm below the metal surface, (ref. 9.12). The data shown was obtained welding steel at a rate of 127 cm per minute, with 1.5 kW from a continuous wave CO_2 laser. It can be seen that a beam focussed on to the surface of the workpiece has a much inferior performance.

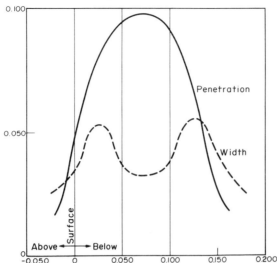

Fig. 9.13 Laser welding - the relationship between weld width and penetration. (Courtesy Laser Focus)

The penetration of the weld is reduced if the weld speed is increased, as shown in Fig. 9.14, these results again being obtained with a 1.5 kW gas laser welding type 302 stainless steel. Upper limits of welding rate occur when the metal resolidifies too quickly to flow and fuse to adjacent metal.

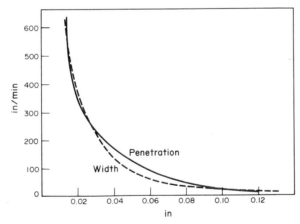

Fig. 9.14 The performance of a 1.5 kW laser in terms of weld rate and penetration. (Courtesy Laser Focus)

Weld preparations are similar to those used in conventional welding, lap and butt joints being most frequently used. However because of the localised heating characteristic of the laser beam, it is important to ensure that tolerances in alignment of the joint are kept to a minimum, otherwise the melting process may not occur at the correct point.

As mentioned in the introductory remarks, the laser is capable of being pulsed. This means that a large number of single weld operations, such as spot welding can be carried out, as well as the continuous welding described above (ref. 9.13). Efficient high power lasers having the ability to carry out repetitive welding operations at the rate of 40 per second are now available. Thus in any automatic spot welding application, such as in the electronics industry, the limiting factors are more likely to be connected with the rate at which the job can be fed to the welding machine, than the process speed of the laser welder itself.

9.5.2 Applications of laser welding

The above discussion cites lasers having powers of 1.5 kW for continuous welding. More powerful units are available, and deep penetration welds can be made in large automotive components (ref. 9.14). Automated welding of miniature electronic or mechanical parts, where weld accuracy, repeatability and quality are necessary is ideally carried out using the pulsed laser, (ref. 9.15). Most other welding techniques are incompatible with automated systems because of the need to replace electrodes, and the less accurate operation possible.

Laser welding of miniature telephone relays at the rate of 1000 per hour is one of the applications of the system described by Aeschlimann and Monnier (ref. 9.15). A similar unit constructed by Alcyon Electronique and Physique SA of Lausanne in conjunction with the Omega Watch Company for welding watch components has also been successfully applied. While laser welders of this type cost five to ten times as much as their conventional counterparts, minimum distortion of the component occurs and quality control requirements are met so easily that checks can be reduced to the level of automatic monitoring and control of the laser output. The cost of each weld was found to be low, only about 0.3p, and the quality of the finished parts was so high that a zero rejection rate could be assumed in cost estimates.

Lasers can be used for spot welds where spot sizes are as small as 0.005 mm, but in some welding applications, such as the joining of plastic film, wide welds without the need to penetrate the surface as in metal laser welding may be required. In these instances the laser may be de-focussed, giving a weld band several millimeters wide.

Lasers can also be used for cutting (cloth can be cut accurately, and this method does not exert any forces on the material), and for heat treatment, which may be required locally and in hitherto inaccessible places.

REFERENCES

9.1 Leslie, P. Dielectric heating; theory and practice. <u>Industrial Process Heating</u> pp 28 - 29, Sept. 1973.

9.2 Anon. Dielectric and microwave heating. The Electricity Council, Publication EC3104, London, May 1973

9.3 Hopwood, J.E. and Neller, W.C. Microwave preheating of conveyor belting. Available from: Dunlop-Angus Belting Group, PO Box No.7, Liverpool L24 1UY (Published 1975)

9.4 Anon. Preheating and vulcanising. The Electricity Council, Publication EC3073, London (1973)

New or Specialised Processes

9.5   Lowry, J.F., Fink, J.H. and Schumacher, B.W. A major advance in high power electron beam welding in air. J. Applied Physics, 47, 1 (1976)

9.6   Leva, M. Fluidization. McGraw-Hill, New York, (1959)

9.7   Ehrlich, S. and McCurdy, W.A. Developing a fluidized-bed boiler. Proc. 9th Intersoc. Energy Conversion Engineering Conf., ASME, Paper 749133, San Fransisco, (1974)

9.8   Thurlow, G.G. Fluidised bed combustion - the present situation. Heating & Air Conditioning Journal, 44, 521 April 1975

9.9   Butler, P. Oil or coal burning fluid bed boiler may help in future crisis. The Engineer, 6 Nov. 1975

9.10 Strimbeck, D.C., Sherren, D.C. and Keddy, E.S. Process environment effects on heat pipes for fluid bed gasification of coal. Proc. 9th Intersoc. Energy Conversion Engineering Conf., ASME, Paper 749108, San Fransisco (1974)

9.11 Moore, B.J. Etched foil elements: A new approach to uniform appliance heating. IEEE Transactions on Industry Applications, 1A-11, 2, March/April, 1975

9.12 Engel, S.L. Kilowatt welding with a laser - Technology update. Laser Focus, 12, 2, Feb. 1976.

9.13 Saifi, M.A. and Vahaviolos, S.J. Laser spot welding and real-time evaluation. Journal of Quantum Electronics, QE-12, 2, Feb 1976

9.14 Schmatz, D.J. and Yessik, M. Laser processing in the automotive industry. Digest of Technical Papers, IEEE/OSA Conference on Laser Engineering and Applications, (1975)

9.15 Aeschlimann, J.P. and Monnier, P. Automated welding of minute parts - Technology Update. Laser Focus, 12, 3, March 1976.

Alternative Sources of Energy

A considerable variety of alternative sources of energy are available to us. Some will require substantial, if not unacceptable, amounts of expenditure, both in manpower and money, to fully exploit. Others, such as solar energy, are currently successfully used on small domestic heating units, but need further development before they can be considered seriously for commercial or industrial use on any scale.

Several of these energy sources are described in the following few pages.

10.1 Wind Power

Of the numerous sources of energy which are available for exploitation on a long term basis, wind power is probably of less significance to industry than others being developed. It may have some limited applications in agriculture and associated activities however, and windmills, albeit of comparatively low power, are commercially available, (see Appendix 4).

The power output of a windmill (now more commonly called wind generators to accommodate the numerous types available) is proportional to the cube of the wind velocity approaching normal to the rotor. If we consider a windmill of diameter D, with an approach wind velocity V, the amount of energy in the wind passing through the disc traced out by the blades is:

$$P = \tfrac{1}{8} \rho \pi D^2 V^3$$

where ρ is the air density

Most wind generators have efficiencies of about 50 per cent, i.e. only 0.5P will be extracted, and this peak efficiency will only occur at the optimum design value used for the ratio of the blade speed to air speed. Thus the power/speed characteristic of a wind generator will be of the form shown in Fig. 10.1.

Apart from the conventional type of windmill, ducted turbines can be used (see Fig. 10.2). In addition a number of other forms of wind generator are being developed, including the Derrieus generator and the vertical axis turbine (see Figs. 10.3 and 10.4). Vertical axis machines seem to offer significant advantages in terms of cost, the Derrieus system, developed by the United States Atomic Energy Commission at their Sandia laboratory costing, in production, about £300/kW. The main disadvantage of this type is the fact that it is not self starting. However the vertical axis machine proposed by Lewis (ref. 10.1) overcomes this drawback. Output of vertical axis machines is of the order of 1 kW for a unit 5 m high in a 32 km per hour wind, (ref. 10.2). Units of this type are also insensitive to wind direction.

The largest windmills constructed to date have had outputs of 1 to 1.25 MW, but in order to achieve this, diameters of the blade disc have approached 60 m.

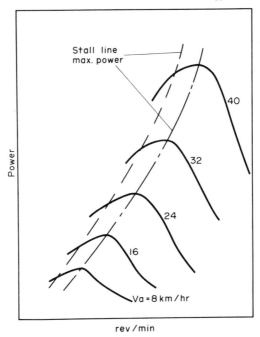

Fig. 10.1 Typical power/speed characteristics of a wind generator

The National Science Foundation are now sponsoring development in the United States of new wind generators in the MW range, with construction of a prototype scheduled for 1976 (ref. 10.3).

It will be of interest to follow United Kingdom activities in wind generator development. Some companies are putting forward proposals to use wind power, converted into heat, for drying agricultural produce. As well as supplying heat, the windmill provides power for the hot air circulating fans.

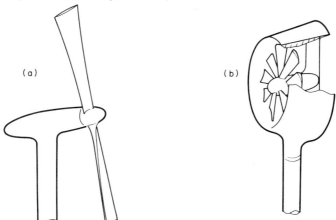

Fig. 10.2 Conventional and ducted windmills

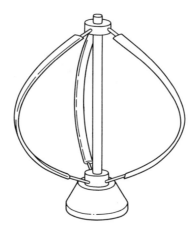

Fig. 10.3 The Derrieus generator

Fig. 10.4 A vertical axis turbine, which is insensitive to wind direction

Because wind power is unreliable as a continuous source of energy, the use of thermal storage systems would be necessary were any application to industrial processes to be considered.

10.2 Geothermal Energy

Geothermal energy, or heat in the earth's interior, is regarded by many as the second most important future source of energy, next to solar power.

The volume of the earth is approximately 10^{12} km^3, and except for the thin crust, all of this volume is believed to be at temperatures sufficiently high to keep the material comprising it in a molten state. While this is obviously a massive potential source of energy, it is to a large extent inaccessible or of only limited use because the technology for recovery and utilisation is not sufficiently advanced, (ref. 10.4). In the United Kingdom, for example, the

Alternative Sources of Energy

crust is about 30 km thick, and the temperature at the base of the crust is estimated to be approaching 1000°C. The temperature below the surface increases at a rate of about 30°C/km (except where the crust is penetrated by hot springs etc.), and current drilling technology will enable sufficient depth to be reached to tap heat at a temperature of about 200°C. In other parts of the world it is possible to find high pressure steam at comparable depths, and other forms of heat energy are potentially useful sources.

There are four basic types of geothermal energy resources; dry steam, wet steam, hot dry rock, and geopressurised zones, (ref. 10.5). Dry steam, the smallest resource, is the only form which has been exploited to any degree to date. It results from the boiling of water in an underground reservoir, the superheated steam passing upwards towards the surface. In order to exploit the energy contained in the dry steam, it is only necessary to feed it through a separator to remove any entrained particles and then pass it through a steam turbine connected to an electricity generator. No fuel handling and converting plant is needed.

Unfortunately dry steam has only been found in three locations, the largest field being near San Francisco. During 1976 it is anticipated that 700 MW will be generated from this field. Other sites being exploited are in Italy and Japan, and Table 10.1 gives data on some existing installations.

TABLE 10.1 Output and Capital Cost of Geothermal Power Plants

Condensing-turbine power plants

Larderello

net capacity 2 x 26 MW: 170 US$/kW;
net capacity 1 x 15 MW: 226 US$/kW;

The Geysers

net capacity 2 x 55 MW: 125 US$/kW;
net capacity 2 x 28 MW: 135 US$/kW;

Back-pressure power plants

Larderello

net capacity 1 x 15 MW: 95 US$/kW;
net capacity 1 x 4 MW: 105 US$/kW;

Namafjall

net capacity 1 x 3 MW: 60 US$/kW.

Wet steam is believed to be much more abundant than dry steam, and its presence is indicated by hot springs. Some systems have been developed for generating electricity, but the resulting hot water is normally so contaminated with dissolved minerals that it is highly corrosive, and contains in some instances up to 30 per cent of dissolved solids. This also hampers the search for this resource, as drill components are attacked. Such attack would also prevent the use of conventional materials for pipelines etc., and the cost of extraction is therefore likely to be high. Most of the reservoirs are also insufficiently pressurised to support extraction of the wet steam, and mechanical pumping is required. Therefore it is unlikely that this form of geothermal energy will be fully exploited for a considerable number of years owing to the economic factors involved.

It has been estimated that the amount of energy contained in the rock forming the earth's crust in the Western United States is equivalent to the total US coal reserves (see Chapter 2). The heat in rocks must be extracted via some heat transfer medium introduced into them, and it has been suggested that water could be pumped from the surface into large subterranean chambers in the 'hot rock' created by explosives. The water heated in the chambers would then be extracted by a different route. Hot rocks are available everywhere in the crust, and thus it is the most prolific energy resource. The technology of exploitation is not developed, but work is being carried out at a number of laboratories throughout the world to improve knowledge of the subject and to formulate plans for exploitation.

The fourth geothermal energy resource is the geopressurised reservoir. This type of reservoir is known to occur in several parts of the world, including along the Gulf Coast of the United States. It consists of a deeply buried sediment containing hot water under very high pressure, with a significant content of dissolved natural gas. The water pressure is sufficiently high to support the rock above it, the rock thickness generally exceeding 4 km, (ref. 10.6). Thus a geopressurised reservoir contains three energy resources, hot water, pressure energy and natural gas. The technology for exploitation of this resource, and indeed the knowledge of the full nature of the resource, are not available at present, and in common with hot rocks, further extensive study is needed before a conclusion can be reached as to the viability of geopressurised energy as a practical energy form.

Volcanoes and lava flows are the most obvious and dramatic demonstration of the potential energy existing below the earth's crust. Proposals have been put forward for extracting energy from lava pools and subterranean lava reservoirs. The technology involved would be more expensive than that needed to extract heat from the 'conventional' geothermal energy resources detailed above, but the phenomenon in itself is a visible example of the impact geothermal energy could make in the future.

10.3 Wave power

The United Kingdom Government Central Policy Review Staff recommended in 1974 (ref. 10.7) that "The first stage of a full technical and economic appraisal of harnessing wave power for electricity generation should be put in hand". It is argued that as far as the United Kingdom is concerned, wave power, if feasible, would have some favourable features. With a coastline of approximately 1500 km which could be used for the location of a practical wave power system, the potential energy generation using this technique is of the order of 30 000 MW, or about 50 per cent of the current installed capacity of the CEGB. A particularly attractive argument in favour of wave power, especially when compared with solar power, is that it provides a seasonal peak in the winter, when demand for electricity is also at a peak.

The above recommendation has now been acted upon, and work at the University of Edinburgh (refs. 10.8, 10.9) and other centres has demonstrated experimentally that more than 80 per cent of the total wave power from water waves could be extracted, using a specially contoured rocking device, illustrated in Fig. 10.5. The device rotates about its centre, O, and absorbs power from waves coming from the direction shown. Its stern is a half cylinder centred at O, but at its lowest point it grows into a surface which is another cylinder centered on O'. This shape continues until it reaches an angle θ to the vertical, at which point it develops into a tangent which is continued above the surface of the sea. In Salter's first model O' was one half radius above O and θ was 15°. The efficiency for wave lengths of about eight times the diameter of the small cylinder is over 80 per cent.

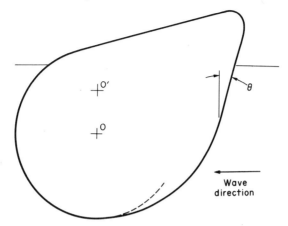

Fig. 10.5 Cross-sectional form of a wave generator device developed at Edinburgh University

When the vane, or 'duck' moves, there is no change in the displacement of the water behind it, and the change in displacement in front of it rises from zero at the base to amounts close to those of the approaching wave. The motion of the vane is best described as 'nodding', and by nodding about its axis the vane will produce useful work. While the rotational velocity of the vane will be far too slow for conventional electricity generation methods to be used, Salter proposes to locate each vane on up to 100 radial piston units which can provide radial and axial location as well as power take-off. These will generate a high pressure hydraulic oil flow which could be used to drive hydraulic swash plate motors at speeds more akin to those needed for electricity generation, (ref. 10.10).

With regard to the likely power output of a string of vanes rotating (or nodding) on a common tubular backbone having a length of about 500 m, as shown in Fig. 10.6, the desirability of acceptable costs may mean a lower efficiency than the 80 per cent plus, demonstrated by Salter, but even at lower efficiencies a 14 m diameter vane will average a power output of 50 kW/m over the year, peaking to 200 kW/m.

As with this, and any other wave power generation system (another three systems are under development in the United Kingdom alone), full output of power could not be guaranteed at any particular time, and stand-by capacity would have to be provided, possibly in the form of gas turbine generator sets. The Central Policy Review Staff estimated that the capital cost of a system based on that developed by Salter would be of the order of £420/kW, and the electricity thus generated would be about 2.5 times more expensive than the expected cost of generation by nuclear fission, and 40 per cent higher than harnessed wind power.

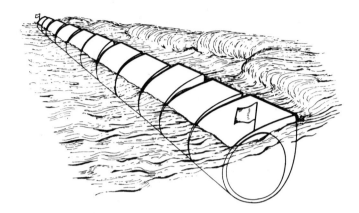

Fig. 10.6 Artist's impression of a string of 'nodding ducks' to recover wave energy

10.4 The Hydrogen Economy Concept

A number of authorities believe that hydrogen could form the basis of energy policy in the period following the downturn in supply of oil and natural gas, i.e. from 1995 onwards. It has been argued that reliance on electricity, be it generated in nuclear power stations or by other means, would be unwise because it cannot be stored and is not particularly portable. Transport using overhead cables is environmentally undesirable, and buried cables are expensive to lay. Even solar energy, wind power and wave power are incapable of producing a reliable, portable form of energy if directed towards the production of electricity. Hydrogen can be stored, although storage costs are currently about three times as high as for natural gas, (engineering developments could reduce the cost), and hydrogen can be used directly as a chemical feedstock or as a fuel for transport. The power density of hydrogen is compared with other fuels in Table 10.2.

TABLE 10.2 Power Density of Fuels

Fuel	Energy/mass MJ/kg	Density g/cm^3	Energy/volume MJ/l
Gasoline	48	0.74	35
Fuel oil	44	0.96	41
Liquid hydrogen	121	0.07	8
Methanol	34	0.83	28

Hydrogen is not a primary fuel, but can be produced from water by the addition of energy. The only direct combustion product of hydrogen is water, so the normal pollution problems associated with other liquid fuels will not be present. The conversion of water to hydrogen can be performed in three ways, (ref. 10.11). It can be generated by using electricity to electrolyse water, freeing hydrogen and oxygen at the electrodes. Alternatively it can be obtained by heating water to a very high temperature (>2500°C), where it spontaneously splits into hydrogen and oxygen. (Because of the high temperaatures needed, well above those in nuclear reactors, this technique is unlikely to be used in the foreseeable future). Thermochemical splitting is the third method of manufacture. A number of materials take part in the production of hydrogen and the reaction temperatures are somewhat less than that required to split water. Ferrous chloride and water reacting at 650°C generates hydrogen (ref. 10.12); and calcium bromide heated to 730°C, forming hydrobromic acid, which can then be added to mercury, yields hydrogen at 250°C (ref. 10.13). In both cases all products with the exception of oxygen and hydrogen can be recycled. The temperatures required for these reactions to take place can be obtained with high temperature nuclear reactors, and development is continuing on a number of thermochemical cycles to obtain acceptable efficiencies and economies.

Production of the hydrogen is only one step in the development of an acceptable new fuel. Storage need not create problems if the hydrogen is kept as a gas, but storage for use in remote locations will probably necessitate keeping the hydrogen in liquid form. A number of hydrides, such as magnesium hydride, can hold hydrogen at room temperature in a 'solid' form. The hydrogen can be released under controlled conditions by heating the hydride, in the case of a magnesium release occurring at temperatures around 260°C.

Transmission of hydrogen by pipeline is believed to be economical, particularly over distances in excess of 500 km (ref. 10.13), and this is advantageous if nuclear reactors are to be used to provide heat for the hydrogen production cycle. This would meet the pressure for remote siting of reactors, with the electricity generation carried out in local centres of demand. An overall concept of a 'hydrogen economy', taking into account the above, is shown in Fig. 10.7.

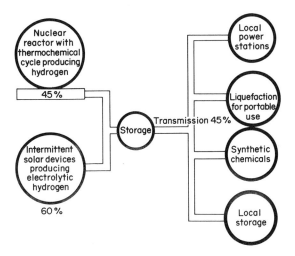

Fig. 10.7 A schematic of the 'hydrogen economy' showing system efficiencies

Of the several potential new energy sources discussed in this chapter, the production of hydrogen within acceptable economic constraints is one of the most difficult to develop. EURATOM, as well as several authorities in the

United States, are engaged in studies to this end, but developments are unlikely to have any impact on industrial energy consumption policy for many years.

10.5 Solar energy

Solar radiation is the most abundant form of energy available to us, and of the energy sources discussed in this chapter solar energy has probably received the most attention from research workers. There are many ways of expressing in an impressive manner the quantity of energy received by the earth from the sun. Eckert, (ref. 10.14) states that the energy received during one or two weeks is equivalent to all the known fossil fuel reserves on earth. Alternatively, the solar radiation falling on $\frac{1}{500}$ th of the area of the United States, if converted with an efficiency of 100 per cent, would meet all the current energy requirements of that country, (ref. 10.14).

There are a number of ways in which solar energy can be used, either directly or indirectly:

(i)   Indirect utilisation through geophysical or meteorological effects such as wind power or ocean thermal gradients.

(ii) Use of biological effects such as photosynthesis, which can yield chemical fuels

(iii) Conversion of solar energy into electricity

(iv) Direct use of solar energy for heating.

Wind power is discussed elsewhere in this chapter, and other indirect utilisation techniques may be studied with the assistance of the Bibliography. In this section we will be concerned with the conversion of solar energy into electricity, and the use of solar energy for heating.

10.5.1 Electricity generation
The generation of electricity using solar energy may be carried out in two ways. Solar radiation may be concentrated using a parabolic mirror to generate a high pressure vapour, which then uses a conventional power cycle via a turbine to drive an electricity generator. Alternatively a photovoltaic cell may be used to directly convert the solar radiation to electricity.

A system utilising a steam turbine is illustrated in Fig. 10.8, (ref. 10.15). The parabolic mirror is more attractive for this application than a flat plate solar collector because much higher temperatures, hence turbine efficiencies, can be achieved. A flat plate collector may heat the fluid to 80 to 110°C, but a parabolic mirror of economic proportions could permit vapour temperatures of 300 to 500°C to be attained. Such fluids as mercury have been investigated for use in the collector, and the fluid may be used directly in the turbine, or used to raise steam in a boiler. A number of other thermal cycles are being studied in an attempt to raise efficiency and enable lower grade heat from diffuse sunlight to be utilised. With current estimates of efficiency ranging from 10 to 20 per cent and with generating costs being up to four times as high as those of a fossil-fuelled power station, a substantial amount of work remains to be done before an economically viable system is obtained.

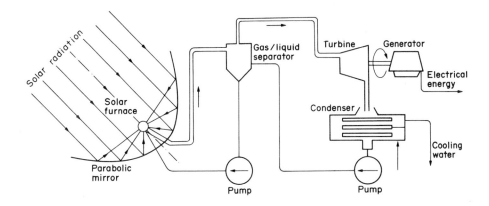

Fig. 10.8 Solar energy concentrators can provide a heat source for turbogenerator sets

Photovoltaic cells, made using semiconductor material, can convert a proportion of the absorbed solar energy directly into electricity, with a conversion efficiency of 13 to 14 per cent (ref. 10.14). The basis of most cells is silicon, which has a maximum theoretical efficiency of 20 per cent, but cadmium sulphide is also receiving attention, and cadmium-tellurium cells give even higher efficiencies (23 per cent). The cost of solar cells is still high, partly because of low production rates, some commercially available cells costing several £ sterling per Watt. On current performances, a 200 MW power station (small by today's standards) would need cells having a total surface area of 5 km^2, (ref. 10.16).

Both of the above techniques provide power only when receiving solar radiation of sufficient intensity, and therefore if substantial reliance is to be placed on these systems, energy storage technology must be extended to cope with very large quantities of heat in the case of the turbine cycle method, or of electrical, mechanical or chemical energy in the case of photovoltaic cells.

10.5.2 Solar heating

The use of solar energy to provide hot water for use in buildings is the simplest and best developed solar conversion system. Solar collectors of the flat plate type are commercially available from a large number of manufacturers, some of which are listed in Appendix 4, and are cheap and reliable. To date the majority of units have been applied in houses but larger buildings are being fitted with solar collectors, which, particularly when used in conjunction with a heat pump, as shown in Fig. 10.9, can provide a large proportion of the total heat energy requirements.

Of course the thermal energy which can be collected is a function of the cloud cover and the latitude of the installation, as shown in Table 10.3, (ref. 10.18), but solar water heaters in some areas of the United States now provide heat more cheaply than electric units (taking into account capital expenditure). Mass production should make these systems competitive with any other current energy form. As hot water storage is already a feature of most domestic hot water services, conversion need not involve too much expenditure.

TABLE 10.3 Solar Energy Received on a Horizontal Surface

| | Watts per square metre | | | | |
| | Clear | | Av clouds | | |
Lat.	June	Dec	June	Dec	Year
60°N	380	30	190	10	100
40	370	130	220	70	150
20	340	220	220	150	210
0	290	310	150	190	190

Reduce by 50 per cent for town
Solar constant 1350 W/m^2

Some estimates suggest that 35 per cent of the energy consumption in buildings in the United States could be met by solar heating by the year 2020, without needing to significantly improve the current level of technology in this area.

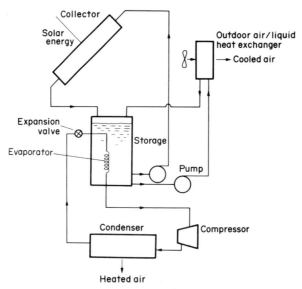

Fig. 10.9 When used in conjunction with a heat pump, flat plate solar collectors can meet thermal demands in buildings

The implications of solar energy for heating in industry are highly speculative at present. Solar collectors of the type just described may provide some of the hot water and air conditioning services in commercial and industrial buildings, but applications in process plant are more difficult to define. One of the few studies in this area concerns the use of solar energy for drying, in this case the removal of moisture from oil shale, (ref. 10.19). This would reduce the energy input needed to carry out the thermal decomposition process leading to liquid fuel production. Solar drying can simply be

Alternative Sources of Energy

effected by spreading the wet material over a suitable piece of ground, having first determined the best thickness of layer, taking into account the ground area and rate of drying required. Other solar dryers used in agriculture, have used air heated up to $100^\circ C$ by solar radiation, but in both cases continuity of drying is determined by the availability of sunshine.

10.5.3 Solar energy in relation to conventional energy resources

It will have become evident to the reader that the utilisation of solar energy on a small scale is entirely possible using commercially available equipment. The area where development is required is in relation to extending the scale to meet the requirements of whole communities, be the commodity finally provided heat or electricity.

The potential of solar energy is illustrated in Fig. 10.10, (ref. 10.16). This represents the future energy supply in the United States, including the likely contribution from solar electricity, photosynthesis and heating in buildings.

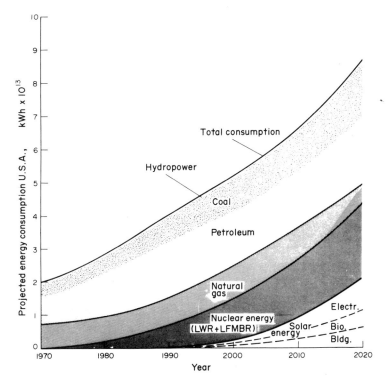

Fig. 10.10 The potential share of solar energy in the future energy supply in the United States.

10.6 Conclusions

Existing sources of energy, particularly those based on fossil fuels, are likely to increase in cost, both as a result of inflation, and in the medium term, because of an increasingly unfavourable supply/demand characteristic.

Solar energy and wind power are, with the present technology available, probably of greater benefit to domestic energy users for home heating etc. rather than to industry. Those who foresee the 'electric economy' may, in the longer term, be proved correct as replacements for fossil fuel will be based largely on electricity generation, either by wave power, geothermal energy or by earth-orbiting solar cell arrays. (A report published early in 1976 estimates that electricity generated by this technique would cost us 5c/kW hr, competitive with the forecast costs of conventionally produced power).

Before these resources, which are essentially inexhaustible, can be applied in industry, many years of conservation and, possibly, energy shortages must be tolerated. Reliance on coal and nuclear energy until the year 2000, when new energy sources may be ready for widespread application, is forecast in the United Kingdom. With some of the alternative energy sources, in particular solar energy, industry may accelerate utilisation by its own initiative in demanding new energy supplies less prone to fluctuating prices and availability.

REFERENCES

10.1   Lewis, R.I. Wind power for domestic energy. Proc. Conf. on Housing and Energy. Univ. Newcastle upon Tyne, 15 - 17 April, 1975

10.2   Booth, D. Derrieus generator spins in the wind for a cheap kilowatt. The Engineer, 13 Feb. 1975.

10.3   Anon. Wind power: A new look at an old draft. Mosaic, 5, 2, Spring 1974.

10.4   Smith, M.C. Geothermal power. Physics and the Energy Problem. Am. Inst. Physics Conf. Proc. No. 19, New York (1974)

10.5   Anon. Geothermal energy: Taking Mother Earth's temperature. Mosaic 5, 2, Spring 1974.

10.6   Hickel, W.J. Geothermal Energy. University of Alaska Publication, (1973)

10.7   Energy Conservation. A study by the Central Policy Review Staff, London, HMSO (1975)

10.8   Salter, S.H. Wave power. Nature, 249, June 21, 1974.

10.9   Salter, S.H. et al. Characteristics of a rocking wave power device. Nature, 254, 504 - 506 (1975)

10.10   Salter, S.H. et al. The architecture of nodding duck wave power generators. The Naval Architect. pp 21 - 24, Jan 1976.

10.11   Gregory, D. The hydrogen economy concept. American Institute of Physics. Conf. Proc. No. 19, 171 - 174, (1974)

10.12   Hampson, P.J. et al. Can hydrogen transmission replace electricity? CEGB Research, 2, 4 - 11 (1975)

10.13   Slesser, M. Why a hydrogen economy? <u>The Chartered Mechanical Engineer</u>, pp 57 - 60, Feb 1975

10.14   Wolf, M. The potential impacts of solar energy. <u>Energy Conversion</u> 14, 9 - 20, (1974)

10.15   Sjoerdsma, A.C. and Over, A.J. Energy conservation - ways and means. Future Shape of Technology Publications, No. 19, The Netherlands, (1974)

10.16   Kelly, B.P., Eckert, J.A. and Berman, E. Investigation of photovoltaic applications. Proc. International Conf: The sun in the service of mankind, Paris (1973)

10.17   Corman, J.C., McGowan, J.G. and Peters, W.D. Solar augmented home heating heat pump system. Proc. 9th Intersoc. Energy Conversion Energy Conversion Engineering Conf., ASME, San Francisco, Aug. 1974.

10.18   Hawthorne, W.R. Energy: A renewed challenge to engineers. <u>Proc. Institute of Mechanical Engineers</u>. 189, 52/75, 446 - 460, (1975)

10.19   Löf, G.O. Use of solar energy for heating purposes: Solar drying. Proc. United Nations Conference, General Report, Agenda Item III C.3, Paper GR/15(S), (1961).

Appendix 1
Energy Management and the Energy Audit

The management of energy in a company should be undertaken using the basic techniques which are applied in assessing the relative merits of, for example, the options for capital expenditure as part of a plant expansion programme. Decisions in this latter area are normally taken at boardroom level, and the ultimate responsibility for, and direction of an energy management programme should also be a function of the directors. If the commitment to the energy management programme is not seen to exist at the top of a company, those responsible for its implementation at plant level will find their job more difficult.

Outside the boardroom there are a number of ways of organising the programme. The most common first step is to appoint an energy manager, who should be given sufficient authority to enlist the support of personnel as required at most staff levels within the company. In a large organisation with factories located in several parts of the country (and overseas), the energy manager will probably advise on the appointment of assistants at each factory, and his function will be to co-ordinate and direct the activities of these assistants, reporting as necessary to the board when significant capital expenditure on equipment is required. Ideally the manager should be allocated financial resources which he can use for 'good-housekeeping' and other tasks, as recommended by his assistants in the field. His budget should be sufficient to permit access to outside resources such as consultants who can assist in audits and more specialised system design.

At plant level it will most often be discovered that the energy used by the various items of equipment is difficult to quantify. While the total costs of electricity, gas and water, for example, will be available, the particular efficiency of individual plant items will be unknown. This data is a necessary part of the energy audit, and in some cases it may take several months to collect and accurately assess this information.

Most companies will have an energy requirement for what may be called 'environmental' tasks, and a requirement for 'manufacturing' tasks. The former includes lighting, space heating and hot water services in offices, production areas and warehouses. The latter includes the energy usage in equipment using heat, electric motors, compressed air supplies and all other plant used in the manufacturing process. (Such an audit should also take into account transport, both internal and external, although this factor is not discussed in this book). As a result of the audit, the amount and cost of energy consumed per unit of production may be assessed.

While it is probable that seasonal factors will not have a significant effect on the energy used in the production process, lighting and space heating (or air conditioning) will be a strong function of the time of year. The seasonal variations may be taken into account using the 'degree-day' concept. A degree-day is the daily difference in °C between a base temperature of

Appendix 1

15.5°C and the 24 hour mean outside temperature (when it falls below the base temperature). Degree-day data is published by the British Gas Corporation and the Meteorological Office, and is used as a matter of course by air conditioning engineers.

Lighting and air conditioning loads are comparatively simple to monitor, but combustion processes and water usage will often need to be monitored over a considerable period commencing at the beginning of the audit. Because of the numerous factors involved, such as air/fuel ratios, oil viscosity, exhaust gas make-up etc., a number of items of measuring equipment may be needed. (Suppliers are listed in Appendix 4. Notes on techniques are given in Chapter 5). Alternatively the company may call in consultants to carry out the audit. Appendix 3 contains information on organisations offering this and similar services.

The energy audit will highlight the most energy-intensive items of plant, and if correctly carried out should enable the energy manager to identify any inefficiencies and wastage which should be corrected. Significant savings will probably be made with comparatively low capital investment, but in the longer term the effectiveness of planning and forecasting of energy demand and prices will determine where future investment in new plant, or modifications to existing equipment, will be best directed.

Some decisions concerning heat recovery from process plant, and similar conservation measures which can obviously benefit a process by enabling re-use of energy, can be made rapidly. The allocation of capital necessary for implementation, if supported by financial analyses which suggest rapid returns (see Appendix 2), then depends upon the wider aspects of company policy. However there are numerous broader implications of energy conservation which should be taken into account by the energy manager, affecting policy over a period as long as ten years.

Among the most important aspects of such a long term analysis is the accurate prediction of changes in processes and technology within the particular industry. This will of course necessitate recourse to data outside the company conducting the study, and the energy audits being undertaken in the United Kingdom by the appropriate Research Associations are of special interest in this respect. As well as technological change associated with a particular product, the impact on energy utilisation of a change in the pattern of production in the company, possibly spurred by the desire to diversify, must be taken into account.

Changes in the primary fuels used, for example a switch from oil to natural gas for boiler firing, may obviously be beneficial. However there may be a number of by-products or waste which could be used as fuels. A common example is the employment of wood shavings and sawdust to raise steam (or provide heat in another form). New types of incinerator designed with energy production in mind are readily available (see Chapter 6), and are but one of several items of plant which can put waste to better use.

Two other aspects of energy conservation are the responsibility of the energy manager. Firstly he should set a realistic target for energy savings, based on the results of the audit and the co-operation he is likely to obtain from the board with respect to capital expenditure. Secondly, he must involve all personnel in the company in an energy conservation programme.

Appendix 1

Both education and motivation are necessary, the latter being implemented in part by participation in suggestion and incentive schemes and other systems, which must be shown to benefit the employee as well as the employer.

The energy manager is likely to become a key figure in industry, as the cost of energy will increase steadily, possibly influencing product cost (including that of raw materials) to an extent similar to the effect of wages. The company which is able to minimise the energy content of its products is, other things being equal, likely to be the more competitive.

Further Reading: The following papers and publications give information on the experience of companies with energy policies, and guidance on implementing energy conservation programmes and audits. (See also Appendix 4).

"A Guide to Energy Management: How to Conduct an Energy Audit" American Society of Executives Association (ASEA), 1101 16th Street, N.W. Washington DC 20036.

"How to Start an Energy Management Program" Superintendent of Documents, US Govt. Printing Office, Washington DC 20402.

"Energy Conservation Program Guide for Industry and Commerce (EPIC)" Superintendent of Documents, US Govt. Printing Office, Washington DC 20402.

"Energy Audits" Fuel Efficiency Booklet 1, UK Dept. of Energy, HMSO, London.

Woodcock, R.B. Energy Management - the Perkins Approach. _Works Study & Management Services_, 19, 7, 242 - 246 (1975)

Spiegel, W.F. Challenge: Engineering and the Conservation Ethic. _ASHRAE Journal_, May 1975.

Blosson, J.S. How to Develop and Implement an Energy Conservation Program. _Heating/Piping/Air Conditioning_, Jan. 1975.

Lowther, D.W. The Outlook for Energy. _Process News_, 2 Sept. 1975.

Appendix 2
Financial Analyses for Evaluating Benefits of Energy Conservation Equipment

Introduction

There has been much use of the phrases 'payback period', 'return on investment' and other similar financial terminology in the main text. Obviously energy conservation is, as far as the company involved is concerned, directed at saving money. Many techniques which may be put into practice involve negligible cost, and the profitability of such 'good housekeeping' activities is generally not open to question, provided of course that they do not adversely affect the quality of the product. With the need to spend capital on energy conservation, accurate assessments of the savings which will result from the expenditure must be made. In times when energy prices are rising with apparent regularity, an energy conservation system which may not be economically viable one year could well prove attractive at a later date. Thus assessments should take into account projected energy costs.

There are always numerous different demands on company capital resources, and the simple rate of return is insufficient by itself on which to base decisions. Risk, cash flow, taxation and company expansion programmes, together with labour costs and numerous other considerations must be taken into account. However, the comparatively straightforward analyses below help in indicating the effectiveness of energy conservation techniques, and it is of course up to individual companies to select additional criteria on which to base their investment programmes.

The payback period:

A first estimate of the effectiveness of an energy conservation measure may be obtained by calculating either the payback period or the return on the investment. The information needed to carry out the calculations is very basic, and generally easy to obtain. The following data is required:

A	Capital cost of energy conservation equipment, including installation cost
B	Annual operating cost
C	Annual fuel savings
D	Fuel price
E	Equipment life.

The fuel price should ideally take into account increases which will occur over the life of the equipment. Current fuel prices may be used, but this will result in a pessimistic figure for the payback period.

Using the above data, the payback period, which is the ratio of the capital cost of the installation to the net annual savings, may be calculated.

Appendix 2

$$\text{Payback period} = \frac{A}{(C \times D) - B}$$

The payback period is then compared to the equipment life, E. In some cases a payback period of less than E/2 is considered profitable where E is less than ten years (ref. A2.1). In the author's experience, industry is interested in payback periods of two years or less, although this may be influenced by the current (early 1976) low level of overall investment.

Calculation of the return on the investment, expressed as a percentage per annum, takes into account equipment depreciation:

$$\text{Return on investment} = \frac{\{(C \times D) - B\} - \text{Depreciation charge}}{A}$$

The depreciation charge may be simply expressed as the ratio of capital cost to equipment life. A return on the investment in excess of 25 per cent is desirable, but more rigorous calculations may be carried out in marginal cases. Discounted Cash Flow (DCF) is one technique which provides a means for comparing projects on a consistent basis (A2.2). Using this method one is able to determine the present value of a future sum of money, assuming a given rate of interest. This allows one to assess whether the cash income, in terms of fuel savings, is sufficient to repay the capital investment needed, taking into account a particular rate of interest over the life of the project. If the interest rate needed to balance income and expenditure is low, it is probably an unattractive investment.

Other financial analyses

The simplest payback period calculations are those applied to investment in capital equipment such as a heat recovery heat exchanger. The study of the economic feasibility of, for example, a total energy plant is a more complicated exercise, as one must take into account projected fuel costs for the total energy plant, the cost of power supplied by the existing utility or national authority, and the amount of power and heat generated by the total energy plant which can be usefully employed, possibly expressed in hours per year.

The total energy gas turbine package manufactured by Centrax is an example of an installation which requires such an analysis. In this case the system is available on payment of an annual rental charge of £24 000, thus spreading the investment over a number of years, easing the burden of capital outflow. Table A2.1 shows the economic appraisal of a 550 kW generator driven by a gas turbine, which also provides 2.2 MW in heat energy. For the first year of operation the costs selected are those prevailing in the United Kingdom in October, 1975:

Purchased electricity: 1.7p/kW h
Natural gas: 10p/therm

For successive years electricity prices are assumed to increase at a rate of 20 per cent per annum, compounded, and natural gas increases at 15 per cent per annum compounded. It is further assumed that uses can be found for all the electricity and recoverable heat.

Appendix 2

TABLE A2.1 Centrax Gas Turbine Generator Economic Analysis

TOTAL COSTS* FOR 4000 HOURS PER ANNUM (NATURAL GAS)

Year	Conventional Method			T.E. SET			Net Savings on Conventional Method
	Purchased Electricity £	Heat £	Total £	Fuel £	Rental £	Total £	
1	37 400	38 480	75 880	50 800	24 000	74 800	1 080
2	44 880	44 252	89 132	58 420	24 000	82 420	6 712
3	53 856	50 889	104 745	67 183	24 000	91 183	13 562
4	64 627	58 523	123 150	77 260	24 000	101 260	21 890

*Costs are approximate

The rental system, which appears to have potential in many other energy conservation areas, eases the capital burden, as mentioned above. However the net savings depend upon the time at which the rental is paid, and its relationship to the payment date of electricity and fuel bills. Thus interest charges or the earning capacity of the rental sum over the year should be accounted for. This may extend the payback period.

---*---

It has been possible to touch on only a few aspects of energy financial analyses. However, while the energy manager is not expected to do the job of the accountants, his case can be improved if he is familiar with the rudiments of financial analyses which, in the end, will be used to support or reject his technical assessments.

REFERENCES

A2.1   Anon. Energy conservation program guide for industry and commerce. NBS Handbook 115, US National Bureau of Standards, Washington, D.C. Sept. 1974.

A2.2   Drummond, W.A. Economics of waste heat recovery from boiler flue gases. Proc. Inst. of Plant Engineers, Waste Heat Recovery Conf., Paper 3, London 25 - 26 Sept. 1974.

Appendix 3
Organisations Offering Services to Industry in the Field of Energy Conservation

A large number of organisations, in both the public and private sectors of industry, are in a position to assist companies in carrying out energy audits, advising on equipment purchases, or managing complete energy conservation programmes. In the United Kingdom most industries are represented by Research Associations (RA) who are able to assist in a variety of ways, and some of these RA's are prepared to give advice to companies outside the scope of their normal activities. A selection of RA's is given in this Appendix.

Government departments, the fuel industries and various private bodies are also well-equipped to give assistance in energy conservation, and some of these organisations, with, where appropriate, the type of service they offer, are listed in Section A3.2.

The organisation of Associations and other energy advisory bodies in the United States closely follows that of the United Kingdom. In general at present they are better equipped to carry out their tasks, and have proved most helpful in providing data on energy conservation studies etc. The third part of Appendix 3 lists a selection of United States organisations fulfilling similar functions to those in the United Kingdom.

A3.1 UK Research Associations

Sira Institute South Hill Chislehurst Kent BR7 5EH	Moisture Measurement and Control Centre
Production Engineering RA Melton Mowbray Leicestershire LE13 0PB	Energy utilisation surveys Industrial energy audits
British Leather Manufacturers RA Milton Park Stroude Road Egham Surrey TW20 9UQ	Specialist advice on energy (normally for members only)
Timber Research & Development Assn. Stocking Lane, Hughenden Valley, High Wycombe Buckinghamshire HP14 4ND	Advice on use of wood waste for heating Advice to members on energy utilisation
Rubber & Plastics RA Shawbury Shrewsbury SY4 4NR	Recovery and recycling of materials Energy content of containers

Appendix 3

British Ceramic RA Queens Road Penkhull Stoke on Trent ST4 7LQ	Projects on energy conservation Reports on energy saving guidelines (some available to non-members)
Machine Tool Industry RA Hulley Road Macclesfield Cheshire SK10 2NE	Advice on reduction of energy losses in machine tools. Equipment is available for measuring machine and gearbox efficiencies
British Cast Iron RA Alvechurch Birmingham B48 7QB	Foundry operation services Energy development work
British Launderers RA Hill View Gardens Hendon London NW4 2JS	Worldwide technical and consultancy services. Data on power, heat, water and materials consumption (target figures)
British Coal Utilisation RA Randalls Road Leatherhead Surrey KT22 7RZ	Long term R & D, primarily on fluidised bed combustion and manufacture of substitute natural gas from coal
Steel Castings Research & Trade Association 5 East Bank Road Sheffield S2 3PT	Member services, particularly in oxy/fuel cutting processes, heat treatment and melting practice
Paint Research Association Waldegrave Road Teddington Middlesex TW11 8LD	Energy savings in painting processes, e.g. stoving, low temperature curing and u/v curing
Lambeg Industrial RA The Research Institute Lisburn Northern Ireland BT27 4RJ	Advice on energy conservation for the textile industry
Wool Industry RA Headingley Lane Leeds LS6 1BW	Energy conservation on behalf of the textile industry
Electrical Research Association Cleeve Road Leatherhead Surrey KT22 7SA	Numerous energy advisory services including audits, management techniques and equipment R & D.

A3.2 Government and Private Companies in the UK

Department of Industry
Abell House
John Islip Street
London SW1P 4LN

Appendix 3

The National Terotechnology Centre
Cleeve Road
Leatherhead
Surrey KT22 7SA

Acoustics & Envirometrics Ltd. Ruxley Towers Claygate Surrey KT10 0UF	Air conditioning, heating and heat recovery project work
A.F. Stobart, Consultancy Services Manor Farm Claydon Banbury Oxfordshire	General consultancy
The Institution of Heating & Ventilating Engineers 49 Cadogan Square London SW1X 0JB	Energy conservation publications
British Gas Technical Consultancy Service 326 High Holborn London WC1V 7PT	Gas utilisation and equipment advisory service
Hughes & Company 33 Lower Park Putney Hill London SW15	General consultancy
Carter Building Engineering Services Ltd Redhill Road Birmingham B25 8EY	Air conditioning maintenance and performance optimisation
British Gas Corporation School of Fuel Management Midlands Research Station Wharf Lane Solihull West Midlands B91 2JW	Fuel management and conservation courses
The Energy Equipment Company Ltd. Energy House Hockliffe Street Leighton Buzzard Bedfordshire LU7 8HE	Energy conservation schemes, including equipment operation and fuel refresher courses
Department of Energy Thames House South Millbank London SW1P 4QJ	Tax allowance scheme for insulation Regional development grants

Appendix 3

National Industrial Fuel Efficiency Service Ltd. Orchard House 14 Great Smith Street London SW1P 3BU	NIFES offer energy management and consultancy services in most areas of energy conservation
Lighting Industry Federation 25 Bedford Square London WC1B 3HH	Advice on improving lighting efficiency
National Utility Services Inc.(UK) Ltd. Carolyn House Dingwall Road Croydon CR9 3LX	Tariff advisers
British Petroleum Group (Area Offices)	Mobile Combustion Laboratories Boiler efficiency Chimney design Advice on efficient furnace operation District heating, refuse incineration with waste heat recovery Lubrication and energy reduction Insulation Transport
Esso Petroleum (Area Offices)	Plant surveys Advice to customers on energy conservation
Shell UK (Area Offices)	Energy conservation service (in certain areas) Advice to customers on optimum fuel utilisation
Computer Aided Design Centre Madingley Road Cambridge CB3 0HB	Computer programs available as design aids for plant integration. Can be used to take into account operating costs etc.
Electron Beam Processes Ltd. Abbot Close Oyster Lane Byfleet Surrey	Electron beam welders
International Research & Development Company Ltd. Fossway Newcastle upon Tyne NE6 2YD	Electron beam welding Heat recovery system design and consultancy
Bell's Asbestos & Engineering Ltd Farnham Road Slough Berkshire	Industrial service covering maintenance, insulation etc.

Appendix 3

Satchwell Control Systems Ltd. P.O. Box 57 Farnham Road Slough Berkshire SL1 4UX	Energy conservation service associated with control systems
IBM United Kingdom Ltd. 17 Addiscombe Road Croydon CR9 6HS.	Power management system for central co-ordination of power-consuming devices. Applicable to large commercial and industrial plant, it runs on an IBM System/7 computer.

A3.3 Selected Organisations Offering Assistance on Energy Conservation in the USA

Portland Cement Association Old Orchard Road Skokie Illinois 60076 USA	Work on energy consumption in types of buildings, including thermal inertia effects
National L-P Gas Association 7 W. Monroe Street Chicago Illinois 60603 USA	Advice on operation of gas-fired appliances
Concrete Reinforcing Steel Institute 180 North La Salle Street Chicago Illinois 60601 USA	Energy aspects of reinforced concrete buildings - future activity
Inter Technology Corporation 100 Main Street Warrenton Virginia 22186 USA	Contract work on solar energy utilisation and energy conservation
American Consulting Engineers Council 1155 Fifteenth Street Northwest Washington DC 20005 USA	Advice concerning consulting engineers able to assist in energy conservation by better design etc.
R.G. Vanderweil Engineers 99 Bedford Street Boston Massachusetts 02111 USA	Consulting engineers, energy audits and improvements to existing facilities, and energy conservation design of new buildings

Appendix 3

Systems Research 379 Wetherell Street Manchester Conn. 06040 USA	Industrial and residential energy conservation work, covering both electrical and fuel energy sources
Hittman Associates Inc. 9190 Red Branch Road Columbia Maryland 21045 USA	Studies on behalf of industry and government in energy conservation
Versar Inc. 6621 Electronic Drive Springfield Virginia 22151 USA	Energy utilisation studies for industry, and development of techniques for energy consumption analysis
Office of Energy Programs US Dept of Commerce Washington DC 20230 USA	Advice on all aspects of energy conservation, backed by handbooks for industry and other sectors
Edison Electric Institute 90 Park Avenue New York 10016 USA	Energy management programmes, and computer optimisation of energy utilisation
Metal Powder Industries Federation PO Box 2054 Princeton NJ 08540 USA	Energy in sintering and allied industries
American Gas Association 1515 Wilson Blvd. Arlington Va 22209 USA	Advice on operation of gas-fired appliances
American Petroleum Institute 1801 K Street, NW Washington DC 20006 USA	Publicity data and advice on conservation
National Oil Fuel Institute 60 East 42 Street New York NY 10017 USA	Publicity data and advice on conservation

Appendix 3

US Energy Research & Development   Energy R & D, advice and consultancy
   Administration
Washington DC 20545
USA

Federal Energy Administration   Energy R & D, advice and consultancy
Office of Oil and Gas
12th and Pennsylvania Avenue, N.W.,
Washington DC 20461
USA

Appendix 4
Manufacturers of Equipment for Energy Conservation

This Appendix lists a selection of companies who manufacture goods of interest to commercial organisations who wish to conserve energy. Data is included on equipment for monitoring energy consumption and conversion efficiency, as well as control equipment, heat recovery devices, and prime movers. While the list is not claimed to be fully comprehensive, it covers companies whose products have been mentioned in the main text, and as such represents a broad cross-section of the energy conservation-type products available.

The addresses of manufacturers are listed in terms of their products as follows:

Air conditioning
Air curtains/doors
Boiler isolators/draught regulators
Burners
Combustion control equipment
Compressors (refrigeration & heat pump)
Condensate recovery systems
Control systems and maximum demand monitors
Economisers
Electron beam welders
Energy from waste/waste recovery
Fluidised bed cleaners
Fluidised bed combustion
Fluidised bed dryers
Heat exchangers (miscellaneous)
Heat pipe recuperators
Heat pumps
Humidity measurement
Infra-red heaters
Laser welders
Lighting controls
Plate heat exchangers
Prime mover heat recovery
Reciprocating engines
Recuperators (static matrix)
Recuperators (tubular)
Rotating regenerators
'Run-around' recuperators
Solar collectors/heaters
Solar control glass and film
Steam traps
Steam/water mixing valves
Temperature control systems
Temperature measurement

Appendix 4

Thermal fluid heaters
Thermal insulation
Turbines
Viscosity measurement
Waste heat boilers
Water conservation systems
Wind generators.

Air Conditioning

Instant Cooling Equipment Ltd.
Kennet House
77-79 Bath Road
Thatcham, Newbury
Berkshire

Coolair Distributors Ltd.
Victoria Works
Howard Street
Stockport SK1 2BJ
Cheshire

UAS (UK) Ltd.
15 Waterloo Place
Leamington Spa
Warwickshire CV32 5LA

Andrews-Weatherfoil Ltd.
Bath Road
Slough
Berkshire SL1 4AP

Air Curtains/Doors

Minikay Ltd.
C/O Rotaire Dryers Ltd.
2 Glebe Road
Huntingdon PE18 7DU

Newman Industrial Sales
Upper Mills Industrial Estate
Stonehouse
Gloucestershire GL10 2BJ

Ream & Co.
308 Elm Park Avenue
Hornchurch
Essex RM12 4DA

Harefield Rubber Company Ltd.
Bell Works
Harefield
Middlesex UB9 6HG

Nationaire Engineering Ltd.
Hutton Close
Crowther Industrial Estate
Washington
Tyne & Wear

Boiler Isolators/Draught Regulators

Energy Equipment Co. Ltd.
The Cottages
Whitchurch
Aylesbury
Buckinghamshire

Burners (Including Recuperative Types)

Stordy Combustion Engineering Ltd.
Heath Mill Road
Wombourne
Wolverhampton WV5 88D

Hotwork Development Ltd.
Little Royd Mill
Low Road, Earlsheaton
Dewsbury
West Yorkshire

Appendix 4

Hamworthy Engineering Ltd.
Combustion Division,
Fleets Corner
Poole
Dorset

Hygrotherm Engineering Ltd.
Whitworth House
115 Princess Street
Manchester M1 6JR

Submerged Combustion Ltd.
Tweedale
Telford
Shropshire TF7 4JZ

Tolltreck Ltd.
Priory House
Friar Street
Droitwich
Worcestershire (integrated in their high efficiency furnace)

Combustion Control Equipment

The Aerogen Company Ltd.
Anstey Mill Lane Works
Newman Lane
Alton
Hampshire GU34 2QW

Shandon Southern Instruments Ltd.,
Frimley Road
Camberley
Surrey GU16 5ET (Combustion testing equipment)

Babcock Product Engineering Ltd.
Woodall-Duckham House
Crawley
Sussex RH10 1UX

Taylor Servomex Ltd.
Crowborough
Sussex
(Gas analysis equipment)

Compressors (Refrigeration & Heat Pump)

J & E Hall
HTI Engineering
Dartford
Kent

Howden Group Ltd.
195 Scotland Street
Glasgow G5 8PJ

York Division
Borg Warner
PO Box 1592
York
Pennsylvania 17405
USA

Condensate Recovery Systems

Girdlestone Pumps Ltd.
Melton
Woodbridge
Suffolk IP12 IER

Crane Ltd.
Pumps Division
Furnival Street
Reddish
Stockport
Cheshire SK5 6LP

Control Systems and Maximum Demand Monitors

Hamworthy Engineering Ltd.
Combustion Division
Fleets Corner
Poole
Dorset BJ17 7LA

Johnson Service Company
507 East Michigan Street
Milwankee
Wisconsin 53201
USA

Appendix 4

Landis & Gyr Billman Ltd.
Victoria Road
North Acton
London W3 6XS

Kristian Kirk Electric Ltd.
London Road
Thrupp, Stroud
Gloucestershire GL5 2AZ

Bryce Capacitors Ltd.
Chester Road
Helsby
Cheshire WA6 0DQ

Clare Instruments Ltd.
Clare Works
Woods Way
Mullberry Industrial Estate
Goring-by-Sea
Sussex BN12 4QY

CSL Energy Management Ltd.
3 Chifford Street
London W1X 1RA

Digicon Electronics Ltd.
26 Portland Square
Bristol BS2 8RZ

Economisers

Rheinstahl ECO GmbH
4010 Hilden
Postfach 660
West Germany

Senior Economisers Ltd.
Otterspool Way
Watford By-pass
Watford
Herts WDZ 8HX

E. Green & Son Ltd.
Calder Vale Road
Wakefield
West Yorkshire WF1 5PF

Stierle Hochdruck-Economiser kg
68 Mannheim
Karl-Ludwig Str 14
West Germany

Electron Beam Welders

Hawker Siddeley Dynamics Ltd.
Electron Beam Division
Manor Road
Hatfield
Hertfordshire

Energy from Waste/Waste Recovery

Combat
Oxford Street, Bilston
West Midlands

Pall Filtration Ltd.
Walton Road
Portsmouth PO6 1TD

UK Waste Materials Exchange
PO Box 51
Stevenage
Hertfordshire SG1 2DT

Topcast Engineering Ltd.
141 The Broadway
Woking
Surrey GU21 5AP

Waste oil combustion systems

Hydraulic and lube oil purification

Industrial waste exchange
system for industry

Solvent extraction

Appendix 4

Crewe Chemicals Ltd. Hall Lane Rookery Bridge Sandbach Cheshire	Solvent recovery, heat transfer fluid and high vacuum fluid recovery
Luco-Engineering Services Old Boundary House London Road Sunningdale Ascot Berkshire	Chemical waste burners for boilers, incinerators and furnaces
Integrated Conveyors Ltd. Dudley Road East Tividale Warley Worcs B69 3HS	Swarf reclamation
Newell Dunford Engineering Ltd. 143 Maple Road Surbiton Surrey KT6 4BD	Swarf reclamation
Liquid Systems Ltd. Stafford Road Newport Gwent NPT 7XR	Waste oil regeneration
Henry Balfour & Co. Ltd. Leven Fife KY8 4YW Scotland	Liquid waste and energy recovery (Nittetu Process)
Robert Jenkins Systems Ltd. Wortley Road Rotherham Yorkshire S61 1LT	Solid waste and energy recovery
Beverley Chemical Engineering Ltd. Billingshurst West Sussex RH14 9SA	All wastes and heat recovery
Geldia (London) Ltd. Holland House Burmester Road London SW17	Water heaters using incinerator heat
Wellman Incandescent Ltd. Cornwall Road Smethwick, Warley West Midlands B66 2LB	Waste gas incinerators

Appendix 4

Fluidised Bed Cleaners

Rolling Stock & Engineering Co. Ltd.
Nestfield Works
Albert Hill
Darlington DL1 2NW

Fluidised Bed Combustion

Combustion Systems Ltd.
Kingsgate House
66/74 Victoria Street
London SW1E 6SL

West's Pyro Ltd.
Dale House
Tiviot Dale
Stockport
Cheshire SK1 1SA

Fluidised Bed Dryers

Anhydro A/S
DK-2860
SØborg-Copenhagen
Denmark

APV-Mitchell(Dryers) Ltd.
Denton Holme
Carlisle CA2 5DU

Arthur White Process Plant Ltd.
Stapeley Manor
Stapeley, Nantwich
Cheshire

West's Pyro Ltd.
Dale House
Tiviot Dale
Stockport
Cheshire SK1 1SA

Heat Exchangers (Miscellaneous)

Eduard Ahlborn, Aktiengesellschaft
D32 Hildesheim
Lüntzelstr 22
Postfach 530
West Germany

Worthington-Simpson Ltd.
Lowfield Works
PO Box 17
Newark
Nottinghamshire

E.I. Du Pont de Nemours
Wilmington
Delaware 19898
USA

Hunt Heat Exchangers Ltd.
Middleton
Manchester M24 1GQ

E.J. Bowman Ltd.
Aston Brook Street East
Birmingham B6 4AP

D.J. Neil Ltd.
PO Box 31
Macclesfield
Cheshire SK10 2EX   (Agents for Lorden
                                     & Co.)

Heat Pipe Recuperators

Q-Dot Corporation
151 Regal Row
Suite 220
Dallas
Texas 75247
USA

Hughes Aircraft Company
Electron Dynamics Division
3100 West Lomita Blvd.
Torrance
California 90509
USA

Isothermics Inc.
PO Box 86
Augusta
New Jersey 07822
USA

Burke Thermal Engineering Ltd.
Nucleus Works
Mill Lane
Alton
Hampshire. (Agents for Q-Dot Corp.)

International Research &
  Development Co. Ltd.
Fossway
Newcastle upon Tyne NE6 2YD

Heat Pumps

York Division
Borg Warner
PO Box 1592
York
Pennsylvania 17405
USA

Westinghouse Electric Corporation
Power Systems Company
700 Braddock Avenue
East Pittsburgh
Pennsylvania 15112
USA

Westinghouse Electric SA
1 The Curfew Yard
Windsor
Berkshire

Lennox Industries Ltd.
Lister Road
Basingstoke
Hampshire

BAHCO Ventilation
S-199 01 ENKÖPING
Sweden

Briton Air Conditioning Ltd.
15 Clarence Street
Staines
Middlesex TW18 4SU

Brown Boveri-York
Kälte-und Klimatechnik GmbH
68 Mannheim 1
Gottlieb-Daimler-Strasse 6
West Germany

International Research & Development
  Co. Ltd.
Fossway
Newcastle upon Tyne NE6 2YD
(Design, procurement etc.)

Humidity Measurement

Shaw Moisture Meters
Rawson Road
Westgate
Bradford

Infra-Red Heaters

SBM (UK) Ltd.
21a Clarence Street
Staines
Middlesex

Laser Welders

International Research & Development Company Ltd.
Fossway
Newcastle upon Tyne NE6 2YD

Appendix 4

Lighting Controls

Photain Controls Ltd.
Unit 18, Hangar No. 3
The Aerodrome
Ford
Sussex

Plate Heat Exchangers

Robert Jenkins Systems Ltd.
Wortley Road
Rotherham
Yorkshire   S61 1LT

The APV Company Ltd.
PO Box 4
Crawley
Sussex RH10 2QB

Marine & Industrial Heat Ltd.
Gazelda Works
Lower High Street
Watford
Herts. WD1 2JN

Topcast Engineering Ltd.
56 Maybury Road
Woking
Surrey GU21 5JD   (Agents for VICARB
                              of France)

Eduard Ahlborn
D32 Hildesheim
Lüntzelstrasse 22
Postfach 530
W. Germany

Prime Mover Heat Recovery

Maxim Silencers
Riley-Beaird Inc.
PO Box 1115
Shreveport
Louisiana 71130
USA

Applied Energy Systems Ltd.
Thermal House
Marlborough Road
Watford WD1 7BR

Reciprocating Engines

Bermotor Ltd. (Renault)
21 London Road
Tunbridge Wells
Kent

Applied Energy Systems Ltd
Thermal House
Marlborough Road
Watford WD1 7BR

Waukesha Motor Company
Waukesha
Wisconsin 53186
USA

W.H. Allen Sons & Co. Ltd.
Queens Engineering Works
Bedford

Onan
1400 73rd Avenue NE
Minneapolis
Minnesota 55432
USA (Generators driven by gas engines)

White Motor Corporation
White Superior Division
1401 Sheridan Avenue
PO Box 540
Springfield
Ohio 45501
USA

Power-Torque Engineering Ltd.
Pickford Brook
Allesley
Coventry CV5 9AN

Appendix 4

Recuperators (Static Matrix)

ITT-Reznor
Barnfield Road
Park Farm Industrial Estate
Folkestone
Kent CT19 5DR

Allied Air Products Co. Inc.
315 E. Franklin
Newberg
Oregon 97132
USA

Fan Installations Ltd.
130 Manchester Road
Kearsley
Bolton
Lancashire   (Recuperator Spa Agents)

Recuperators (Tubular)

Lee Wilson Engineering Co. Inc.
20005 Lake Road
Cleveland
Ohio 44116
USA

Fluidfire Waste Heat Systems Ltd.
Unit 10
Washington Street
Netherton, Dudley
West Midlands DY2 9RE

Stein Atkinson Stordy Ltd.
Dorking Surrey
Surrey RH5 4BA

Thermal Efficiency Ltd.
Otterspool Way
Watford By-pass
Watford
Herts WD2 8HX

Rotating Regenerators

The Wing Company
Division of Aeroflow Dynamics Inc.
Linden
New Jersey
USA

Uni-Tubes Ltd.
189 Bath Road
Slough
Berks SL1 4AR (Domestic units)

Corning (UK Office)
Corning Glass International SA
1a Cumberland House
Kensington Court
London W8 5NP

Curwen and Newbery Ltd.
Westcroft House
Westcroft Works
Alfred Street
Westbury
Wilts BA13 3DZ

James Howden & Co. Ltd.
195 Scotland Street
Glasgow G5 8PJ

Acoustics & Envirometrics Ltd.
Ruxley Towers
Claygate
Surrey KT10 0UF (Agents for
         Munter of Sweden)

Cargocaire Engineering Co.
6 Chestnut Street
Amesbury
Massachusetts 01913
USA

Appendix 4

Run-Around Recuperators

AB Svenska Fläktfabrieken,
Fack
S-104 60 Stockholm
Sweden

SF Air Treatment Ltd.
Staines House
158 High Street
Staines
Middlesex   (Agents for AB Svenska Fläkt)

Solar Collectors/Heaters

Air Distribution Equipment Ltd.
64 Whitebarn Road
Llanishen
Cardiff CF4 5HB
Wales

Asahi Trading Co. Ltd.
Asahi House
Church Road
Port Erin
Isle of Man

PPG Industries Inc.
One Gateway Center
Pittsburgh
Pennsylvania 15222
USA

Solar Control Glass & Film

Thomas Bennett Ltd.
Goodman Street
Leeds LS10 1QN

3M United Kingdom Ltd.
90 Mitchell Street
Glasgow G1 3NJ

Steam Traps

Armca Specialities Ltd.
19/20 Cowcross Street
London EC1

Gestra (UK)
9/11 Bancroft Court
Hitchin
Hertfordshire SG5 1PH

Steam/Water Mixing Valves

Meynell Valves Ltd.
Bushbury
Wolverhampton WV10 9LB

Temperature Control Systems

Control Devices Inc.
670 N River Street
Wilkes-Barre
PA 18705
USA

Satchwell Control Systems Ltd.
PO Box 57
Farnham Road
Slough
Berks SL1 4UH

Landis & Gyr-Billman Ltd.
Victoria Road
North Acton
London W3 6XS

Appendix 4

Temperature Measurement

Martron Associates Ltd.
81 Station Road
Marlow
Buckinghamshire

Thermal Fluid Heaters

Hygrotherm Engineering Ltd.
Botanical House
1 Botanical Garden
Talbot Road
Manchester M16 OHL

Wanson Company Ltd.
7 Elstree Way
Borehamwood
Hertfordshire WD6 1SA

Thermal Insulation

Morganite Ceramic Fibres Ltd.
Neston
Wirral
Merseyside L64 3TR

The Carborundum Co. Ltd.
Mill Lane
Rainford, St. Helens
Merseyside WA11 8LP

ICI Ltd.
PO Box 6
Billingham
Cleveland TS23 1LD

Capricorn Industrial Services Ltd.
49 St. James Street
London SW1A 1JY   (Allplas balls)

Micropore Insulation Ltd.
1 Arrowe Brook Road
Upton
Wirral
Merseyside L49 1SX

Fibreglass Ltd.
Insulation Division
St. Helens
Merseyside WL10 3TR

Turbines

Chemical Construction Corporation
Regal House
Twickenham TW1 3QJ (Sofrair
   furnace energy recovery system)

Ishikawajima-Harima Heavy Industries
   Ltd.
2-1, 2-chome, Otemachi
Chiyoda-ku
Tokyo 100
Japan

Ruston Gas Turbines Ltd.
PO Box 1
Lincoln LN2 5DJ

Centrax Ltd. (Gas Turbine Division)
Shaldon Road
Newton Abbot
Devon TQ12 4SQ

Peter Brotherhood Ltd.
Peterborough PE4 6AB (Steam
   turbines and waste heat recovery)

Viscosity Measurement

EUR-CONTROL GB Ltd.
222a Addington Road
Selsdon
South Croydon
Surrey CR2 8YH

Appendix 4

Waste Heat Boilers

Engineering Controls Division
Pott Industries Inc.
611 E Marceau Street
St. Louis
MO 63111
USA (Vaporphase system)

Combustion Engineering Inc.
Windsor
Connecticut 06095
USA

Stone-Platt Crawley Ltd.
PO Box 5
Gatwick Road
Crawley
Sussex RH10 2RN (Agents for
  Conseco Inc.)

ETS (Lonertia) Ltd.
Prudential House
Wellesley Road
Croydon
Surrey (Agents for Pott Industries
  Inc.)

Schmidt'sche Heissdampf-Gesell-
  schaft GmbH
35 Kassel
Postfach 103 429
West Germany

Hamworthy Engineering Ltd.
Combustion Division
Fleets Corner
Poole
Dorset BH17 7LA

Hygrotherm Engineering Ltd.
Whitworth House
115 Princess Street
Manchester M1 6JR

ABCO Industries Inc.
2675 East Highway 80
PO Box 268
Abilene
Texas 79604
USA

Eclipse Inc.
PO Box 4756
Chattanooga
Tennessee 37405
USA

Hawthorn Leslie (Marine)
St. Peters Works
Newcastle upon Tyne NE99 1PD

Deltak Corporation
6950 Wayzata Blvd.
Minneapolis
Minnesota 55426
USA

Water Conservation/Recovery Systems

Perflex
Energy Systems Division
500 West Oklahoma Avenue
Milwaukee
WI 53207
USA

Simon Hartley Ltd.
Etruria
Stoke-on-Trent ST4 7BH

Babcock Water Treatment
Unifilter House
25 Raleigh Gardens
London Road
Mitcham
Surrey CR4 3UP

Vickers Ltd.
South Marston
Swindon
Wiltshire

Appendix 4

Seitz-Werke GmbH
D-6550 Bad Kreuznach
PO Box 1049
West Germany

International Research &
  Development Co. Ltd.
Fossway
Newcastle upon Tyne NE6 2YD

Chemviron Ltd.
34 Union Street
Oldham
Lancashire

Biomechanics Ltd.
Smarden
Ashford
Kent

PCI Reverse Osmosis Divn.
Laverstoke Mills
Whitchurch
Hampshire RG28 7NR

Water Management Ltd.
Stourport Road
Kidderminster
Worcestershire DY11 7QF

Mather & Platt
Anti-Pollution Systems
14 Buckingham Palace Road
London SW1W 0QP

Capital Controls Company
Advance Lane
Colmar
Pennsylvania 18915
USA

Morgett Electrochemicals Ltd.
27 Tower Road South
Heswall
Wirral
Merseyside L60 7SY

Elga Products Ltd.
Lane End
Buckinghamshire HP14 3JH

Ames Crosta
Heywood
Lancashire OL10 4NF

Foster Wheeler Energy Corporation
110 South Orange Avenue
Livingston
NJ 07039
USA

Aluminium Company of America
727 Alcoa Building
Pittsburgh
PA 15219
USA

Aminoden A/S
9990 Skagen
Denmark

Wind Generators

Control Technology Ltd.
Bolney Avenue
Peacehaven
Sussex

Grumman International Inc.
64-65 Grosvenor Street
London W1X DB

Wind Energy Supply Co. Ltd.
South Block
Redhill Aerodrome
Surrey

Appendix 5
Bibliography

The Bibliography contains data on energy conservation publications, most of which are additional to those listed in the main text. References are listed in random order, but are classified under the following headings:

    Air conditioning and environmental control
    Diesel and gas engines
    Energy conservation - general books and conference proceedings
    Energy conservation - hints and good housekeeping
    Heat pumps in industry
    Industrial energy conservation
    Industrial gas turbines
    Total energy systems
    Waste heat boilers
    Use of natural resources
    Useful periodicals on energy conservation

Air conditioning and environmental control

Bowlen, K.L. Energy recovery from exhaust air for year round environmental control. Trans. ASHRAE, 1, 314 - 321, (1974)

Miller, R.R. Trends in in-plant environment. Trans. ASHRAE, 1, 211 - 215 (1974)

Schoenberger, P.K. Energy saving techniques for existing buildings. Heating, Piping and Air Conditioning, Jan. 1975

Energy conservation in building design. PSA Library Service Bibliography, PSA, London (1976)

Systems and systems control in buildings. PSA Library Service Bibliography, PSA, London (1976)

Bernstein, H.M. and McCarthy, P.M. Analysis of factors related to energy use in the commercial sector. Proc. 57th Annual Conference, Planning 75: Innovation and Action. American Institute of Planners, San Antonio, Texas, (1975)

Stoecker, W.F. (Ed). Procedures for simulating the performance of components and systems for energy calculations. 3rd Edition, ASHRAE, New York, (1975)

ASHRAE Standard 90-75. "Energy conservation in new building design" ASHRAE, New York, (1975)

Ambrose, E.R. Heat reclaiming systems - a review. Heating, Piping & Air Conditioning, 47, 55 - 58, (1975)

Appendix 5

Diesel and gas engines

Conrad, J.C. Natural gas engine drive for centrifugal compressors. *Heating, Piping and Air Conditioning*, 34, 111 - 116, (1962)

Barrangon, M. The gas engines: On-site prime movers. *ASHRAE Journal*, 12, 50 - 53, (1970)

McLure, C.J.R. Reciprocating engine heat recovery systems. *Heating, Piping & Air Conditioning*, 39, 165 - 170, (1967)

Marsh, R.C. The gas engine is alive. *Heating, Piping, Air Conditioning*, 39, 142 - 149, (1967)

Energy conservation - General books and conference proceedings

Waste heat recovery. Proc. Institute of Fuel Conference, Bournemouth, 1961, Publ. Chapman & Hall, London, (1963).

Energy use and conservation in the metals industry. Proc. AIME Energy Symposium, New York, Feb. 1975.

Technology of efficient energy utilisation. Report of a NATO Science Committee Conference, Les Arcs, France, Oct. 1973. Publ. Scientific Affairs Division, NATO, Brussels (1974)

Chapman, P. *Fuel's Paradise*. Penguin Books, Harmonsworth, (1975)

Ion, D.C. *Availability of World Energy Resources*. Graham & Trotman, London, (1975)

Waste Heat Utilisation. Proc. National Conf. Oct. 27 - 29, 1971, Gatlinburg, Tennessee, Yarosh, M.M. (Editor), Oak Ridge National Laboratory, May 1972

Dryden, I.G.C. (Editor). *The Efficient Use of Energy*. Institute of Fuel/ IPC, Guildford, (1975)

Richardson, H.W. *Economic Aspects of the Energy Crisis*, Lexington Books, Hampshire (1975)

Energy Technology II. Proc. 2nd Energy Technology Conference, May 12 - 14, 1975. Publ. Government Institutes Inc., Maryland USA, 1975. (Energy Technology III. Proc. 3rd Energy Technology Conference, March 29 - 31, 1976, to be published 1976)

Efficient use of fuels in process and manufacturing industries. Proc. Inst. of Gas Technology Symposium, April 16 - 19, 1974. Published IGT, Chicago, (1975)

Efficient use of fuels in the metallurgical industries. Proc. Institute of Gas Technology Symposium, Dec. 9 - 13, 1974. Published IGT, Chicago (1975)

Energy research and development. Reports for Energy Policy Project of Ford Foundation, by Hollomat, H. et al. Ballinger Publishing Co. USA, (1975).

Crawley, G.M. *Energy*. Collier-Macmillan, London (1975).

Appendix 5

Energy R & D - Problems and Perspectives. OECD, Paris, (1975)

Gyftopoulos, E.P. and Lazaridis, L.J. Potential Fuel Effectiveness in Industry Ballinger Publishing Company, USA, (1975)

Energy consumption in manufacturing. Prepared by National Science Foundation. Published Ballinger, USA (1975)

Energy conservation in commercial, residential and industrial buildings. Proc. Conference Ohio State University, May 5 - 7, 1974. Publ. National Science Foundation, NSF RA N74 123, PB-240 306, May 1974

Energy recovery in process plants. Proc. Conference in London, 29 - 31 Jan, 1975. Proceedings publ. Institution of Mechanical Engineers, London (1975)

Energy conservation - hints and good housekeeping

Anon. Exhibits at ASHRAE/ARI 1975 Expo stress energy conservation ideas. ASHRAE Journal, 17, 35, (1975)

Anon. Energy conservation handbook, US Dept. Commerce, Washington D.C., May 1974

Anon. Industry's vital stake in energy management. US Dept. Commerce, Washington D.C., (1974)

Anon. 33 Money-saving ways to conserve energy in your business. US Dept. Commerce, Washington D.C., April 1974

Anon. The truth about the first law of thermodynamics. Light Reading, Kodak Ltd., No. 4, (1975)

Anon. Energy conservation and a common sense approach. Southern California Gas Company, Los Angeles, (1973)

Anon. Energy saving in industry. Department of Energy, London, June 1975

Anon. Gas energy conservation hints for the food service industry. Southern California Gas Company, Los Angeles.

Saving energy in commercial gas kitchens. American Gas Association, Publication R01025, (1975)

How to save energy in commercial buildings. Southern California Gas Company, Publication 7436, Los Angeles, (1974)

Bataille, G.S. Stop wasting energy - here are some practical ideas. Plant Engineering, pp 137 - 139, June 13, 1974.

Beatson, C. Good housekeeping is the key to more power conservation. The Engineer. pp 32 - 35, 31 July/7 Aug. 1975.

Weston, H.B: Energy: A way of thinking. Maintenance Engineering, April 1975

Lyle, O. The efficient use of steam, HMSO, London, 1956

Appendix 5

Anon. Energy conservation right now? Grumman International Inc. Publication, London, (1975)

Anon. Recommended guidelines for retail food store energy conservation. US Commercial Refrigerator Manuf. Assn., Rept. CRMA-EC-1, Washington D.C., (1975)

'Energy conservation - 2'. (Tables of waste heat content in exhaust gases). British Ceramic Research Association, Stoke on Trent, (1975)

Energy audits. Fuel efficiency booklet 1. Department of Energy, HMSO, Feb, 1976.

An outline guide to electric space and water heating. The Electricity Council, London, (1975)

Frick, E.T. Guidelines for energy survival. (Heat Processing). Metal Progress 108, 4, 62 - 67 (1975)

I.H.V.E. Energy notes for offices. The Institution of Heating and Ventilating Engineers, London, (1975.).

I.H.V.E. Energy notes for factories. The Institution of Heating and Ventilating Engineers, London, (1975)

Four guides to fuel economy. ESSO Oilways International, 22, 4, (1976)

Anon. I.H.V.E. Energy notes for factories. The Institution of Heating and Ventilating Engineers, London, August 1975

Anon. The sensible use of latent heat. Fuel Efficiency Booklet 2. Department of Energy, HMSO, London, (1976)

Heat pumps in industry

Aktuelle Wege zu verbesserter Energie anwendung. VDI-Berichte 250. Publ. VDI-Verlag GmbH, Düsseldorf, (1975)

BSRIA Bibliography 103 (Heat Pumps). BSRIA, Bracknell, (1975)

Trenkowitz, G. Energy saving by using heat pumps. Ki Klima + Kälte-Ingenieur 4/74, 155 - 162, (1974) (in German)

Ross, P.N. The Templifier for process heat. Proc. Annual Conference EEI Conservation and Energy Management Division, Atlanta, Georgia, March 16 - 18, 1975.

Bowen, J.L. Energy conservation in the 'seventies. Refrigeration and Air Conditioning, April 1975

Leidenfrost, W. and Eisele, E.H. Rotating heat exchangers and the optimisation of a heat pump. IEEE Trans. on Industry Applications. 1A-8, 3, (1972)

Applied heat pump systems. In: ASHRAE Systems Handbook, Chapter 11, New York, (1973)

Appendix 5

Anon, Industrial heat pump cuts the fuel bill. Electrical Review, pp 404 - 5, 28 March - 4 April, 1975.

Bridgers, F.H. How new technology may save energy in existing buildings. Heating, Piping and Air Conditioning, 47, 9, 50 - 55, (1975)

Anon. Heat pump refinements. New Scientist, p.180, 22 Jan. 1976.

Macadam, J.A. Heat pumps - the British experience. Building Research Establishment Note N117/74, (1974)

Juttemann, H. Heat pumps in large buildings. OA-Trans-939, Electricity Council, London, (1974)

Yanagimachi, M. Air source heat pump/heat storage system of Hiroshima regional station of Japan Broadcasting Corporation. Trans. SHASE, Japan, 3, 23 - 30, (1965)

Vicktor, H. Regenerative heat exchange and the heat pump. Heiz. Luft. Haustech., 21, 10, 363 - 368, (1970) (in German)

Villaume, M. Centralised heat pumps and waste heat recovery. Chaud. Froid. Plomb., 27, 73 - 84, 113 - 119, 320 - 321, Jan & Feb. 1973 (in French)

Trenkowitz, G. Use of heat pumps for heat recovery. Elektrowärme Int., 30, 4, A180 -A187, (1972)

Industrial energy conservation

Woods, S.E. Heat conservation in zinc and lead extraction and refining. Metals and Materials, 8, 3, March 1974.

Gray, P.M.J. Conservation in primary extraction processing. The Metallurgist and Materials Technologist, 7, 2, Feb 1975.

Beatson, C. Save energy with electric heating in manufacturing. The Engineer 26 Feb. 1976.

Kern, W.I. Increasing heat exchanger efficiency through continuous mechanical tube maintenance. Combustion (US), pp 18 - 27, Aug. 1975.

Mortimer, J. Exchanging heat to save fuel. The Engineer, 2 - 9 August 1973.

Anon. Industrial energy study of the petroleum refining industry. Report PB-238671/2, Federal Energy Authority FEA/EI-1656 (USA), (1975)

Pettman, M.J. et al. Energy in plant. Hydrocarbon Processing (USA), 54, 1, (1975)

Kratochvil, J.A. and Rayner, H.M. Minimising fuel costs. Trans. ASME, Paper 75-WA/Pwr-13, Dec. 1975.

Anon. Planned maintenance pays off in foundry. Dept. of Industry, Committee for Terotechnology, Case History No. 1, HMSO, London (1975)

Appendix 5

Anon. Planned maintenance cuts food factory down time. Dept. of Industry, Committee for Terotechnology, Case History No. 3, HMSO, London (1975)

Balmer, I.R. Effective use of fuels and resources in the metals industries. Metallurgia and Metal Forming, pp 319 - 321, Sept. 1975

White, W.C. Energy problems and challenges in fertilizer production. Proc. Fertilizer Institute Round Table Meeting, Washington D.C., 4 Dec. 1974.

Day, C.E. How we specify fuels for our plants. Trans. ASME, Paper 69-FU-7, New York, (1969)

Becker, H.P. Energy conservation analysis of pumping systems. ASHRAE Journal pp 43 - 51, April 1975.

Schwindt, H.J. Utilisation of waste heat from inductive melting installations. Electrowärme International, 32, B2, (1974)

Fulton, R. and Strindehag, O. Ecoterm - A method of waste heat recovery. Fläkt/SF Air Treatment Ltd., Staines, (1976)

Robson, B.G. Heat recovery from exhaust air with rotary heat exchangers. Australian J. Refrigeration, 24, 3, 16 - 20, (1970)

Waste heat recovery systems (Processing Report) Processing (UK), p 46, May 1975.

Homfeld, E.W. Utilising gas engine waste heat in a cracking plant. Gas Wärme International, 19, 8, (1970) (In German)

Anon. Spiral heat exchangers. Heating and Air Conditioning Journal, 43, 510, (1974)

Goodell, P.H. Conserving energy with gas infra-red (heating). Industrial Gas (US), pp 9 - 13, Aug. 1971

Berg, C.A. Energy conservation through effective utilisation. National Bureau of Standards NBSIR 73 102, Washington D.C., Feb. 1973.

Cook, C.S. Evaluation of a fossil fuel fired ceramic regenerative heat exchanger. Rept. PB-236 346, General Electric Co. Philadelphia, Oct. 1974.

Sacton, J.C. et al. Industrial energy study of the industrial chemicals group. Int. Research and Technology Corp., Washington D.C., Report PB-236 322, Aug. 1974.

Rogers, L.M. The application of thermography to plant condition monitoring and energy conservation. British Steel Corporation, TB/TH/71. Report PB-236 360, Aug. 1974.

Anon. Industrial energy study of selected food industries. FEA Report EI 1652, PB-237 316, July 1974.

Anon. Fuel and energy conservation in the coal industries. Hittman Associates Inc., Columbia. FEA Report EI 1659, PB-237 151, May 1974.

Queen, D.M. Industrial energy study of the hydraulic cement industry. FEA Report EI 1665, PB-237, 142, August, 1974.

McQuade, F.E. The prospects of energy demand scheduling. National Science Foundation Report NSF RA N-74 152, 8, PB-239, 763, (1974)

Bunt, B.P. The energy used by packaging and its minimisation. J. Society of Dairy Technology, 28, 3, (1975)

Quartulli, O.J. Stop wastes: re-use process condensate. Hydrocarbon Processing. 54, 10, 94 - 99, (1975)

Anon. Feed pump and fan efficiency and reliability. Power (US) pp 25 - 29, Nov. 1975.

Edwards, J.V. et al. Computation of transient temperatures in regenerators. Int. J. Heat and Mass Transfer, 14, 1175 - 1202, (1971)

Loyd, S. and Starling, C. Heat recovery from buildings. An annotated bibliography with a survey of available products and their suppliers. BSRIA Bibliography 104/76, April, 1976.

Boyen, J.L. Practical Heat Recovery   John Wiley, New York, (1975)

Dutch Patent 7403678. Heat pumps in bottle washing plant. Milpro N.V. Brussels, 19 March 1974.

Industrial gas turbines

Booth, D. Plastics shave time off industrial gas turbine development. The Engineer, 6 Feb. 1975.

Stansell, J. Powerful future for gas turbines. Electrical Review, 28 Nov. 1975.

Ford, E. Prospects for natural gas fuelled total energy systems. Steam & Heating Eng., 40, 472, (1971)

Koch, H. and Sharan, H.N. Low emission power systems. Sulzer Technical Review, 4, 205 - 211, (1975)

Mottram, A.W.T. The gas turbine - recent improvements and their effect on the range of applications. Proc. Eighth World Energy Conference, Bucharest, June 28 - July 2, 1971.

Total energy systems

Butler, P. A showpiece to save money by promoting waste heat recovery. The Engineer, 23 Oct. 1975.

Wittner, B.R. and Culp, R.E. Turbine total energy for offshore rigs. Gas Turbine International, Nov - Dec, 1973.

Klein, S. Total on-site power. Machine Design, 42, 2, (1970)

Appendix 5

Aulbach, R.E. When does the 'all-fuel' concept of energy make sense? Trans. ASME, Paper 69-FU-6, (1969)

Anon. Total energy: An annotated bibliography. Department of the Environment, PSA, London, ISBN 0 900014 51 2, (1975)

Fleming, W.S. Total energy concepts for refrigeration and air-conditioning engineers. ASHRAE Journal, Nov. - Dec. 1970.

Gogia, J.K. Controls for total energy systems. Heating, Piping & Air Conditioning, April 1967.

Dukelow, J.S. Operation and maintenance of total energy systems. Heating, Piping & Air Conditioning, April 1967

Obrecht, M.F. Water treatment in total energy. Heating, Piping & Air Conditioning, April, 1967.

Sage, D. General procedure for economic analysis of onsite energy systems. Heating, Piping & Air Conditioning, April 1967.

Waste heat boilers

Clay, P.E. Ins and outs of heat recovery equipment. Air Conditioning, Heating and Ventilating, 65, 1, (1968)

Van den Hoogen, B. Designing gas turbine heat recovery boilers. Gas Turbine Int., pp 32 - 35, Nov. - Dec. 1973.

Wright, R. Finned tubes for waste heat recovery. Chem. Proc. Engineering. Heat Transfer Survey, (1968)

Waterland, A.F. Energy conservation in an industrial plant. Trans. ASME Power Division, Paper 73-IPWR-9, (1973)

Use of natural resources

Senior, J. Coal utilisation in industry. Heating and Air Conditioning Journal, July 1974.

Hammond, A.L. et al. Energy and the future - gasification. American Association for Advancement of Science, pp 11 - 16, (1973)

Grainger, L. Future trends in utilisation of coal energy conversion. Energy Digest, pp 2 - 7, Jan/Feb, 1974.

Reserves and resources of uranium in the United States. Printing and Publishing Office, National Academy of Sciences, Washington D.C., (1975)

Enhancement of recovery of oil and gas - Progress Review No. 1. ERDA Energy Research Centre, Washington D.C., (1975)

Fifteenth International Symposium on Combustion. Proceedings: Combustion Institute, Pittsburgh, Penn., (1975)

Strimbeck, D.C. Process environment effects on heat pipes for fluid-bed gasification of coal. Proc. 9th Intersoc. Energy Conversion Engineering Conf. ASME, San Fransisco, (1974)

Bainbridge, G.R. UK prime energy resources. Proc. Conference on Housing and Energy, University of Newcastle upon Tyne, 15 - 17 April, 1975.

Tostevin, G.M. and Luxton, R.E. Solar energy in industrial thermal energy systems. Proc. Thermo. Fluids Conf. '75 on Energy, pp 32 - 36, Brisbane, Australia, 3 Dec. 1975.

Field, A.A. Geothermal energy rediscovered. Heating Piping & Air Conditioning pp 113 - 116, Jan. 1975.

Anon. Geothermal power station. US Army Foreign Science and Technology Center FSTC-HT-23-1674-73, AD-785948, March 1974.

Kirchhoff, R.H. et al. Hot side heat exchanger for an ocean thermal difference power plant. Proc. 9th Intersoc. Energy Conversion Engineering Conference, ASME, San Fransisco, (1974)

Wilson, E.M. Energy from the tides. Science Journal, pp 50 - 57, July 1965.

Anon. The new energy sources. Industrial Research, Nov. 15, 1974.

Chapman, P.F. The energy cost of materials. Energy Policy, pp 47 - 57, March 1975.

Rayment, R. Energy from the wind. Heating and Air Conditioning Journal, April, 1976.

Kaplan, G. For solar power: sunny days ahead? IEEE Spectrum, 12, 12, (1975)

Useful Periodicals on Energy Conservation

Actual Specifying Engineer. Cahners Publ. Co. Inc., 5, S. Wabash Avenue, Chicago, IL 60603, USA.

Consulting Engineer. St. Joseph, MI 49085, USA.

Diesel Engineering. Whitehall Press, Earl House, 27 Earl Street, Maidstone, Kent ME14 1PE.

Electrical Review. IPC Press, Dorset House, Stanford Street, London SE1 9LU.

Electrowaerme International. Vulkan-Verlag Dr. W. Classennachf. GmbH, Haus der Technik, D-43 Essen, West Germany, (Summaries in English)

Energy. Pergamon Press, Headington Hill Hall, Oxford, OX3 OBW.

Energy Conversion. Pergamon Press, Headington Hill Hall, Oxford, OX3 OBW.

Energy Developments. International Review Service, 15 Washington Place, New York, NY 10003, USA.

Appendix 5

<u>Energy Info</u>. Robert Morey Assoc., Box 98, Dana Point, California 92629, USA.

<u>Energy Report</u> Microinfo. Ltd., The Post House, High Street, Alton, Hampshire, GU34 1EF, England.

<u>Energy Sources</u> Crane, Russak & Co. Inc. 52, Vanderbilt Avenue, New York, NY 10017, USA.

<u>Energy Systems & Policy</u> Crane, Russack & Co. Inc., 52, Vanderbilt Avenue, New York, NY 10017, USA.

<u>Fuel</u> IPC Science & Technology Press, IPC House, 32 High Street, Guildford, Surrey.

<u>Heating, Piping & Air Conditioning</u>. 25 Sullivan Street, Westwood, NJ 07675, USA.

<u>Maintenance Engineering</u>. Mercury House Publications Ltd., Mercury House, Waterloo Road, London SE1 8UL.

<u>Natural Gas For Industry</u>. Benn Brothers Ltd., London EC4.

<u>Oil & Gas Journal</u>. Petroleum Publishing Co. 211 S. Cheyenne Avenue, Tulsa, Oklahoma 74101, USA.

<u>Power Generation Industrial</u>. Fuel and Metallurgical Journals, Queensway House, 2 Queensway, Redhill, Surrey.

<u>Power and Works Engineering</u>. Fuel and Metallurgical Journals, Queensway House, 2 Queensway, Redhill, Surrey.

<u>'Spectrum'</u>. Alternative Sources of Energy, Route 4, PO Box 90, Golden, CO 80401, USA.

Appendix 6
Useful Conversion Factors and Fuel Properties

A6.1 Basic Physical Quantities

Length	1 ft	=	0.3048 m
Area	1 ft^2	=	0.0929 m^2
	1 in^2	=	6.451 cm^2
Volume	1 ft^3	=	0.0283 m^3
	1 gallon	=	4.546 litres
	1 US gallon	=	0.833 Imperial gallons
Volume Rate of Flow	1 gal/min	=	0.0758 litre/s
Mass	1 lb	=	0.4536 kg
	1 ton	=	1.016 tonnes
	1 tonne	=	1000 kg
Mass Flow Rate	1 lb/hr	=	1.259×10^{-4} kg/s
Density	1 lb/ft^3	=	16.019 kg/m^3
Force	1 lbf	=	4.448 N
Pressure	1 lbf/in^2	=	6.894 kN/m^2
	1 bar	=	10^5 N/m^2
	1 atm.	=	101.325 kN/m^2
Dynamic Viscosity	1 Poise	=	0.1 Ns/m^2
	1 lbf.s/ft^2	=	0.047 Ns/m^2
Energy	1 kW h	=	3.6×10^6 J
	1 hp h	=	2.684×10^6 J
	1 Btu	=	1.055 kJ
	1 Btu	=	0.251 k cal
Power	1 hp	=	0.745 kW
	1 hp	=	1.013 metric hp
Temperature	$(^\circ F - 32) \times 5/9$	=	$^\circ C$
Quantity of Heat	1 Btu	=	1.055 kJ
	1 k cal	=	4.186 kJ
Heat Flow Rate	1 Btu/h	=	0.293 W
	1 k cal/h	=	1.163 W

Appendix 6

Density of Heat Flow	1 Btu/ft^{2}h	=	3.154 W/m^2
Thermal Conductivity	1 Btu/ft h $^\circ$F	=	1.730 W/m $^\circ$C
Coefficient of Heat Transfer	1 Btu/ft^2 h $^\circ$F	=	5.678 W/m^2 $^\circ$C
Specific Heat Capacity	1 Btu/lb $^\circ$F	=	4.186 x 10^3 J/kg $^\circ$C
Enthalpy	1 Btu/lb	=	2.326 J/g
	1 k cal/kg	=	4.186 J/g
Calorific Value (Volume basis)	1 Btu/ft^3	=	0.037 J/cm^3
	1 therm/gal	=	2.32 x 10^4 J/cm^3
Light-Illumination	1 lux	=	1 lumen/m^2
	1 foot candle	=	1 lumen/ft^2

A6.2 Fuel Equivalents and Calorific Values

<u>Coal</u>   1 million tons:
                                      260 x 10^6 therms
                                      585 x 10^3 tons petroleum
                                      25 000 x 10^6 ft^3 natural gas
                                      2200 GWh electricity generated

            Gross calorific value (CV): 27 500 MJ/tonne

<u>Natural Gas</u>   1 million therms:
                                      2.75 x 10^6 m^3
                                      4000 tons coal
                                      2300 tons petroleum
                                      9 GWh electricity generated

            Gross calorific value (CV): 37.3 - 41.0 MJ/m^3

<u>Liquified Petroleum Gas</u>

        Propane CV: 93.3 MJ/m^3
        Butane CV: 124 MJ/m^3

<u>Oil & Petroleum</u> 1 million tons:
                                      7.5 x 10^6 barrels
                                      430 x 10^6 therms
                                      1.7 x 10^6 tons coal
                                      3800 GWh electricity generated

        1 barrel: 34.9726 gallons
                                        42 US gallons
                                        0.159 m^3

        Gross calorific values:
            Fuel oil: 43.3 - 44.4 MJ/kg
            Diesel fuel: 45.1 - 46.1 MJ/kg
            Heating oil: 46.1 - 47.0 MJ/kg
            Turbine fuel: 46.3 - 46.8 MJ/kg

Index

ABCO Industries, 330
AB Svenska, 207, 328
Absorption cycles, 256
Accumulators, 263 et seq.
    Ruths, 264
Acid pickling plant, 123
Acoustics & Envirometrics, 314, 327
Additives, 120, 147 et seq.
Adhesive curing, 277
Aerogen Co., 321
Aeschlimann, J.P., 290
Aggregate drying, 128, 172
Agriculture, 22, 293
Air/fuel ratios, 119, 307
Air cleaners, 112
Air conditioning: 45, 110 et seq.,
  152 et seq.
  computers, 114
  control systems, 113
  filters, 112
  heat recovery, 113, 152, 203, 209, 320
  heat redistribution, 152 et seq.
  insulation, 112
  luminaires, 139 et seq., 157
Air coolers, 25
Air curtains, 138, 143, 320
Air Distribution Equipment, 328
Air preheaters, 73, 119, 162
Air registers, 135
Albright & Wilson, 222
Alcoa system, 62, 341
Alcyon Electronique, 290
Alkyd resin plant, 272
Allen, W.H., 326
Allied Air Products, 327
Allied Chemical, 245
Allplas system, 269
Alternative energy sources: 292 et
  seq.
  geothermal energy, 294
  hydrogen, 298
  solar energy, 300
  wave power, 296
  wind power, 292

Aluminium: 59 et seq.
  Alcoa systems, 62, 331
  Bayer process, 62
  energy usage, 60
  furnace, 170
  Hall-Heroult, 60
  storage media, 271
  Toth process, 61
American Ceramic Society, 89
American Consulting Engnrs., 316
American Gas Association, 257, 317
American Petroleum Institute, 317
Ames Crosta, 331
Aminodan A/S, 331
Ammonia manufacture: 63 et seq.
  catalysts, 64, 66
  compressors, 65
Analysis:
  cost, 306 et seq.
  flue gas, 145
Andrews-Weatherfoil, 320
Anhydro A/S, 324
Annealing:
  glass, 86
  heat recovery in, 86 et seq.
Applied Energy Systems, 326
APV Co., 221, 326
APV-Mitchell (dryers), 324
Aquaculture, 22
Arc furnace, 55, 183
Armca Specialities, 328
Arrow Chemicals, 149
Arthur White Process Plant, 177, 324
Artos, 102
Asahi Trading, 328
ASHRAE, 48, 102, 218
ASME, 24
Atomic Energy Commission, 292
Audits, 306 et seq.
AXCESS program, 114
Azeotropic mixtures, 245

Babcock Product Engng., 321
Babcock Water Treatment, 330
Babcock & Wilcox, 264

Index

Back pressure turbine, 21
  total energy, 24 et seq., 81
Bacteria destruction, 94
BAHCO Ventilation, 325
Balfour, Henry, 188, 323
Ball blankets, 269
Bark, 78 et seq.
  calorific value, 79
Barometric control, 118
Basiulis, A., 206
Batch dryers, 179
Bayer process, 62
BCIRA, 183
Bell's Asbestos, 315
Benzene production, 8
Bermotor (Renault), 326
Beverley Chemical Engng., 323
Bibliography, 332 et seq.
Biomechanics Co., 331
Blast furnace, 54
  energy recovery, 56
Blowdown recovery, 117
  cost of, 168
  heat exchanger, 167
Boilers, 117 et seq., 162 et seq.
  air preheaters, 162, 217
  blowdown, 117, 166
  condensate, 85, 117, 165
  corrosion, 148
  draught regulators, 320
  efficiency, 228
  firetube, 228
  fluidised bed, 283
  good housekeeping, 85, 117
  heat recovery, 117, 163, 224
  isolators, 118, 320
  maintenance, 117
  steam traps, 146
  thermal storage, 118, 264
  utilisation, 118
  waste heat, 65, 227 et seq., 330
Borg-Warner, 321, 325
BOS, 54 et seq.
Bottles:
  washing, 95 et seq.
  weight, 91
Bowman, E.J., 324
Brayton cycle, 256
Breweries, 98
Brick kilns, 175, 252
Bricks, 175
British Cast Iron RA, 183, 313
British Ceramic RA, 313
British Coal Utilis. RA, 4, 313
British Food RA, 93

British Gas Corporation, 169
  Fuel Management School, 314
  Gas engine research, 39
  Tech. Consultancy Serv., 307, 314
British Launderers RA, 313
British Petroleum, 119, 135, 315
British Steel Corporation, 57
Briton Air Conditioning, 325
Brown-Boveri, 325
Bryce Capacitors, 322
Buell Ltd., 173
Buildings:
  air conditioning, 110, 143, 152
  Chester, 153
  heat balance, 111, 153
  heating, 143
  heat storage, 162
  lighting, 139, 157
  structure, 111, 153
Burke Thermal Engineering, 325
Burners:
  control, 119, 123
  flat flame, 121, 185
  fuel additives, 148
  heat recovery, 169 et seq.
  high velocity, 90, 121
  nozzle-mixing, 169
  packaged, 170
  preheating, 121
  recuperative, 169, 320
  tip alignment, 135

Calorific values, 189, 343
Cameras, infra-red, 145
Canned foods, 93 et seq.
Cannon Air Engineering, 174
Capenhurst, 21, 90, 178
Capital Controls, 331
Capital costs, 143, 168, 221
Capricorn Industrial Services, 269, 329
Carborundum Co., 234, 329
Carburettors, 40
Cargocaire, 327
Carnot cycle, 239
Carter Building Eng. Serv., 314
Case hardening, 184
Catalysts:
  in ammonia plant, 64, 66
Catalytic combustion, 105, 207
Catalytic incinerator, 186
Catalytic rich gas, 11
Catering equipment, 99
Cells - solar, 301

Central Electricity Generating Board, 19 et seq., 271
Central Policy Review Staff, 20, 296
'Central Waste Heat Recovery', 69
Centrax Ltd., 39, 48, 310, 329
'Centrax' gas turbine, 33
 economics of, 34, 310 et seq.
Ceramic fibres, 132
Ceramic recuperator, 57
Ceramic regenerator, 57, 213
Cereal dryers, 176
Chemical Construction Corp., 57, 329
Chemical industry: 62 et seq.
 ammonia, 63 et seq.
 energy usage, 63
 ethylene, 69 et seq.
 methanol, 66 et seq.
Chemviron, 331
Chillers, 45, 244
Chimneys: 118 et seq.
 barometric control, 118
 contamination, 118
 draught, 118
Chlorination, 98
Chrysler Corp., 238
Clare Instruments, 322
Clay drying, 175
Cleaning:
 fluidised bed, 284, 324
Coal: 4 et seq.
 calorific value, 5
 costs, 10
 extraction, 6
 gasification, 7, 63, 284
 liquefaction, 8, 63
 rank, 6
 reserves, 5, 304
'Coefficient of Performance', 21, 240
Coke, 6, 183
Combat Co., 322
Combined cycles: 7, 36 et seq.
 efficiency, 36, 81
 examples, 36 et seq., 81
 total energy, 36
Combustion:
 control, 123, 321
 fluidised bed, 282 et seq., 324
Combustion Chemicals, 149
Combustion Engineering, Inc., 330
Combustion Publ. Co., 79
Combustion Systems Ltd., 324
Combustion systems: 119 et seq.
 automatic control, 123, 321
 catalytic, 105
 efficiency, 88, 147
 'Environ', 130
 improvements, 117, 119 et seq.
 in glass furnaces, 88
 new burners, 121 et seq.
 noise in, 130
 submerged, 121, 147
 test kits, 119
Company audit, 306 et seq.
Compressed air: 124
 dryers, 125
 filters, 125
 freeze drying, 88
 in glass moulds, 88
 maintenance, 125
Compressors:
 air, 73, 89, 124
 centrifugal, 45, 245
 gas turbine drive, 29
 heat pump, 245
 manufacturers, 321
 refrigeration, 245
 screw, 245
Computers:
 air conditioning, 114, 153
 refineries, 73
Computer-Aided Design Centre, 315
Concrete Reinforcing Steel Inst., 316
Condensate:
 in engines, 48
 return of, 117, 165, 321
Condensers:
 double-bundle, 161
Consumat, 189
Consumption:
 natural resources, 4 et seq.
 UK, 5, 9, 51 et seq.
 USA, 4 et seq.
 World, 4 et seq.
Control:
 air conditioning, 113
 automatic, 119
 combustion, 119, 321
 cooling towers, 126
 flow, 119
 lighting, 142, 326
 of engines, 48
 systems, 321, 328
 temperature, 93, 146, 328
 voltage, 321
Control Devices Inc., 328
Control Technology, 331
Convection:
 recuperators, 233
Conversion factors, 342
Conveyor belt curing, 278
Coolair Distributors, 320

Coolers:
  air, 25
  fluidised bed, 284
  oil, 254
Cooling:
  ebbulient, 48
Cooling towers: 125 et seq., 161, 252
  carry-over, 125
  control, 126
  fouling, 125
  good housekeeping, 125 et seq.
Corning, 327
Corrosion control, 119, 148
Costs:
  analyses, 27, 164, 188, 221
  capital, 143, 168
  energy, 10, 146
  heat exchanger, 102
  maintenance, 27, 47, 143, 148
  running, 27, 47, 96, 102
Courtaulds, 99
Crane Co., 321
Crewe Chemicals, 322
Csathy, D., 187
CSL Energy Management, 322
Cupolas, 182
Curwen & Newbery, 213, 327
Cycles: 238 et seq.
  combined, 36 et seq.
  gas turbine, 28 et seq.
  Rankine, 27, 240, 256
  steam, 19
  Stirling, 256
Cyclone separators, 190

Dairies: 95 et seq.
  plate heat exchangers, 22 et seq.
  pasteurisation, 222
Davy Powergas, 67
Degree day, 306
Deltak Corporation, 188, 330
Demand:
  changes in, 24
DEODO System, 187
Department of Energy, 308, 314
Department of Environment, 114
Department of Industry, 313
Department of Interior, 7
Derrieus generator, 292
Desulphurisation, 186
Dielectric heating, 276
Diesel engines: 38 et seq.
  dual fuel, 38
  efficiencies, 38, 120
  maintenance, 46 et seq.

  manufacturers, 326
  total energy, 42 et seq.
  waste heat, 42 et seq., 230
Digicon Electronics, 322
Dimmers, 142
Dishwashers, 99, 258 et seq.
Distillation, 67, 73
District heating, 21
Divided blast, 183
Domestic:
  heat pumps, 241
  hot water, 145
Doors, 143
Dowtherm, 194
Draught:
  control, 118
  regulators, 266, 320
Dryers: 126 et seq., 171 et seq.
  aggregate, 128, 172
  air volume, 128
  batch, 126, 179
  cereal, 176, 293
  compressed air, 125
  energy costs, 104
  filtration in, 180
  fluidised bed, 179, 284, 324
  food, 176
  gas, 126
  heat pumps in, 178, 249
  microwave, 83, 277
  paper, 78
  regenerative, 176
  r.f., 106
  rotary, 172, 174
  solar, 302
  textile, 100 et seq., 126
  turbo-tray, 173
Dual firing, 38, 137
Dunham Bush, 46
Du Pont, 149, 243
Dye-houses, 99 et seq.

Ebullient cooling, 48, 254
Eclipse Inc., 330
Economic boiler, 264
Economics:
  analyses, 27
  payback period, 57
  total energy, 22 et seq.
Economisers: 224 et seq.
  applications, 104, 224
  performance, 119, 226
  types, 224, et seq., 322
Econovent EX, 234
ECOTERM, 208

Edison Electric Inst., 114, 317
Eduard Ahlborn, 324, 326
E.E.C., 1, 6
Efficacy, 139
Efficiency:
  burner, 119
  furnace, 88, 134
Effluent treatment, 77, 96, 100
E.I. Dupont, 324
Electric Energy Association, 152
Electric equipment, 130 et seq., 238 et seq.
Electric heating: 90, 285
  furnaces, 55, 183
Electrical Research Association, 313
Electricity Council, 21, 83, 139, 152
Electricity:
  costs, 47, 131, 292, 297, 310
  power factor, 131
Electricity generation:
  costs, 23, 45, 47, 310
  efficiency, 20, 54
  in-plant, 22, 79, 106, 130, 310
  national, 15, 19
  power factor, 84, 131
  solar, 300
  transformers, 131
  transmission loss, 131
Electroflotation, 98
Electron Beam Processes, 315
Electron beam welding, 279 et seq.
  conventional, 279, 322
  in air, 281
Electrostatic fibre laying, 83
Elga Products, 331
Emhart Corporation, 89
Endothermic reactions, 137
Energy:
  geothermal, 294
  solar, 300
  wave, 296
  wind, 292
Energy audit, 306 et seq.
Energy consumption:
  aluminium industry, 59
  chemical industry, 62
  coal, 5
  food industry, 92
  gas, 8
  glass industry, 85
  laundries, 102
  national, 4 et seq., 51
  oil, 13
  paper industry, 76
  steel industry, 51 et seq.
  textile industry, 52, 99

uranium, 16
Energy content, 24, 97
Energy costs, 10, 15, 38 et seq., 310 et seq.
Energy Equipment Co., 314, 320
Energy manager, 1, 306 et seq.
Energy policy:
  management, 306 et seq.
Energy storage: 263 et seq.
  accumulators, 264
  boilers, 264
  chemical, 263
  electromagnetic, 263
  in lehrs, 90
  kinetic, 263
  Metro-Flex system, 265
  potential, 263
Engelhard, 105, 187
Engines:
  cooling systems, 326
  diesel, 38 et seq.
  gas, 38 et seq.
  gas turbine, 32
  manufacturers, 326
  multiple units, 231
  rental, 28 et seq.
  reciprocating, 38 et seq.
ERDA, 318
Escher Wyss, 80
ESSO, 269, 315
Etched foil, 285
Ethylene plant, 69 et seq.
  cracking furnace, 70
  heat recovery, 69 et seq.
  pyrolysis, 70
  reliability, 71
ETS (Lonertia), 330
EURATOM, 299
European Economic Community, 1, 6
EUR-Control, 120, 329
Evaporators:
  heat pump, 240
Examples:
  air conditioning, 116
  air preheaters, 163, 217
  blowdown, 166
  bottle washers, 96
  condensate recovery, 165
  conveyor belts, 279
  cooling towers, 125
  cupolas, 183
  dairies, 96, 222
  dish washers, 99, 258 et seq.
  dryers, 81, 128, 173
  drying oven (heat pipes), 205
  evaporating plant, 221

Index

furnaces, 132, 183, 238
heat treatment, 184
hospital, 209
incinerators, 186
injection moulding, 252
insulation, 132, 136
laser welding, 286
lighting, 142
paper manufacture, 84 et seq.
printing machines, 218
recuperative burners, 170
sausage skin factory, 209
steam traps, 146
Templifier, 250
textile dryers, 104
total energy, 25 et seq., 33, 42
Exhaust:
  diesel, 38 et seq.
  gas engine, 38 et seq.
  gas turbine, 28 et seq.
  heat recovery, 154 et seq., 229
Exothermic Ladle Cover, 136
Exothermic reactions, 137

Fan Installations, 327
Federal Energy Administration, 318
Feedwater: 24
  cost, 165
  quality, 165
  recovery, 117, 165
Feedwater heating: 70, 224
  methanol plant, 68
Fibreglass Ltd., 268, 329
Fibrous insulation, 132, 267
Filters:
  air conditioning, 112
  dryers, 125
Finance Act, 113, 267
Financial analyses, 309, et seq.
Firebricks, 130
Fission, 15
Fixation, 102
Flat flame burners, 185
Flow control, 119
Flowmeters, 145
Flue gases, 117, 224
Flues: 118, 182
  balanced draught, 118
  furnace, 134, 226
  gas analysis, 119, 145
  heat losses, 118
Fluidfire Systems, 327
Fluidisation, 282
Fluidised bed: 95, 282 et seq.
  boiler, 283

cleaners, 284, 324
combustion, 7, 283, 324
coolers, 284
dryers, 179, 284, 324
'fast', 7
Fluids:
  thermal, 191 et seq.
Fluorescent lighting, 140
Fly ash, 191
Foil heating element, 285 et seq.
Food industry: 92 et seq., 219
  bacteria destruction, 94
  canned foods, 93
  energy usage, 52, 92
  fluidised beds, 95
  good housekeeping, 93
  heat pumps, 97 et seq.
  milk products, 95 et seq.
  packaging, 92
Ford engines, 40 et seq.
Fossil fuels, 4 et seq.
Foster Wheeler Energy Corp., 331
Foundries, 182
Freeze drying:
  compressed air, 88
Freons: 73
  in turbines, 27
  properties, 244
Fuel:
  additives, 120, 147
  coke, 6
  engine, 38 et seq.
  equivalents, 298
  feed control, 49
  fossil, 4, 343
  hydrogen, 298 et seq.
  injection, 38
  waste as, 77, 185 et seq.
Fume incinerators, 184
Furnaces:
  arc, 55
  batch, 181
  BOS, 54 et seq.
  continuous, 181
  cracking, 70
  cupola, 182
  divided blast, 183
  electric arc, 55, 183
  flues, 134, 182
  glass melting, 86 et seq.
  good housekeeping, 120, 132 et seq.
  heat flow in, 121, 181
  heat load, 135
  heat recovery, 34, 56, 70, 132, 182
  heat treatment, 132, 183
  idling, 134, 184

induction, 183
  ladle, 136
  monitoring, 134
  rails, 136
  reheat, 183

Garrett R & D Co., 195
Gas: 9 et seq., 133
  analysis, 88, 145
  from coal, 6
  natural, 9
  synthesis, 65, 69
Gas-coupled heat exchanger, 211
Gas engines: 38 et seq.
  applications, 39, 46, 125, 246
  costs, 47
  governing, 38
  heat balance, 43, 254
  manufacturers, 326
  reliability, 40, 46 et seq.
  total energy, 42 et seq.
  waste heat, 42 et seq., 230
Gasification, 6, 63, 284
Gas turbines: 28 et seq., 329
  Centrax, 33, 310
  compressor drive, 29
  costs, 34, 310 et seq.
  cycle, 29
  dryers, 81
  efficiency, 30, 33
  industrial, 32 et seq., 69, 81, 310
  injection water heating, 226
  intercoolers, 31
  reheat, 31
  Ruston, 32
  temperatures, 30
  theory, 29
  total energy, 32, 74, 310
  waste heat, 32, 42, 214, 226
  waste heat boilers, 33, 229
Geldia (London), 323
Generation:
  electricity, 21 et seq., 47, 81, 310 et seq.
Generators:
  Derrieus, 292, 330
  wave, 297
Geothermal energy: 294 et seq.
  power plants, 295
Gestra (UK) Ltd., 167, 328
Gillies, J., 100
Girdlestone Pumps, 321
Glass:
  double glazing, 112
  solar control, 112, 328

Glass industry: 85 et seq.
  annealing, 89
  combustion efficiency, 88
  energy consumption, 85
  forming, 88 et seq.
  furnace recuperators, 87
  furnace regenerators, 87
  melting process, 86
Glass Industry Res. Assn., 87
Good housekeeping: 2, 110 et seq., 306 et seq.
  air conditioning, 110
  boilers, 117
  combustion systems, 119
  compressed air units, 124
  cooling towers, 125
  dryers, 83
  electricity generation, 130
  electric equipment, 131
  food industry, 94
  furnaces, 132
  general process heating, 136
  hot water systems, 145
  incinerators, 138
  lighting, 139
  paper industry, 83 et seq.
  plant buildings, 143
  steam traps, 146
  textile industry, 106
  vats, 147
Governing, 41
Grasso Compressors, 245
Gray, A.G., 54
Green, E. & Son, 322
Greenhouses, 21
Grumman International, 331
Gulf Coast, 296

Hall, J. & E., 321
Hall-Heroult process, 60
Hamworthy Engineering, 123, 321, 330
Harefield Rubber Co., 320
Hawker Siddeley, 322
Hawthorn Leslie (Marine), 330
Heat:
  high grade, 254
Heating:
  dielectric, 276
  district, 21
  foil element, 285 et seq.
  infra-red, 144, 325
  microwave, 276
  solar, 328
  thermal fluid, 191 et seq., 329

Index

Heat balance: 42
  buildings, 154
Heat exchangers:
  air-to-air, 104, 154, 172
  economiser, 224, 322
  gas-coupled, 211
  heat pipe, 104, 154, 199 et seq.
  liquid-coupled, 154, 207 et seq.
  Lordan, 257 et seq.
  miscellaneous, 167, 257, 324
  plate, 219 et seq., 326
  recuperative, 56, 232 et seq., 324, 327
  rotating, 57, 211 et seq., 327
  'run-around' coil, 154, 207 et seq., 328
  static matrix, 154, 234 et seq., 327
  suppliers, 319 et seq.
Heat pipes:
  control of, 206
  costs, 104, 204
  heat exchangers, 104, 154, 180, 199 et seq.
  manufacturers, 324
  performance, 202 et seq.
  variable conductance, 206
Heat pumps: 238 et seq.
  absorption cycle, 256
  air-to-air, 160, 241, 249
  applications, 21, 65, 73, 159, 244 et seq.
  chilling/space heating, 159, 252
  coefficient of performance, 21, 240
  compressors, 245, 321
  cycles, 238 et seq.
  dehumidifiers, 178, 249
  domestic, 241, 302
  electric drives, 73, 97, 249
  examples, 160
  gas engine drives, 39, 65, 246
  gas turbine drives, 65
  in bottle washers, 97 et seq.
  in dish washers, 99
  in dryers, 178, 249
  in refineries, 73
  performance of, 97
  suppliers, 321, 325
  Templifier, 245 et seq.
  thermoelectric, 255
  water-to-water, 97, 125, 247
  working fluids, 244
Heat recovery from:
  air conditioning plant, 113, 154
  blowdown, 167
  boilers, 85, 117, 163

  bottle washers, 96 et seq.
  burners, 169 et seq.
  catalytic combustors, 105
  dishwashers, 259
  dryers, 79, 85, 171 et seq.
  ethylene plant, 69 et seq.
  furnaces, 56 et seq., 132, 329
  gas turbines, 33, 44, 329
  incinerators, 139, 185 et seq.
  lehrs, 89
  lighting, 157 et seq.
  methanol plant, 66 et seq.
  oil coolers, 254
  reciprocating engines, 43, 326
  steam turbines, 21, 28
Heat recovery systems:
  double duct, 161
  heat exchangers, 154
  heat pipes, 154, 324
  heat pumps, 159, 238 et seq.
  recuperators, 154, 324, 328
  suppliers, 319 et seq.
Heat storage: 90, 147, 270 et seq.
  accumulators, 270
  buildings, 145, 162
  heat exchangers, 272
  in resin plant, 272
  latent, 273
  sensible, 162, 191, 271
Heat transfer salts, 191, 226
Heat treatment: 147
  furnaces, 132, 183, 238
High Temperature Reactor, 59
Hirt/Hygrotherm, 121, 185
Hittman Associates, 317
HMSO, 225
Hot springs, 295
Hot water supply, 145
Hotwork Development, 122, 320
Howden Group, 163, 217, 321, 327
Hughes & Co., 314
Hughes Aircraft, 324
Humidity:
  dryers, 107, 130
  measurement, 126, 325
Hunt Heat Exchangers, 324
Hydraulic flow control, 119
Hydrogen economy, 298 et seq.
Hydrogen production, 8, 298
Hygrotherm Engineering, 321, 329, 330

IBM Chicago, 116
IBM (UK), 316
ICI, 66, 125, 146, 243, 329

Idle operation:
  furnaces, 184
IEEE, 155
IHI Co., 28, 329
IHVE, 314
Illuminating Engineering Society, 139
Immersion heaters, 147
Impco, 40
Incandescent lights, 140
Incinerators: 184 et seq.
  catalytic, 186
  Consumat, 189
  DEODO, 187
  fume, 139, 184
  good housekeeping, 138 et seq.
  liquid, 188 et seq.
  solid, 189 et seq.
  thermal, 184
  waste heat, 139, 226
Indirect heat exchangers:
  gas-coupled, 211
  liquid-coupled, 154, 172, 192
Induction melting, 183
Industries:
  chemical, 62 et seq.
  fibreboard, 76
  food, 91, 92 et seq.
  glass container, 86
  'heat supplied' to, 51
  paper, 76 et seq.
  steel, 52 et seq.
Industrial Services, 149, 132 et seq.
Infra-red:
  cameras, 145
  heaters, 144, 325
  paper heating, 82 et seq.
Installations:
  BCIRA, 136
  British Gas, 46
  Dunlop, 279
  Electricity Council, 272
  ICI, 146
  Isothermics, 205
  James Howden, 163
  Long Eaton, 46
  Omega Watch, 290
  Petbow, 42
  Standard Oil, 25 et seq., 146
  Sunbeam, 184
Instant Cooling Equipment Ltd., 320
Institute of Gas Engineers, 169
Instrumentation:
  flow, 145
  moisture content, 126, 325
  oxygen analysis, 88, 145
  radiotelemetry, 94

  temperature, 94, 145
Insulation: 267 et seq.
  air conditioning, 112
  fibrous, 133, 267
  ladles, 136
  performance, 133, 268 et seq.
  thermal, 132 et seq., 267, 329
Integrated Conveyors, 323
Integrated waste heat recovery, 69
International Research & Development
  Co. Ltd., 201, 279, 288, 315, 325,
  331
Inter Technology Corp., 316
Investment criteria:
  returns, 309 et seq.
Iron & steel industry: 52 et seq.
  energy conservation, 56 et seq.
  energy usage, 51 et seq., 182
  fuel substitution, 54
  Japan, 56
  Netherlands, 52
  nuclear heat, 59
  UK, 53
  USA, 55
  West Germany, 57
Irrigation, 22
Isothermics Inc., 205, 325
ITT-Reznor, 235, 327

James Howden, 163
Japan, 56, 295
Jenkins, Robert, 189, 220, 323, 326
Johnson Service Co., 321

Kaowool, 132
Kawasaki Industries, 183
Kilns:
  drying, 249
  insulation, 133
  recuperators, 175
K-lite A, 136
Kolbusz, P., 21
Konus-Kessel, 191
Kristian Kirk Electric, 321
Krypton gas, 112

Ladles (furnace), 136
Lagging, 137, 267
Lambeg Industrial RA, 313
Lancashire boiler, 264
Landis & Gyr Billman, 321, 328
Larderello, 295
Laser Focus, 289

Index

Laser welders, 286 et seq., 325
Latent heat storage, 273
Laundries, 102
Leather Manufac. RA, 312
Lee Wilson Engineering, 327
Lehrs:
  energy storage, 90 et seq.
  heat recovery, 86, 89
Lennox Industries, 325
Lighting:
  air-cooled, 157
  controls, 142, 326
  dimmers, 142
  efficacy, 139
  fluorescent, 140, 157
  good housekeeping, 139 et seq.
  heat recovery, 142, 153, 157
  incandescent, 140
  luminaires, 141, 157
  maintenance costs, 143
  maintenance factors, 141
  water-cooled, 159
Lighting Industries Fed'n., 139, 315
Link Plastics, 252
Liquefaction, 8, 63
Liquid-coupled heat exchangers, 192, 207 et seq.
Liquid natural gas, 11
Liquid Systems, 323
Liquified petroleum gas, 14 et seq., 343
Ljungstrom wheel, 211 et seq.
Loading bays, 143
Lordan & Co., 258
Long Eaton, 46
LTM refractories, 132
Luco-Engineering Services, 323

Machine Tool Industry RA, 313
Maintenance: 46 et seq., 134
  boilers, 117
  costs, 27, 47
  factors, 141
  lights, 141
  regularity, 112, 141
Manufacturing industry:
  energy usage, 51 et seq., 59, 62, 71, 76, 85, 92
Marine & Industrial Heat, 219, 326
Martron Assoc., 329
Mather & Platt, 331
Maxim Silencers, 326
Maximum demand: 90
  monitors, 321

Measurement: 307
  combustion, 119, 321
  flow, 119, 145
  moisture, 107, 126, 146, 325
  temperature, 94, 329
  viscosity, 307, 329
Melting: 135
  electric, 183
  glass, 86 et seq.
Mercury, 27
Mercury vapour lights, 140
Metabolic heat, 153
Metal halide, 140
Metal Powder Industries Fed'n., 317
'Metal Progress', 184
Methanation plant, 207
Methane, 8, 65
Methanol manufacture: 66 et seq.
  distillation, 67
  gas use, 66
  ICI plant, 66 et seq.
  preheating, 66
  reboiler heat, 67
  water savings, 69
Metro-Flex, 266
Meynell Valves, 328
Micropore Insulation, 329
Microwave heating: 276 et seq.
  drying, 277
  in paper industry, 83
  oven, 278
  welding, 278
Middle East, 13
Milpro NV, 97
Minikay, 138, 320
Minnesota Mining (3M), 328
Mixing valves, 328
Mol, A. 69
Monnier, P., 290
Morganite, 329
Morgett Electrochemicals, 331
Motors, 25
Munter wheel, 211 et seq., 327

Nalfloc, 149
Namafjall, 295
NASA, 196
Nationaire Engineering, 320
National Asphalt Pavement Assn., 128, 173
National Coal Board, 6
National Engineering Laboratory, 269
National LP Gas Assn., 316
National Oil Fuel Inst., 317

National Science Foundation, 8, 207, 293
National Utility Services, 132, 315
NATO, 76
Natural gas: 9 et seq., 38
  costs, 10
  reserves, 9
  shortages, 7
Neil, D.J., 257, 324
Netherlands:
  energy usage, 9, 51 et seq., 71, 76, 92
Newell Dunford, 323
Newman Ind. Sales, 320
NIFES, 148, 315
Nittetu Chemical Co., 188
North Sea, 10, 13
Nozzle mixing burner, 169
Nuclear power, 15
  costs, 15
  heat, 59
  programmes, 304

Obermaier & Cie., 101
Office of Energy Programs, 317
'Off-peak' energy, 162, 270
Oil: 12 et seq.
  coolers, 254
  demand, 12
  injection, 120
  resources, 13
  shale, 13, 302
  tar sands, 13
Oil refineries: 25 et seq., 71 et seq., 168
  capacity, 71
  distillation, 72
  gas turbines in, 35
  heat pumps, 73
  operating costs, 25 et seq., 72
Omega Watch Co., 290
Onan, 326
Ore:
  thorium, 15
  uranium, 15
Osram-GEC, 158
Ovens: 137
  microwave, 278
  paint, 185, 205
Over, A.J., 63
Oxygen analysis equipment, 88

Package dyeing, 99 et seq.
Packaged burner, 170

Paint ovens, 185
Paint Research Assn., 313
Pall Filtration, 322
Paper manufacture: 76 et seq.
  drying, 78 et seq., 278
  energy consumption, 52
  heat balance, 80 et seq.
Paraflow heat exchanger, 221
Pasteurisation, 176, 222
Payback period, 57, 98, 164, 204, 309 et seq.
PCI Reverse Osmosis, 331
Pelapone, 46
PERA, 194, 312
Perflex Inc., 330
Perkins, 41, 46
Petbow, 42
Peter Brotherhood, 329
Pharmaceutical drying, 126
Phase change materials, 263
Photain Controls, 326
Photosynthesis, 300
Photovoltaic cells, 301
Plant:
  buildings, 143 et seq.
  utilisation, 131
Plate heat exchangers: 219 et seq.
  efficiencies, 222
  examples, 221 et seq.
  pressure drops, 220
  types, 219, 326
Plost, M., 206
Pollution, 6, 37, 71, 105, 119, 139, 190
Portland Cement Assn., 316
Potential energy storage, 263
Pottery kiln, 133
Potts Industries, 231, 330
Power factor, 131
Power generation: 19 et seq.
  by industry, 23 et seq., 80, 130
  costs, 25 et seq.
  methods, 19 et seq., 80
  policy, 19, 23
  reliability, 38 et seq.
Power/heat ratio: 25, 80
  gas turbines, 28 et seq.
  reciprocating engines, 43
Power plant:
  good housekeeping, 130
  types, 21 et seq.
Power stations:
  efficiency, 20 et seq.
  thermal discharge, 22
  waste heat, 22
Power Torque, 40, 326

Index

PPG Industries, 328
Preheating:
  boiler air, 117, 162
  dryer air, 171 et seq.
  incinerator air, 185 et seq.
  injection water, 226
  scrap, 57
Prestcold, 252
Prices: 10 et seq., 15, 27
Prime movers:
  gas turbines, 28 et seq.
  heat recovery, 22, 35
  reciprocating engines, 38 et seq.
  steam turbines, 19 et seq.
Printing industry, 218
Property Services Agency, 114
Pulp & paper industry: 76 et seq.
  drying, 78
  good housekeeping, 83 et seq.
  total energy, 81 et seq.
Pumps (heat), 238 et seq.
Pyrolysis: 195
  coal, 7
  ethylene, 70

Q-Dot Corporation, 200, 324

Radiant heating, 144
Radiation recuperators, 235 et seq.
Radiotelemetry, 94
Rankine cycle, 27 et seq., 240
Reactions:
  endothermic, 137
  exothermic, 137
Ream & Co., 320
Reciprocating engines: 38 et seq.
  condensate removal, 48
  control systems, 48
  diesel, 38 et seq
  dual fuel, 38
  gas, 38 et seq., 254 et seq.
  heat recovery, 43 et seq., 228, 254, 326
  maintenance, 46 et seq.
  manufacturers, 326
  oil consumption, 47
  spark ignition, 38
  turbo-charged, 39
  waste heat boilers, 228
Recuperative burners, 130, 169 et seq.
Recuperators:
  ceramic, 56, 88, 233
  composite tube, 234
  convection, 233

Econovent, 234
flue tube, 234
heat pipe, 105, 154
in cupolas, 183
radiation, 235 et seq.
static matrix, 156, 327
tubes, 233
tubular, 173, 327
Reding, J.T., 88
Refractory brick, 132
Refineries, 71 et seq.
Refrigeration: 91, 98, 159 et seq., 238
  compressors, 245, 321
  fluids, 27, 73, 244
Regenerators:
  rotating, 57, 154, 211 et seq., 327
Regenerative dryers, 176
Reheat, 20
Reheating furnace, 183
Reliability:
  gas engines, 40, 46 et seq.
  plant, 71, 146
  power generation, 23
Rental of equipment, 35, 310 et seq.
Research Associations, 76, 312 et seq.
Reserves & Resources: 4 et seq., 292 et seq.
  coal, 5
  gas, 9
  oil, 13
  uranium, 16
Reverse osmosis, 98
R.F. heating, 83
R.G. Vanderweil Engineers, 316
Rheinstahl ECO, 322
Rolling Stock & Engng. Co., 324
Rooke, D.E., 10
Rotary dryers, 130, 174
Rotating regenerators: 211 et seq.
  air conditioning, 154
  boiler air preheating, 162, 217
  ceramic, 57, 213
  costs, 164
  cross-contamination, 213 et seq.
  manufacturers, 327
  performance, 155, 163, 212
  types, 212 et seq.
Rubber & Plastic RA, 312
'Run-around' coils, 154, 172, 192, 207 et seq., 328
Running costs, 88
Ruston Co., 32, 39, 329
Ruths accumulator, 264

Saffil fibre, 133
Salavat (USSR), 87
Salter, S.H., 297
Salts:
  heat storage, 274
  heat transfer, 191, 226
Sandia Laboratory, 292
Satchwell Control Systems, 316, 328
SBM (UK), 325
Scheduling, 133
Schmidt'sch. Heiss., 330
Scrap preheating:
  aluminium, 62
  glass, 87
  iron, 55 et seq.
Seitz-Werke, 331
Semiconductor materials, 274
Senior Economisers, 322
Sensible heat storage, 162, 191, 270
Separators:
  cyclone, 190
Services (Industrial), 312 et seq.
SF Air Treatment, 208, 328
Shale, 13, 302
Shandon Southern Instrum., 321
Shaw Moisture Meters, 325
Shearer, A., 49
Shell Centre, 142
Shell (UK), 142, 315
Shepherd, B., 88
Simon-Hartley, 330
Sira Institute, 126, 312
Sjoerdsma, A.C., 63
Soil heating, 22
Solar cells, 300
Solar collectors, 300, 328
Solar concentrators, 300
Solar drying, 302
Solar energy: 243, 300 et seq.
  costs, 301 et seq.
  radiation, 112
Solar film, 112, 328
Solar glass, 112, 328
Solar heating, 301 et seq., 328
Space heating, 73, 144, 211, 252
Spiezman, 104
Standard Oil Company, 25 et seq., 146
Stand-by plant: 26
  reliability, 38 et seq.
  types, 19 et seq.
Start-up controls, 48, 113
Static matrix heat exchangers, 155, 234, 327
Steam:
  process, 81, 131, 138, 165, 229, 264
  traps, 117, 146 et seq., 328
Steam cycle, 19
Steam turbines: 24 et seq., 329
  back-pressure, 21, 81
  condensing, 24
Steel Casting R. & T. Assn., 132, 313
Stein Atkinson Stordy, 327
Stenters, 104
Sterilisation, 93 et seq.
Stierle Hochdruch kg., 322
Stirling cycle, 256
Stobart, A.F., 314
Stone-Platt Crawley, 330
Storage:
  gas, 14
  heat, 162, 263 et seq.
  in lehrs, 91
  potential energy, 263
Storage tanks, 147, 162
Stordy Combustion, 121, 130, 320
Submerged combustion, 121, 147, 321
Sulzer Review, 81
Sump pits, 102
Sunbeam Equipment Co., 184
Superheating, 226
Supplementary heating, 162
Svenska Fläktfabriken, 307, 328
Synthesis gas, 68, 69
Systems Research, 317

Tariffs, 131
Tar sands, 13
Taylor-Servomex, 321
TDS level, 166
Teisen Furnaces, 88
Temperature measurement, 94, 126, 145, 329
Templifier: 245 et seq.
  applications, 250
  performance, 248
Tennessee Valley Authority, 22
Terotechnology Centre, 314
Textile industry: 99 et seq.
  dryers, 104 et seq., 126
  dye-houses, 99 et seq.
  energy consumption, 52
  fixation, 102
  good housekeeping, 100
  stenters, 100
Textile Institute, 106
The Engineer, 218, 252
The Geysers, 295
Thermal Cleaning Bath, 284
Thermal Efficiency Ltd., 233, 327

Thermal efficiency:
  boilers, 228
  burners, 119, 169 et seq.
  dryers, 104
  furnaces, 88, 134
  turbines, 19 et seq.
Thermal fluids:
  heaters, 191, 329
  properties, 193
Thermal incinerators, 184 et seq.
Thermal insulation: 132 et seq., 267 et seq.
  ball blanket, 269
  blanket, 133, 267
  performance, 133, 268 et seq.
  suppliers, 329
Thermal storage, 42, 118, 264 et seq.
Thermex, 191
Thermoelectrics:
  heat pumps, 255
Thermostats, 147
Thomas Bennett, 328
Thorium reserves, 16
Timber drying, 249
Timber R & D Assn., 312
Tolltreck, 321
Topcast Engineering, 322, 326
Total energy:
  combine cycle, 36
  diesel engine, 42 et seq.
  examples, 42, 46, 79, 106
  gas engine, 45
  gas turbine, 32 et seq.
  steam turbine, 22 et seq.
Transformers, 131
Transmission losses, 131, 299
Traps:
  good housekeeping, 146 et seq.
  suppliers, 328
  steam, 146
Tribol, 149
Triple-E (UK), 149
Tubular recuperators, 173, 327
Turbines:
  back pressure, 21
  condensing, 21
  Freon, 27 et seq.
  gas, 28, 69
  manufacturers, 329
  mercury, 27
  on blast furnace, 56
  steam, 19 et seq., 300
  vertical axis, 294
Turbo-charging, 39
Turbo-generators: 19 et seq.
  efficiencies, 26

Turbo-Tray dryers, 173
Tuyere, 56

UAS (UK), 320
United Kingdom:
  energy usage, 5, 9, 51 et seq., 76, 85, 92
United Lubricants, 149
United States, 76, 92
UniTubes, 327
Universities:
  California, 54
  Edinburgh, 296
  Manitoba, 218
Uranium: 15 et seq.
  costs, 15
  resources, 16
US Bureau of Mines, 57
USSR, 183
Utilities, 9, 26

Valves (mixing), 328
Van Vechten, J.A., 274
Vaporphase system, 231
Vats, 147
Velocity measurement, 119, 145
Ventilation, 113, 144
Versar Inc., 317
Vickers Ltd., 330
Viscosity: 120, 148
  measurement, 329
Voltage stabilisation, 131

Wanson Company, 329
Warren Spring Lab., 194
Washing:
  bottles, 95 et seq.
  heat recovery, 96 et seq.
Waste:
  heating value, 78, 189, 195
  Materials Exchange, 194, 322
  organic, 195
  recovery, 193 et seq., 322
  recycling, 77, 193
  treatment, 22, 100, 138, 187, 322
Waste heat:
  air conditioning, 113, 152, 203, 209
  boilers, 85, 330
  from incinerators, 138, 185 et seq., 226
  power stations, 22, 28
  recovery, 28, 33, 85, 100, 199 et seq.

Waste heat boiler: 227 et seq.
  applications, 65, 67, 87, 106, 229
  classification, 227
  in gas turbines, 33, 229
  manufacturers, 330
  multiple installations, 231
  performance, 228 et seq.
Water conservation: 69, 125, 145
  dairies, 96
  equipment manufacturers, 330
  in laundries, 102 et seq.
  in methanol plant, 69
  sump pits, 102
Water jackets, 255
Water Management Ltd., 331
Waukesha Co., 39, 326
Wave power, 296 et seq.
Welding:
  dielectric, 278
  electron beam, 279, 322
  laser, 286 et seq., 325
Wellman Incandescent, 182, 323
Westinghouse Electric, 245, 325
Westinghouse Research Labs., 281
West Germany, 57
West's Pyro, 324
White Superior, 39, 326
Wind Energy Supply Co., 331
Wind generators:
  suppliers, 331
  types, 292 et seq.
Windows, 112
Wind power. 292 et seq.
  Derrieus generator, 292
Wing Company, 212, 327
Wool industry:
  Research Assn., 106, 313
Work loads, 100, 133
Worthington-Simpson, 324

-------oOo-------